Structural Renovation in C

T0264772

Structural Reform in Japan

Structural Renovation in Concrete

Zongjin Li, Christopher Leung
and Yunping Xi

Routledge
Taylor & Francis Group

LONDON AND NEW YORK

First published 2009
by Taylor & Francis

This edition published 2013 by Routledge
2 Park Square, Milton Park, Abingdon, Oxon OX14 4RN
711 Third Avenue, New York, NY 10017, USA

Routledge is an imprint of the Taylor & Francis Group, an informa business

© 2009 Zongjin Li, Christopher Leung and Yunping Xi

Typeset in Sabon by
RefineCatch Limited, Bungay, Suffolk

British Library Cataloguing in Publication Data
A catalogue record for this book is available from the British Library

Library of Congress Cataloging-in-Publication Data
 Li, Zongjin, Dr.
 Structural renovation in concrete / Zongjin Li.
  p. cm.
 Includes bibliographical references and index.
  1. Concrete construction—Maintenance and repair. 2. Buildings,
Reinforced concrete—Maintenance and repair. I. Title.
 TA683.L48 2009
 B624.1′8340288—dc22   2008032399

ISBN 10: 0–415–42371–6 (hbk)
ISBN 10: 0–203–93136–X (ebk)

ISBN 13: 978–0–415–42371–7 (hbk)
ISBN 13: 978–0–203–93136–3 (ebk)

Contents

Figures

Tables

Preface

In the past few decades, the deterioration of buildings and infrastructures has been occurring at an ever increasing rate. This has resulted in a worldwide need for the maintenance, repair and rehabilitation of degraded structures. To meet the needs, available knowledge and technologies must be collected, organized and disseminated to engineers and practitioners. Poor understanding of the deterioration problem and the inability to adopt proper remediation approaches not only lead to a waste of natural resources, but also have a negative impact on the economy of the whole world.

The maintenance, repair and rehabilitation of concrete structures involve several broad issues which encompass technical, social, and economic aspects, such as the fundamental knowledge of materials and structures as well as basic understanding of economy and sociology. In this book, we focus on the technical aspects. From material and structural points of view, a project for maintenance, repair and rehabilitation of a concrete structure usually requires knowledge and expertise in the areas of structural materials, repair materials, structural inspection, material testing, repair and/or strengthening techniques. To the best of the authors' knowledge, a book that systematically covers all these issues is currently not available. The present book can be considered the first attempt to fill such a void.

The book is divided into six chapters. Chapter 1 gives a brief introduction to renovation engineering. Chapter 2 explores the causes of deterioration of concrete structures, covering durability issues and disaster factors. Chapter 3 discusses the techniques of inspection, including destructive and non-destructive methods. Chapter 4 focuses on the traditional repair and strengthening methods such as crack repair. Chapter 5 covers the application of glass fiber reinforced plastics components for bridge deck replacement, and presents the fundamental mechanics essential for component analysis and design. Chapter 6 provides updated knowledge on the strengthening of concrete structures with fiber reinforced polymers.

The book is designed and written primarily to meet the teaching needs for undergraduate students at senior level and graduate students at entry level. It can also serve as a reference or a guide for professional engineers in their practice.

In the process of writing this book, the authors received enthusiastic help and invaluable assistance from many people, which was deeply appreciated. The authors would like to express their special thanks to Dr. Xiangming Zhou, Dr Garrison C. K. Chau, and Mr. Xiangyu Li. In particular, Dr. Xiangming Zhou made a sterling effort in collecting information and drafting some parts of the book.

The support from China Basic Research Grant, Basic Research on Environmentally Friendly Contemporary Concrete (2009CB623200) is greatly acknowledged.

Finally, we would like to thank our wives, Xiuming Cui, Garlin Lee and Xiaoxin Cui. This work is dedicated to their understanding and support.

Zongjin Li, Christopher Leung (Hong Kong, China),
Yunping Xi (Boulder, CO, USA)

1 Introduction

Renovation is a carefully designed series of activities to recover the load-carrying capability, enhance performance, and extend the service life of existing buildings and infrastructures with a satisfactory quality. Such activities include the repair, strengthening, and rehabilitation of aged or damaged structures. Renovation engineering is a combination of maintenance, inspection, and rehabilitation including repair and strengthening.

Over the past few decades, there has been considerable interest in the development of renovation techniques and renovation activities mainly due to the deterioration and durability problems of concrete structures. Many new techniques and a wealth of knowledge in renovation engineering have been developed and promulgated at various conferences, in papers, in press reports, and in class notes. This book aims to summarize the state-of-the-art knowledge in this field and form a systematic tool for teaching and practicing renovation activities.

In the past, outmoded and functionally obsolete buildings and infrastructures were normally demolished. However, in recent years the amount of repair and refurbishment of all types of structures has increased significantly. Owners, engineers and architects of structures need to consider economic aspects as well as the historical significance and long-term serviceability by choosing either demolition and rebuilding or renovation. The owners and the public often share an ethos of conservation and adaptive reuse and their preference is usually renovation rather than demolition. Moreover, as zoning and environmental regulations make it ever more difficult to construct new buildings, renovation has become the most practical course of action. Besides, recycling buildings can be viewed as a way to convert resources and reduce landfill demand (Newman 2001). Maintaining and repairing existing building stock, and repair and replacement of the infrastructure, have been a feature of the construction industry for nearly half a century, first in Europe and then in North America. It should be noted that preparation of specification for renovation work is quite different from the design of a new structure.

1.1 Building and infrastructure degradation

Recently, renovation engineering has attracted increasing international attention because of the frequent occurrence of serious degradation of buildings and infrastructures. Constructed infrastructure is essential for the development and progress of commerce and industry in modern society. The gravity of infrastructure degradation can be seen from the following facts. For example, ASCE's 2005 Report Card for America's Infrastructure assessed the condition and capacity of US public works with an overall grade of D. ASCE estimates that US$1.6 trillion is needed over a five-year period to bring the nation's infrastructure up to a good condition. As of 2003, 27.1 percent of the nation's bridges (160,570) were structurally deficient or functionally obsolete; it would cost US$9.4 billion a year for 20 years to eliminate all bridge deficiencies. According to the results of a study by the Association of State Dam Safety Officials, the total investment to bring US dams into safety compliance or to remove obsolete dams tops $30 billion. About 75 percent of schools need extensive repair or replacement and the repair bill for this is as high as $268 billion according to ASCE's 2005 Report Card for America's Infrastructure. In 1999, the European Union set a requirement that all European highways must be able to carry 44-ton vehicles. In the UK, about 40,000 bridges cannot fulfil this requirement and need to be strengthened. Building and infrastructure degradation has become a serious social and financial problem. It can be seen that the cost for infrastructure rehabilitation has become a huge burden on the national economy of the developed countries and soon it will be the same in developing countries. Structural deterioration, together with the need to increase load-carrying capacity, has created a big market for renovation engineering. In China, according to the report of the China Academy of Engineering, the loss caused by corrosion in reinforced concrete structures reached $140 billion per year.

Hence, evaluation and rehabilitation of existing infrastructures have become more and more important in the past few decades and will be more critical in the future. It is predicted that in the new century, fewer new designs and more rehabilitation work will be seen in civil engineering. More funds have to be used on inspection, maintenance, and management of existing infrastructure. More new technologies need to be developed for application in the rehabilitation of infrastructures. And, of course, there is an urgent need for a new book regarding this new branch of structural engineering.

1.2 Common causes of structural degradation

It is important to understand the basic causes and mechanisms of the various forms of deterioration that degrade construction material and infrastructure made of reinforced concrete. Only with this understanding, is it possible to undertake realistic assessments of the current condition of concrete structures, and to design and implement the appropriate renovation work.

Although deterioration of structure is usually a medium- to long-term process, the onset of deterioration and its rate may be influenced by the presence of defects which have their origin at the time of construction, or in the very early stages of the life of the structure (Kay 1992). Structural degradation can be divided into the following categories: (1) progressive structural failure, e.g. collapse of bridges due to repeated traffic loading and gravity loading; (2) sudden damage due to extreme loading such as fire, high speed wind or earthquake; (3) serviceability deficiencies, e.g. excessive deflections and vibrations; and (4) materials degradation, i.e. slow interaction with the environment. Deterioration of concrete can be caused by chemical attack from external sources or between the internal materials of which the structure is built, or by physical deterioration due to climatic changes, abrasion, fire, impact, explosion, earthquake, foundation failure or overloading. Specifically, the common causes responsible for structural degradations are:

- repeated loading, including:
  - traffic loading on bridges and highways;
  - wind-induced vibrations in bridges/buildings;
  - machine-induced vibrations in industrial plants;

- overloading:
  - heavy materials and equipment on floors designed for light live loading;
  - change of use resulting in higher loading than was allowed for in the original design;

- non-uniform dimensional changes:
  - shrinkage of constrained concrete;
  - differential thermal expansion of layered system (e.g. asphaltic pavement on a bridge deck);
  - expansion of internal phases (e.g. rusting steel in concrete);

- severe loading or unexpected hazards:
  - earthquake;
  - hurricane;
  - impact;
  - explosion;
  - fire which can result in some weakening of parts of the structure, as well as physical damage to columns, beams, slabs, etc.;

- loss of foundation support:
  - scouring at bridge piers which may topple after loss of support;
  - cyclic desiccation and re-hydration of clay soil;

- soil pumping under concrete pavement, with a poorly designed sub-base layer;
- abrasion/erosion of concrete surfaces:
  - wear of pavement surface by tires of trucks;
  - abrasion caused by steel-wheeled trolleys;
  - abrasion of a floor slab in a factory;
  - abrasion of marine structures by sand and shingle;
  - erosion of hydraulic structures;
- external chemical attacks:
  - acid rain;
  - sulfate attack;
  - chloride diffusion;
- internal chemical attack:
  - corrosion of reinforcing steel;
  - alkali–aggregate reaction;
  - stress corrosion coupled with chemical/stress effect;
  - phase changes;
- indirect effects of bacteria:
  - in warm temperatures, bacteria in sewage can convert sulfur compounds into sulfuric acid. Deterioration of metallic and concrete sewage pipes can then occur.

Besides these causes due to serviceability, structural deficiency can also result from errors in design and defects in construction. It is noted that a significant proportion of the problems associated with concrete structures can be traced back to design or to construction defects (Rasheeduzzafar *et al.* 1989). For instance, a design consideration of inadequate concrete cover may maximize the chance of oxygen and moisture penetrating the reinforcement, thus increasing the chance of corrosion. As far as construction is concerned, one main problem is oversight of curing and this causes early age cracking that permits external agencies, such as air and moisture, to enter the concrete and attack the cement matrix and the reinforcement. Other common construction errors may include failure to place the reinforcement in the right position, and failure to provide sufficient cover for the reinforcement, or inadequate compaction for concrete. These common flaws, occurring in design, construction and serviceability, of structural degradation may cause the following defects in the structure (Chandler 1991, p. 21):

- excess deflection in beams and floors due to weak design/unforeseen loading;

- inadequate/insufficient fixing between precast and in-situ concrete components;
- lack of sufficient load-carrying packing between precast units;
- misalignment of precast concrete panels;
- inadequate movement joints between claddings and structure;
- inadequate insulation leading to internal condensation;
- surface finishes spalling or flaking;
- distortion of wall panels.

1.3 The objectives and scope of renovation engineering

Renovation is a process of substantial repair or alteration that extends a building's useful life (Newman 2001). Renovation engineering is a very young subject in civil engineering for concrete structures. The missions of renovation engineering are:

1 to develop a better understanding of the degradation process by identifying major parameters governing the deterioration process;
2 to develop effective structural evaluation techniques; these techniques should be non-destructive in nature, fast and reliable;
3 to develop economic, functional, and effective repair, strengthening, and rehabilitation techniques;
4 to develop reliable maintenance procedures;
5 to develop the codes and specifications for repair and rehabilitation so that public safety and health are not jeopardized.

Unfortunately, systematic studies on structural renovation of concrete are scarce and there are only a few textbooks. Other limited information has been presented only in journal papers, or special conference proceedings. So far there are no comprehensive textbooks available addressing issues on renovation of concrete structures. A confluence of several factors usually establishes the need for building renovation. Some of the common ones are:

1 change in use;
2 upgrading of mechanical and electrical systems;
3 deterioration of building envelope;
4 structural damage and failure;
5 upgrading of buildings for lateral loads;
6 reducing serviceability problems.

Renovation engineering normally covers various technologies related to: (1) repair of degraded structures to recover initial load-carrying capacity; and (2) strengthening of structures to increase load-carrying capacity for current needs. The proper renovation of structures requires: (1) a good understanding of the degradation mechanisms for proper action to be taken and to avoid the recurrence of problem in the future, such as rusting when placing steel to strengthen the concrete; (2) reliable evaluation techniques for the existing condition, including a framework for structural appraisal and maintenance and non-destructive testing methods; (3) effective techniques for repair/strengthening with practical guidelines and specifications.

The focus of this book will be on the rehabilitation of concrete structures, on materials and structural aspects, not on architectural features and not on utilities, and less on interactions with other engineers. On the other hand, renovation engineering is more material-oriented. Deterioration problems are basically a materials problem, especially for concrete structures. Only in the final stage of progressive failure do structural problems become significant. Renovation engineering is very practical and requires heavy field work. Inspection and field evaluation are very important in preparing the renovation work since they provide the current condition of the structure and the suggestion for remedial work for the structure. So far, not many specifications and design codes are available for renovation. Though the specific renovation work depends on the type of the structure and its condition, the following steps are generally required for a renovation job:

1 decision on the details of the investigation;
2 investigation (preliminary and detailed) of the structure;
3 diagnosis of the causes of the deterioration and evaluation of the overall condition of the structure;
4 preparation of report to the client to suggest either renovation or rebuild;
5 if renovation is recommended, preparation of specification and contract documents;
6 conducting the designed renovation work;
7 inspection of the renovation work;
8 regular post-contract inspection and monitoring and advising on a practical program of maintenance.

1.4 Useful definitions

The following common definitions are used for various terms in this book:

Assessment – Systematic collection and analysis of data, evaluation, and recommendations regarding the portions of an existing structure which would be affected by its proposed use (ASCE 2000).
Evaluation – The process of determining the structural adequacy or the infrastructure or component for its intended use and/or performance. Evaluation, by its nature, implies the use of personal and subjective judgment by those functioning in the capacity of experts (ASCE 2000).
Infrastructure – In general, the basic economic, social, or military facilities

and installations of a community, including highways, bridges, parking lots, dams and tunnels (ASCE 2000).

Inspection – The activity of examining, measuring, testing, gauging, and using other procedures to ascertain the quality or state, detect errors, defects, or deterioration and otherwise appraise materials, components, systems, or environments (ASCE 2000).

Rehabilitation – The process of repairing or modifying a system to a desired condition. It is an upgrade (of a damaged structure) required to meet the present needs; it implies sensitivity to building features and a sympathetic matching of original construction (Newman 2001).

Repair – To replace or correct deteriorated, damaged, or faulty materials, components, or elements of a system to regain strength, density and durability.

Restoration – The process of re-establishing the materials, form, and appearance of a system to those of a particular era of the system.

Retrofitting – The process of increasing the load-resistance capacity or improving the performance of a structure or portion of the structure. (An example of performance improvement is to retrofit a damper into a structure to reduce its vibration.)

2 Degradation of reinforced concrete structures

Concrete is the most widely used construction material in the world, up to 10 billion tons per year worldwide consumption. Deterioration of concrete structure has become a world-wide problem and a huge burden on human society and the economy. For instance, in the UK, £500m is spent on concrete repair per year. In the USA, US$300–400m dollars are needed for the renovation of bridges and car parks alone where de-icing slats are commonly used in practice and cause severe concrete deterioration and steel corrosion. In China, the annual economic loss due to corrosion in concrete structures has reached 100 billion RMB. Deterioration is any adverse change of normal, mechanical, physical, and chemical properties either in the surface or in the body of concrete, generally due to the disintegration of its components. Degradation processes of concrete usually start from the materials level and then proceed to the structural level. They can be classified as *physical* (caused by natural thermal variations such as freeze–thaw cycles) *artificial* (such as those produced by fires), or by *natural disasters* (such as earthquakes and typhoon), by *abrasion* (erosion), *chemical* (attack by acids, sulfates, ammonium and magnesium ions, pure water, salts or alkali–aggregate reactions), *biological* (fouling, biogenic attack), and *mechanical* (impact, explosion, overloading, settlement, cyclic loading) (Bertolini *et al.* 2004).

In practice, these processes may occur simultaneously, which makes things even worse. Among various degradation causes, steel corrosion is found to be the most severe problem for reinforced concrete structures that can create cracks, cause concrete cover spalling, reduce the effective cross-section area of reinforcement, and lead to collapse. The corrosion of reinforcing steel can be induced either by carbonation or chloride diffusion. Degradation of concrete occurring within the first hours to months after casting can do significant damage to mature concrete structures. As we know, cast in-situ concrete structures are hardly ever built under ideal conditions so defects may occur as the concrete is being cast or very soon afterwards, such as early-age cracking due to plastic settlement, plastic or drying shrinkage, creep, or thermal shrinkage. These defects permit the atmosphere and other environmental agencies to penetrate the body of the concrete and to take part in the chemical and physical processes which give rise to deterioration. There are three basic visual symptoms of deterioration in a concrete structure – cracking, spalling, and disintegration, each occurring in several different forms (Mailvaganam 1992). In a given concrete structure, the three basic indicators of deterioration may occur not only in combination, but also with several forms of each symptom being manifested simultaneously. Before any rehabilitation and renovation work can be done, the cause of the damage must be identified as clearly as possible. As far as concrete structures are concerned, degradation of reinforced concrete structures can be caused by many factors, including non-uniform dimensional changes, repeated loading, lack of durability, natural or human disasters such as typhoon, earthquake, and fire.

2.1 Degradation caused by non-uniform dimensional changes

Deformation in concrete occurs mainly as the material's response to the external load and environment. When freshly hardened concrete is exposed to the ambient temperature and humidity, it generally undergoes thermal shrinkage (shrinkage strain associated with cooling), chemical shrinkage (shrinkage associated with hydration product formation), and drying shrinkage (shrinkage strain associated with moisture loss). When the shrinkage strain is restrained, tensile stress will be built up in structural members. The degradation caused by non-uniform dimensional changes is usually associated with the difference in thermal expansion and volumetric instability of concrete. Normal concrete is very liable to dimensional changes as internal and external conditions change. This situation arises because concrete responds to both temperature and humidity effects and is almost always in a state of dynamic disequilibrium with its environments. Those dimensional changes which may be important in concrete are: (1) thermal expansion; (2) bleeding; (3) plastic shrinkage in fresh concrete; (4) drying shrinkage and cyclic swelling and shrinkage; and (5) creep.

2.1.1 Influence of non-uniform thermal expansion

Non-uniform thermal expansion can be caused by the material's different coefficients of the thermal expansion under the same heating conditions or similar materials under different thermal conditions. The coefficient of linear thermal expansion is a measure of the length change occurring in a material when it is subjected to a temperature change. Let us take a bridge deck heated by the sun as an example (Figure 2.1). During the heating process, the temperature at the top of pavement rises much faster than that at the bottom, resulting in a tendency for the pavement to bend upwards. Consequently, the concrete deck restrains the upward movement, which leads to interfacial shear stresses. However, the process is reversed during

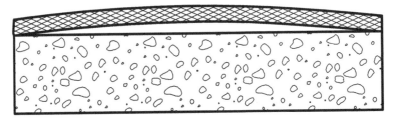

Figure 2.1 A bridge deck heated by the sun.

the cooling process. Due to creep, only part of the deformation is recoverable. Thus, under repeated heating/cooling processes, the pavement may finally debond from the deck and buckle. Traffic loading may also enlarge the debonded region and, eventually, the debonded and hence unsupported part of pavement cracks under traffic loading.

Another example of uneven thermal expansion is thermal effects generated during the hydration process of concrete. It is well known that hydration of cement is a heat-releasing chemical reaction. The heat of hydration of cement raises the temperature of the concrete, so that the concrete is usually slightly warmer than its surroundings when it hardens, and in thick sections and with rich mixes the temperature rise may be quite considerable. For a large volume of a concrete member, the temperature distribution in the member might be quite different, higher inside and lower outside. This may cause uneven expansion or uneven dimensional changes, which could lead to cracking in concrete structures. For example, a concrete dam's surface cools down faster than the interior. When the interior cools down, it "pulls away" from the hardened exterior surface, which may result in tensile stress, thus cracks, in different layers. As a common serviceability problem, the contraction and expansion that result due to seasonal temperature fluctuations can cause preexisting cracks to open and close. In general, thermal changes which cause damages to a structure are the rapid change when the concrete surface temperature changes quickly and the temperature in the core of the member changes slowly. This condition produces a curved temperature gradient with the steepest portion of the curve at the surface. As explained at the above, the upward buckling and debonding of pavement are typical examples of distress due to thermal stresses.

Restrained thermal contraction is a fairly frequent cause of cracking, and often designers do not make adequate provision for thermal movements. Non-uniform thermal expansion is at first sight a material problem, but actually it is better considered as a structural problem. Temperature differences within a concrete structure result in differential dimensional changes. When the contraction or expansion is restrained, the resultant tensile stresses exceed the tensile strain capacity of the concrete and damage (cracking) can be built up.

2.1.2 Effects of bleeding

Fresh concrete is a fluid mixture consisting of water, cementitious materials, sand, and coarse aggregate (gravel or crushed rock). The mixture remains plastic until the development of the cement hydrates starts to be connected each other. During concrete placing, compaction and subsequent plastic state, water or moisture tend to immigrate from bottom to top due to its smaller density. Meanwhile, aggregates tend to move downwards due to the equilibrium lost. The movements of the water or moisture will be obviously blocked by large size aggregate and reinforcing steel, resulting in a water film at the lower surface of these obstacles. Eventually, some water or moisture will be able to reach the surface of the specimen and form a layer of water film there. This phenomenon is called bleeding (see Figure 2.2). The concentration of water at the boundary of large size aggregate and reinforcing steel created by bleeding will form a weak interface in concrete. The characteristics of the interface include: large size of calcium hydroxide, less calcium silicate hydroxide, more porosity, and general weak nature. Moreover,

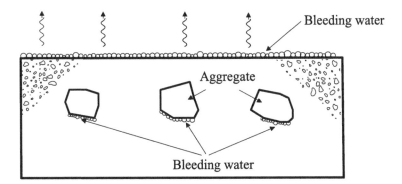

Figure 2.2 Phenomenon of water bleeding.

microcracks may form at the interface, sometimes even a microscopic crack. When cracks are formed in this way, their pattern on the surface tends to mirror that of the reinforcement. The water concentration on the surface of the specimen will result in more calcium hydroxides there, creating weak abrasion resistance on the surface.

Water also concentrates due to its upward movement when fresh concrete is compacted. Bleeding water may be trapped at the bottom of the reinforcing bar and aggregates. In the former case, the concrete may separate from the lower surfaces of the bars and a void forms due to bleeding. If this occurs in the plastic rather than the fluid state, the concrete may crack. In such a condition, the transition zone forms after hardening and unprotected steel becomes potential site for corrosion. Due to water bleeding and sedimentation (downward movement) of coarse aggregates, paste volume increases near the concrete surface and the evaporation of water eventually slows down. The fresh concrete surface then re-absorbs water to give a higher water/cement ratio and, after hardening, resistance to surface wear and abrasion is reduced. The movements associated with the reduction in volume are resisted by the concrete which lies immediately below and which is not subject to volume change. The restraint from the lower concrete causes tensile stresses to build up in the surface layer and, because the material still remains in the fresh state or plastic and has very low strength, cracking can result. Excessive bleeding can be avoided by improving the cohesiveness of the mix through reduction of water/cement ratio, using a better aggregate grading and/or increasing the cohesiveness of fresh concrete.

2.1.3 Effects of plastic shrinkage

In the fresh state, the top surfaces of concrete pours are subject to evaporation and consequent loss of the mix water. The rate of evaporation depends upon ambient conditions such as temperature, exposure to sun, wind speed and relative humidity. The water lost by evaporation is usually replaced by water rising to the surface from the bottom of the concrete by the action of bleeding.

In both cases, there are local reductions in volume when the rate of removal of water from the surface exceeds the rate at which it can be replaced by bleeding. From this point of view, a concrete mixture which provides some bleeding is helpful in reducing deterioration caused by plastic cracking. Protection of concrete surface from drying winds by the use of barriers and the earliest possible application of covering to the surface may be helpful in reducing bleeding effect. ACI 305R (1999) indicates that precautions against surface drying out and cracking should be taken if evaporation is likely to exceed 1 kg m^{-2} h^{-1}.

Shrinkage is caused by the surface tension of water within the capillary pores formed in the cement paste during evaporation. As a capillary pore starts to dry out, the remaining water forms a meniscus between the adjacent cement particles, and the forces of surface tension pull the particles together. The loss of water from concrete may cause mainly two kinds of shrinkage: plastic shrinkage and drying shrinkage. Plastic shrinkage is the phenomenon when the surface layer is hardened and starts to shrink due to water loss while the inside concrete is still wet and plastic. So, plastic shrinkage is normally caused by rapid drying of fresh concrete at the surface (Allen *et al.* 1993). The water lost by evaporation is usually replaced by water rising to the surface of the concrete by the action of bleeding as discussed in Section 2.1.2. Where the rate of water evaporation from the surface exceeds the rate at which it can be replaced by bleeding, there is local reduction in volume. The upward bleeding of water may be accompanied by a downward movement of the solid and heavier ingredients. This downward movement may be resisted by the top layer of reinforcement or by the formwork. In the former case, the layer of concrete above the reinforcement tends to become draped over the bars. If this occurs in the plastic rather than the fluid state, the concrete may crack. In addition, the concrete may separate from the lower surfaces of the bars creating a void, as shown in Figure 2.3. When cracks are formed in this way, their pattern on the surface tends to mirror that of the reinforcement. The surface profile tends to be undulating with high points over the bars. Under other conditions, the downward movement of concrete can be restrained by the shape of the formwork. Plastic shrinkage cracks tend to propagate predominantly through the matrix rather than through the aggregate (Kay 1992). The process leading to plastic shrinkage cracking is shown in Figure 2.3.

These plastic shrinkage cracks provide a path for water and other chemicals to reach the steel reinforcement, which can greatly affect the durability of concrete structures, for instance, corrosion is easily generated. In general,

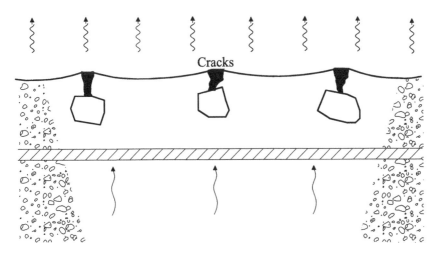

Figure 2.3 Formation of plastic shrinkage crack.

if cracks appear in an exposed concrete surface very soon after it has been finished or even before finishing is complete, they are termed "plastic shrinkage cracks" (Allen *et al.* 1993). These cracks form when wind or heat causes the concrete to lose water rapidly, usually from 30 min. to 6h after the concrete placement. They are often wide and deep and are usually discontinuous. Plastic shrinkage cracks are usually of shallow depth, generally 38–50 mm, 300–450 mm long and usually perpendicular to the wind, which cause the water loss by evaporation. Plastic shrinkage cracks typically run parallel to one another and, because of their size, they can be structurally significant. The most effective way of preventing their occurrence is by sheltering the surface from wind and sun during construction and by covering the concrete surface immediately after finishing, measures which are all directed toward reducing the rate of evaporation. BS 8110 has recommended minimum periods of curing and protection (shown in Table 2.1) for fresh concrete to reduce plastic shrinkage. Changes in concrete mix design, and especially the use of air entrainment, may also be helpful in reducing plastic shrinkage. Remedial measures after the cracks have formed usually consist of sealing them against ingress of water by brushing in cement or low-viscosity polymers (Allen *et al.* 1993).

The magnitude of plastic shrinkage may in extreme cases be as large as 10,000 microstrain (Troxell *et al.* 1968) and has been shown by L'Hermite (1960) as over 6000 microstrain for paste and 2000 microstrain for concrete. As in the plastic state, no great stress is induced in concrete and further working of the concrete can generally be applied to eliminate consequential cracks.

2.1.4 Influence of drying shrinkage

The volume change and water loss of a concrete member do not stop after its initial setting and hardening. The loss of moisture is accompanied by the reduction in volume known as shrinkage. Drying shrinkage is a slow process

Table 2.1 Minimum periods of curing and protection

Type of cement	Ambient conditions after casting	Min. period of curing and protection	
		Average concrete surface temperature	
		5 to 10 °C	*Above 10 °C*
Portland cement, SRPC	Average	4 days	3 days
	Poor	6 days	4 days
All except Portland cement and SRPC, and all with GGBS or PFA	Average	6 days	4 days
	Poor	10 days	7 days
All	Good	No special requirements	

in thick members, so it may lead to a gradual build-up of tensile stress if it is restrained and, further, cracking of concrete. Cracks due to restrained shrinkage are often noticed soon after construction, but when slow drying occurs they may not be apparent until much later. The drying shrinkage crack can vary from fine hairlines to as wide as 3 mm ($^1/_8$ in.), and often serve as ports of entry for moisture, carbon dioxide, and other injurious salts (Mailvaganam 1992).

The formation of drying shrinkage can be explained as follows. In normal practice, in order to produce workable concrete, nearly twice as much water that is theoretically needed to hydrate the cement is usually added to the concrete mix. After concrete has been cured and begins to dry, the excessive water that has not reacted with the cement will begin to migrate from the interior of the concrete mass to the surface. As the moisture evaporates, the concrete volume shrinks. From cement chemistry, it is known that the drying shrinkage of concrete is caused by the contraction of the hydrated hardened calcium silicate gel during the withdrawal of water from the concrete. Even for hardened concrete, loss of water (e.g. during a hot season) can lead to drying shrinkage. The magnitude of the drying shrinkage is influenced by many factors, including:

1 the stiffness and amount of aggregate;
2 water to binder ratio;
3 the total amount of paste;
4 the types and amounts of chemical admixtures;
5 the curing regime and the age of the concrete member at which it is exposed to air;
6 the types and amounts of mineral additives;
7 the theoretical length of the concrete member that is defined as the ratio of section area of the member to its semi-perimeter in contact with the atmosphere;
8 the diameter, amount and distribution of reinforcing steel;
9 the relative humidity and its change rate;
10 the carbonation.

Among various factors, the most important ones affecting drying shrinkage are water to binder ratio and the total amount of the paste that determines the total amount of water contained in a unit volume of the concrete. Concrete with a wetter consistency will shrink more than one with a drier consistency because the former is obtained by the use of a higher water/cement ratio, by a greater quantity of paste, or by a combination of the two. The loss of moisture from the concrete varies with distance from the surface. Drying occurs most rapidly near the surface because of the short distance the water must travel to escape and more slowly from the interior of the concrete because of the increased distance from the surface. A nearly linear relationship exists between the magnitude of the shrinkage and water content of the mix for a particular value of relative humidity. Also, the shrinkage of concrete decreases as the relative humidity increases. When concrete is exposed to 100 percent relative humidity or submerged in water, it will actually increase in volume slightly as no water immigrates out. The gel continues to form because of the ideal conditions for hydration. If concrete is exposed to relative humidity less than 90 percent, it will normally shrink.

Drying shrinkage strain, ε_{sh}, is time dependent. Approximately 90 percent of the ultimate shrinkage occurs during the first year. The magnitude of the ultimate shrinkage is primarily a function of the initial water content of the concrete and the relative humidity of the surrounding environment. For plain concrete members, drying shrinkage ranges from 400 to 700 micro-strain under normal conditions. For reinforced concrete members, the shrinkage strain values are between 200 and 300 micro-strain. Because concrete adjacent to the surface of a member dries more rapidly than that in interior, shrinkage strains are initially larger near the surface than in the interior. As a result of the differential shrinkage, a set of internal self-balancing forces, i.e. compression in the interior and tension on the outside, is built up. The stresses induced by shrinkage can be explained by imagining that the cylindrical core of a concrete cylinder is separated from its outer shell and that the two sections are then free to shrink independently in proportion to their existing water content.

Since deformations must be compatible at the junction between the core and the shell, shear stresses must be created between the core and the shell. If free-body diagrams of the upper half of the cylinder are considered, it is clear that vertical equilibrium requires the shear stresses to induce compression in the core and tension in the shell. In addition to the self-balancing stresses set up by different shrinkages, the overall shrinkage creates stresses if members are restrained in the direction in which shrinkage occurs. Tensile cracking due to shrinkage will take place in any structural element restrained by its boundaries. Drying shrinkage cracks are often straight or ragged. Surface crazing on walls and slabs is an example of drying shrinkage and it is usually shallow and cosmetic in nature. Non-uniform environmental conditions produce moisture gradients which may cause differential shrinkage with resulting warping or cracking depending upon the degree of restraint experienced by the concrete – for example, curling of slabs on grade is caused by drying shrinkage and by moisture and temperature gradients.

Shrinkage must be controlled since it permits the passage of water and other chemicals, is detrimental to appearance, reduces shear strength, and exposes the reinforcement to the atmosphere. For large concrete surface, joints need to be provided to prevent such cracking. Alternatively, shrinkage compensation concrete can be used to reduce drying shrinkage. The cement, used for making shrinkage compensation concrete, contains significant amounts of calcium sulfate and calcium sulfoaluminate. On hydrating, the reaction products occupy a larger volume than the original reactants. The volume of concrete keeps increasing during the first few days. When water loss occurs later, shrinkage occurs which reduces the volume to roughly the original value. Precautions in its application include: (1) expansive reaction stiffens up the concrete rapidly; (2) at high temperatures, above 27–29 °C, slump loss and quick setting may be a problem, so ice is normally needed to reduce the temperature; (3) quick stiffening makes plastic cracking easy and extra care is required to prevent rapid evaporation.

2.1.5 Influence of creep

Creep is a continuous deformation which occurs in concrete under sustained load, and its consequences are only evident after years. Creep can thus be defined as the increase in deformation under a sustained load. The deformation increased by creep can be several times as large as the deformation under loading. Creep is thus of considerable importance in structures. Inadequate design which fails to consider the influence of creep of the structural elements of a building, for instance, shortening of columns or deflection of floors and beams, may result in load being transferred to non-structural elements such as partition walls or cladding panels. In prestressed concrete, in flat slabs and in slender members liable to instability and buckling, creep may be harmful and the advantages of low creep concrete should be considered by the designer in these circumstances.

When a sustained load is removed from a concrete member, the strain decreases immediately by an amount equal to the elastic strain at the given age, generally lower than the elastic strain on loading. This instantaneous recovery is followed by a gradual decrease in strain, called creep recovery, but the recovery of creep is not complete. Creep is not a simply reversible phenomenon, so that any sustained load results in a residual deformation. In concrete, only the hydrated cement paste undergoes creep while the aggregate is relatively hard and considered creep-free. In fact, due to its relatively high stiffness, aggregates can restrain the creep of cement paste. Therefore, creep can be considered as a nonlinear function of the volumetric content of cement paste in concrete. The volumetric content of unhydrated cement paste, the volumetric content of aggregate, the grading, maximum size, shape and texture of the aggregate and certain physical properties of aggregate will affect the magnitude of creep. Among various physical properties of concrete, the modulus of elasticity of aggregate is probably the most important factor which influences creep of concrete and the influence of other aggregate characteristics may be indirect. The higher the modulus of the aggregates, the greater the restraint provided by the aggregate to the potential creep of the hydrated cement paste. Normally aggregates with a higher porosity have a lower modulus of elasticity, thus a lower restraint to creep of concrete.

Also, the porosity of aggregate plays a direct role in the transfer of the moisture within concrete, which is closely associated with creep of concrete.

As a result, some lightweight aggregates batched in a dry condition exhibit high initial creep. What's more, the rate of creep of lightweight aggregate concrete decreases with time more slowly than in normal weight concrete. The type of cement affects the creep of concrete since cement influences the strength of the concrete at the time of application of the load. Experimental data show that within a wide range of concrete strengths, creep is inversely proportional to the strength of concrete at the time of application of load (Mehta and Monteiro 2006). Fineness of cement affects the rate of hydration and the strength development at early ages and thus influences creep. Extremely fine cements, with a specific surface up to 740 kg/m^2, lead to a higher early creep but to a lower creep after one or two years under load (Bennett and Loat 1970). Strength increases in order of: low heat, ordinary, and rapid-hardening cements, so that for a constant applied stress at a fixed (early) age, creep increases in order of rapid-hardening, ordinary, and low heat cements. Creep of concrete made with expansive cement is larger than that of concrete with Portland cement only.

The mechanism of creep in concrete can be related to thermally activated creep. It assumes that the time-dependent strains are the result of thermally activated processes that can be described by rate process theory. Creep strains will originate through deformation of a micro-volume of paste, called a "creep center". The creep center will undergo deformation to a lower energy configuration under the influence of energy added to the system by external sources. This deformation can only occur by going through an energy barrier in the form of an intermediate, high-energy state. The most prevalent view involves slip between adjacent particles of C–S–H under a shear stress. If there is a sufficient amount of water between layered C–S–H, which reduces the van der Waals' forces sufficiently, slippage is ready to occur and hence the creep. Creep can also result from the diffusion of micro water under stress (Mindess *et al.* 2003). Water-reducing and set-retarding admixtures lead to pore refinement in the hydration product and have been found to increase the basic creep in many, but not all cases (Hope *et al.* 1967; Jessop *et al.* 1967).

The environmental humidity of concrete is also an important factor influencing creep. An increase in the atmospheric humidity is expected to slow down the relative rate of moisture flow from the interior to the outer surfaces of concrete. Taking a broad view, for a given concrete, the lower the relative humidity, the higher the creep. The strength of concrete has a significant influence on creep: within a wide range, creep is inversely proportional to the strength of concrete at the time of application of the load. The size and the shape of a concrete element also have some influence on the magnitude of creep since the rate of water loss is obviously controlled by the length of the path travel by the water and the resistance of water escaping from the interior of the concrete is closely related to creep.

Other factors influencing creep include the curing history of concrete, the temperature of exposure and the applied stress. Creep strains can be significantly different when concrete elements are cured in different histories. For instance, drying cycles can enhance micro-cracking in the transition zone and thus increase creep. The exposure temperature of concrete can have two counteracting effects on creep. On the one hand, if a concrete member is exposed to a higher than normal temperature as part of the curing process before it is loaded, the strength will increase and the creep strain would be less than that of a corresponding concrete stored at a lower temperature. On the other hand, exposure to high temperature during the period under load can increase creep. As far as the applied stress is concerned, there is a direct proportionality between creep and the applied stress for hardened concrete. It appears safe to conclude that, within the range of stresses in structures in service, the proportionality between creep and stress holds good, and creep expressions assume this to be the case. Also, creep recovery is also proportional to the stress previously applied (Yue and Taerwe 1992).

Creep is usually determined by measuring the change with time in the strain of a specimen subjected to a constant stress and stored under appropriate conditions. ASTM C 512-94 describes a spring-loaded frame to measure the creep of a concrete sample which maintains a constant load on a concrete test cylinder despite any change in its length. Since creep may develop over as long as 30 years, the measurement can only cover a short period of the age of concrete. Numerous mathematical expressions relating creep and time have been suggested, among which includes the modified Ross expression of ACI 209R-92 (1994a) and those suggested by Bazant and his co-workers (1992).

Creep affects strain, deflection and, often, stress distribution. In concrete structures, creep reduces internal stresses due to non-uniform shrinkage, so that there is a reduction in cracking (Neville *et al.* 1983). In reinforced concrete columns, creep results in a gradual transfer of load from the concrete to the reinforcement. In an eccentrically loaded column, creep increases the deflection and can lead to buckling. In statically indeterminate structures, creep may relieve stress concentrations induced by shrinkage, temperature changes, and so on. However, in mass concrete, creep may cause cracking when a restrained concrete mass undergoes cyclic temperature changes. Creep can also lead to an excessive deflection of structural members and cause other serviceability problems, especially in high-rise buildings and long bridges.

2.2 Degradation caused by repeated loading

Components of reinforced concrete structures such as machine foundations and bridges are frequently subjected to repeated loading (cyclic loads), and the resulting cyclic stresses can lead to microscopic physical damage in the materials. The damage can accumulate and further lead to the strength reduction and then structural degradation. The trend of strength reduction

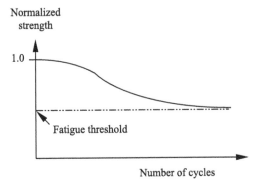

Figure 2.4 Normalized fatigue strength as a function of number of loading cycles.

can be seen from the so-called S–N curve as shown in Figure 2.4. The cyclic fatigue can be defined as a failure caused by a sufficient repeated application of loads that are not large enough to cause failure in a static application. This implies that some internal progressive permanent structural damage must be accumulated in the reinforced concrete structure under the repeated stress.

A typical cyclic loading is shown in Figure 2.5. Some useful definitions and basic concepts for cyclic loadings can be introduced, referring to Figure 2.5:

Constant amplitude stressing – cycling between maximum and minimum stress levels that are constant (see Figure 2.5(a));
Stress range, $\Delta\sigma$, is the difference between the maximum and the minimum values;

$$\Delta\sigma = \sigma_{max} - \sigma_{min} \tag{2.1}$$

Mean stress, σ_m, is the average of the maximum and minimum values;

$$\sigma_m = \frac{\sigma_{max} + \sigma_{min}}{2} \tag{2.2}$$

Stress amplitude, σ_a, is the half of stress range;

$$\sigma_a = \frac{\Delta\sigma}{2} = \frac{\sigma_{max} - \sigma_{min}}{2} \tag{2.3}$$

Completely reversed stressing – mean stress equal to zero with constant amplitude;
Stress ratio, R, is $\sigma_{min}/\sigma_{max}$;
Amplitude ratio, A, is σ_a/σ_m

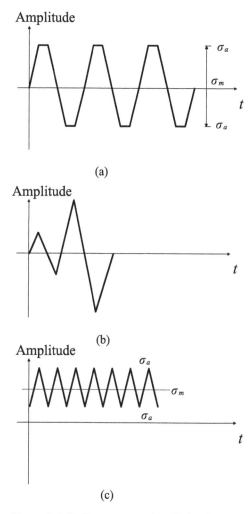

Figure 2.5 Different types of cyclic loadings.

In cyclic fatigue, the symbol "S" is usually used to represent nominal or average stress, which is different from the true stress at a point, σ. Nominal stress distribution is determined from the load or the moment using formulae in mechanics of materials as a matter of convenience while true stress is determined according to the real materials state (stress concentration, yielding). S is only equal to σ in certain situations. The fatigue strength of a material is largely influenced by the maximum stress applied, the difference between maximum and minimum stress (stress range), and the number of cyclic loading. It should be noted that for one cycle of loading and unloading, the concrete stress–strain curve is a closed cycle (see Figure 2.6). The area enclosed is proportional to hysteresis and represents the irreversible

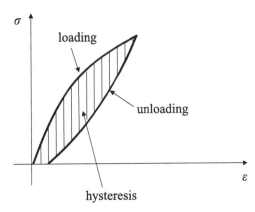

Figure 2.6 Hysteresis loop during loading and unloading.

energy of deformation, i.e. energy due to crack formation or irreversible creep.

The fatigue life of a material is usually plotted as a figure of nominal stress versus cyclic number, S–N diagram. To get an S–N diagram, fatigue tests have to be conducted. Each test deals with a fixed stress amplitude and mean stress. The test continues until the specimen fails at a cyclic number of N. Each experiment result will generate one point on the S–N diagram. The S–N diagram should have sufficient data points to make the empirical analysis meaningful. To make things simple, usually a completely reversed stressing, mean stress equalling to zero with constant amplitude, is adopted first to build up the S–N diagram. For the cases of mean stress not equalling zero, fatigue life can be estimated by using the S–N diagram of completely reversed stressing as stated in the following section.

When sufficient experimental data are obtained from completely reversed stressing fatigue test, the S–N diagram can be plotted in a linear-linear coordinates, or linear-log coordinates, or log-log coordinates. If S–N data are found to be a straight line on a log-log plot, the relationship between stress amplitude and fatigue cycles can be written as:

$$\sigma_{ar} = A(N_f)^B \tag{2.4}$$

where σ_{ar} is the stress amplitude for completely reversed stressing corresponding to N_f. The A and B are material constants; and N_f is the cycle to failure.

For the cases where $\sigma_m \neq 0$, the relationship between the stress amplitude where $\sigma_m \neq 0$ and the stress amplitude for completely reversed stressing can be expressed by the empirical modified Goodman law,

$$\sigma_a = \sigma_{ar}\left(1 - \frac{\sigma_m}{\sigma_\mu}\right) \tag{2.5}$$

where σ_a is the stress amplitude that $\sigma_m \neq 0$ for a given fatigue life, σ_{ar} the stress amplitude of completely reversed stressing at fatigue failure (N_f), σ_m the mean stress, and σ_μ is the static strength of the material. This equation provides the base for estimating the fatigue life for a case that $\sigma_m \neq 0$ by utilizing the *S–N* diagram for completely reversed stressing.

Goodman's law can also be given in the form of stress range. To keep the same number of cycles to failure:

$$\Delta\sigma = \Delta\sigma_r(1 - \sigma_m/\sigma_u) \tag{2.6}$$

where $\Delta\sigma$ is the stress range when $\sigma_m \neq 0$; and $\Delta\sigma_r$ is the stress range when mean stress $= 0$.

The fatigue life of any (σ_m, σ_a) combination can be estimated from the following procedures. First, substitute Eq. (2.4) into (2.5), we get:

$$\sigma_a = \left(1 - \frac{\sigma_m}{\sigma_\mu}\right) A N_f^B \tag{2.7}$$

It can be seen that this equation reduces to $\sigma_a = A(N_f)^B$ as it should, if $\sigma_m = 0$. On a log–log plot, Eq. (2.7) produces a family of *S–N* curves for different values of mean stress, which are all parallel straight lines. In general, let the *S–N* curve for completely reversed loading be

$$\sigma_a = \left(1 - \frac{\sigma_m}{\sigma_\mu}\right) f(N_f) \tag{2.8}$$

which is the corresponding family of *S–N* curves. From Eq. (2.7), we can get:

$$N_f = B\sqrt{\frac{\sigma_a}{A(\sigma_\mu - \sigma_m)}} \tag{2.9}$$

If more than one amplitude or mean level occurs in a fatigue test, fatigue life may be estimated by summing the cycle ratios, called the Palmgren-Miner Rule.

$$\Sigma(N_i/N_{fi}) = 1 \tag{2.10}$$

where N_i is the number of applied cycles under $\Delta\sigma_i$ or σ_{ai} and N_{fi} the number of cycles to failure under $\Delta\sigma_i$ or σ_{ai}.

Often, a sequence of variable amplitude loading is repeated a number of times. Under these circumstances, it is convenient to sum cycle ratios over one repetition of the history, and then multiply this by the number of repetitions required for the summation to reach unity.

$$B_f\left[\sum {}^{N_i}\!/_{N_{fi}}\right]_{\text{one repetition}} = 1 \tag{2.11}$$

where B_f is the number of repetitions to failure.

Another approach for fatigue life prediction involves fracture mechanics concepts. Consider a growing crack that increases its length by an amount Δa due to the application of a number of cycles ΔN. The rate of growth with cycles can be characterized by da/dN.

Assume that the applied loading is cyclical with constant values of the loads P_{max} and P_{min}, hence also with constant values of the nominal stresses S_{max} and S_{min}. For fatigue crack growth, it is conventional to use the nominal stresses that are generally defined based on gross area to avoid the change of stress values with crack length.

The primary variable affecting the growth rate of a crack is the range of the stress intensity factor. This is calculated using the stress range ΔS:

$$\Delta K = F\Delta S\sqrt{\pi a} \tag{2.12}$$

The value of F depends only on the geometry and the relative crack length, $a = a/b$ just as if the loading was not cyclic. Since K and S are proportional for a given crack length, the maximum, the minimum, and the range for K during a loading cycle are given by

$$K_{max} = FS_{max}\sqrt{\pi a} \tag{2.13}$$

$$K_{min} = FS_{min}\sqrt{\pi a} \tag{2.14}$$

$$\Delta K = K_{max} - K_{min} \tag{2.15}$$

For a given material and set of test conditions, the crack growth behavior can be described by the relationship between cyclic crack growth rate da/dN and stress intensity range K. The empirical fitting curve suggests that the following relationship can be used,

$$\frac{da}{dN} = C(\Delta K)^m \tag{2.16}$$

where C and m are curve fitted constant (from log–log plot).

It should be indicated that Eq. (2.16) is obtained empirically and valid for an intermediate crack growing rate or ΔK range. At low growth rates, the curve generally becomes steep and appears to approach a vertical asymptote denoted K_{th}, which is called the fatigue crack growth threshold. This quantity is interpreted as a lower limiting value of K, below which crack growth does not ordinarily occur. At high growth rates, the curve may again become steep. This is due to rapid unstable crack growth just prior to the final failure of the test specimen.

The number of cycles to failure during a fatigue test can be calculated using the following equation:

$$N_f = \int_{a_i}^{a_f} \frac{da}{C[\Delta K]^m}$$

(2.17)

where a_i = initial crack size obtained from inspection (if no crack is found, take a_i as the crack detection threshold); and a_f = final crack size obtained from $K_{max}(a_f) = K_c$. By substituting the parameter expressions in Eq. (2.17), we can obtain:

$$N_{if} = \left[\frac{1 - \left(\frac{a_i}{a_f}\right)^{\left(\frac{m}{2}-1\right)}}{C(F\Delta S\sqrt{\pi})^m \left(\frac{m}{2} - 1\right)} \right] \frac{1}{a_i^{\left(\frac{m}{2}-1\right)}}$$

(2.18)

where N_{if} is the cyclic number for material fail in fatigue from crack growing from a_i to a_f; F is a geometrical function that depends on loading pattern and ratio of crack size to specimen size.

2.3 Degradation caused by lack of durability

Almost universally, concrete has been specified principally on the basis of its compressive strength at 28 days after casting. In many cases, the specified strength can be achieved with a high water/cement ratio and hence permeable concrete. R.C structures, on the other hand, are almost always designed with a sufficiently high safety factor. Thus, it is rare for concrete structures to fail due to lack of intrinsic strength. However, gradual deterioration caused by the lack of durability makes reinforced concrete infrastructures fail to survive their specified service lives in ever increasing numbers. The extent of the problem is such that concrete durability has been described as a "multi-million dollar opportunity" (Anonymous 1988). In a report by the National Materials Advisory Board of the National Research Council, USA, it suggests that there is a lack of proper application of the durability knowledge-base by practitioners, as evidenced by the continuing problem of inadequate durability of concrete in service. The report also notes that the magnitude of durability problem is great enough to merit national action, and seeks reasons for the lack of proper implementation of available data by engineers.

It finds that available concrete technology is inadequately used because of the fragmentation of the industry, confused responsibility for the training and skill of the workers, lack of financial or other incentives for the industry, and poor management and dissemination of available technical information. It further finds that two reasons may contribute to the poor durability of national infrastructure. The first is that the contractors, who can most directly provide it, do not make durability their responsibility. The second is a product of economic management and tax structure. Most developers have no intention of maintaining the ownership of new buildings for the life of those buildings because it is advantageous to sell them within a few years of construction. The initial owner has therefore little incentive to spend money to ensure continued durability when the benefits will not be realized during the period of his ownership. The report argues that for many construction projects neither the owner nor the contractor is motivated to ensure long-term durability of the structures, yet there is potentially a severe financial loss to the community. Restoration of durability is by far the most common repair work of concrete structures.

According to ACI Committee 201, the durability of Portland cement concrete is defined as its ability to resist weathering action, chemical attack, abrasion, or any other process of deterioration to keep its original form, quality and serviceability when exposed to its intended service environment (Mehta and Monteiro 2006). Durability is most likely to relate to long-term serviceability of concrete and concrete structures. Serviceability refers to the capability of the structure to perform the functions for which it has been designed and constructed with exposure to a specific environment. The structure should be able to resist or withstand, during its service life, all the intended loads and environmental conditions without excessive deterioration, wear or failure. Thus, durability in the broadest sense will depend on the nature of the concrete and the aggressiveness of the in-service environment. Durability of concrete is very dependent on the proportioning of the concrete mix. Good concrete quality and adequate reinforcement cover are important factors for the durability of concrete structures. Permeability is another key factor in durable concrete and concrete structures. Concrete permeability is governed by the water/cement ratio, the cement content, the curing and the degree of compaction. Over-batching of water or under-batching of cement are construction errors which can lead to problems early on, particularly in aggressive environments, because they cause more pores in the matrix and it is thus more permeable.

The requirement for the reinforcement cover is usually specified with respect to exposure conditions that the concrete is supposed to serve. Normally design codes have reasonable specifications for the minimal cover thickness, which is important to prevent corrosion of reinforcement and ensure durability of concrete structures. It is now known that, for many conditions of exposure of concrete structures, both strength and durability have to be considered explicitly at the design stage. Therefore, the provision of durability consists mainly of a prescription of maintaining the required margins and factors with time (Keyser 1980). There are four main methods that ensure adequate durability of concrete in service as follows:

1 compliance with current standards of good practice during construction;

2 the use of new and improved materials and innovative construction systems designed for increased durability at competitive cost;

3 provision of protection to existing undamaged structures against adverse environments;

4 use of materials and procedures that incorporate the best available standards of good practice in repair, replacement, and subsequent protection of already damaged structures.

2.3.1 Causes of deterioration and main durability problems

The causes of degradation of Portland cement concrete can be classified as different groups of factors that can affect the performance of a building material, component, or system. For the sake of clarity, the classification of causes of concrete deterioration can be roughly grouped into three categories in this book: (1) physical causes; (2) chemical causes; and (3) mechanical causes. *Physical causes* of deterioration and durability problems include the effects of high temperature or of the differences in thermal expansion of aggregate and of the hardened cement paste, such as alternating freezing-thawing cycle and the associated action of de-icing salts, surface wear or loss of mass due to abrasion, erosion, and cavitation, and cracking which appear widely due to volume changes, normal temperature and humidity gradient, crystallization of salts in pores, structural loading, restrained shrinkage, and exposure to fire. The most common *chemical causes* of problems with concrete durability are: (1) hydrolysis of the cement paste component; (2) carbonation; (3) action-exchange reaction; and/or (4) reaction leading to expansion (such as sulfate expansion, alkali–aggregate expansion, and steel corrosion) (Mehta and Gerwick 1982; Mehta and Monteiro 2006).

Chemical degradation is usually the result of an attack, either internal or external, on the cement matrix. Portland cement is alkaline, so it will react with acids in the presence of moisture and, in consequence, the matrix may become weakened and its constituents may be leached out. *Mechanical causes* include impact and overloading. In reality, major durability problems of concrete structures include corrosion of the reinforcing steel, freeze/thaw damage, salt scaling, alkali–aggregate reactions, and sulfate attack. The common result of these attacks is that all of them can result in cracking and spalling of the concrete. It should be noted that the physical and chemical processes of deterioration usually act in a synergistic manner and the deterioration and durability problem of concrete is rarely due to one isolated cause.

2.3.2 Basic factors influencing durability

Concrete is a permeable and a porous material. The capillary pore structure allows water under pressure to pass slowly or ions or moisture under concentration gradient to migrate gradually through concrete. The property that governs the rate of flow of a fluid into a porous solid under pressure is defined as permeability. The property that governs the rate of migration of ions in a porous solid under concentration gradient is defined as diffusivity. The durability of concrete depends, to a large extent, on permeability and diffusivity. In fact, with the exception of mechanical damage, all the adverse influences on durability involve the transport of fluids through the concrete, which is closely related to concrete permeability and diffusivity.

From a microstructure point of view, permeability is greatly affected by the nature of the pores, both their size and the extent to which they are interconnected. The average size of capillary pores in concrete is about 0.1 μm while the gel pores are very much smaller. In general, concrete permeability is affected by: (1) the quality of cement and aggregate; (2) the quality and quantity of the cement paste, including the amount of cement in the mixture, water/cement ratio and degree of hydration of cement; (3) the bond developed between cement paste and aggregate; (4) the effectiveness of compaction of the concrete; (5) the extent of curing; (6) the presence or absence of cracks; and (7) the characteristics of admixtures used in the concrete. Normally aggregates also contain pores, but these are usually discontinuous. Moreover, aggregate particles are enveloped by the cement paste so that the pores in the aggregate do not contribute to the permeability of concrete (Neville 1996). For steady-state, the coefficient of permeability (K_1) can be determined by Darcy's law that describes a volume flow rate of water under the equilibrium flow condition

$$\frac{dq}{dt} = K_1 \frac{HA}{L\mu} \tag{2.19}$$

where dq/dt is the volume flow rate (m^3 s^{-1}); H is the pressure head loss across the thickness of the medium (m); A is the surface area of the medium (concrete) normal to the direction of flow (m^2); L is the thickness of the medium (m); μ is viscosity of the fluid; and K_1 is the coefficient of permeability depending on the properties of the medium and of the fluid, normally water in the case of concrete, (m s^{-1}).

Obviously the permeability is a function of the pores inside materials. This includes two concepts, percentage of porosity and size distribution of pores. In comparison, porosity is a measure of the proportion of the total volume of concrete occupied by pores and it determines the quantity of liquid or gas that can be contained by a concrete. The size distribution of the pores decides the inter-connecting properties of pores. If the pores are inter-connected, the transport of fluids through concrete is easier, leading to high permeability. On the other hand, if the pores are discontinuous or ineffective with respect to fluid transport, the permeability of the concrete is low, even if its porosity is high. The permeability of naturally hardened paste, that had never been allowed to dry, was found by Powers *et al.* (1954) in a range from 0.001×10^{-12} to 1.20×10^{-12} m s^{-1} for water/cement ratios ranging from 0.3 to 0.7. Drying was found to increase the permeability. As cement hydration proceeded, the permeability was significantly reduced and in 24 days was found to be only one-millionth of its initial value. Table 2.2, after Powers *et al.* (1954), shows this effect. For cement containing higher proportions of C_3S, the time needed to produce complete closure of the capillary pores can be less, leading to lower permeability at an early age. Since water flows more easily through the capillary pores than through the much smaller gel pores, the permeability of the cement paste as a whole is 20–100 times greater than that of the gel itself (Neville 1996).

Different from permeation that refers to flow under pressure differential, diffusion is the process in which a medium moves under a differential in concentration and the relevant property of concrete is referred to as diffusivity. In other words, diffusivity is the rate of moisture migration at the equilibrium diffusion condition and it is defined by Fick's law. Fick's first law is:

$$J = -D \frac{dc}{dx} \tag{2.20}$$

where J is the diffusion flux or mass transport rate per unit area per unit time of a medium in kg/m^{2}s. C is concentration in kg/m^3, D is diffusion coefficient in m^2/s, x coordinate or thickness of the sample in diffusion direction, and dC/dx is concentration gradient in kg/m^4. Even though diffusion takes place only through the pores, the values of J and D refer to the cross-section of the concrete sample; thus, D is actually the effective diffusion coefficient.

Fick's second law is:

$$\frac{\partial C}{\partial t} = D \frac{\partial^2 C}{\partial x^2} \tag{2.21}$$

Table 2.2 Reduction of permeability of cement paste (water/cement = 0.7) with the progress of hydration

Age (days)	Permeability coefficient ($\times 10^{-12}$ ms^{-1})
Fresh	2,000,000
5	400
6	100
8	40
13	5
24	1
Ultimate	0.6

Source: After Powers *et al.* (1954).

The difference between the two parameters, K_1 and D, is that permeability is the parameter characterizing water flow when the pores inside the concrete are filled with water, while diffusivity is the parameter describing the diffusion of water vapor before saturation is reached in the pores. K is a function of pressure difference and D is concentration difference. If one of the parameters is known, the other can be deduced indirectly. As far as the diffusion of gases in concrete is concerned, carbon dioxide and oxygen are of primary importance: the former leads to carbonation of hydrated cement paste, and the latter makes possible the progress of corrosion of embedded reinforcement steel. Like permeability, diffusion is low at lower water/cement ratios, but the influence of the water/cement ratio on diffusion is much smaller than on permeability.

2.3.2.1 Permeability test

Permeability tests measure the rate at which a liquid or gas passes right through the test specimen under an applied pressure head. Concrete is a kind of porous material which allows water under pressure to pass slowly through the concrete, but the rate of flow through dense, good quality concrete is extremely difficult. Till now, testing the permeability of concrete has not been generally standardized (Ludirdja *et al.* 1989). There are two common practices to evaluate the permeability of concrete using water as permeation liquid: the steady flow method and the depth of penetration method. The steady flow method is performed on a saturated specimen and applies a pressure head to one end of the sample. When a steady flow condition is reached, the measurement of the outflow enables the determination of the coefficient of permeability by using Darcy's law:

$$k_1 = \frac{\frac{dq}{dt} L}{HA} \tag{2.22}$$

where:

k_1 = coefficient of permeability (m/sec)
dq/dt = rate of the steady flow (m³/sec)
L = thickness or length of the specimen (m)
H = drop in hydraulic head across the sample (m)
A = cross-sectional area of the sample (m²).

To evaluate the coefficient of permeability by the steady flow method, water should be absorbed into all pores of the sample so that the pore surfaces do not provide friction nor capillary attraction to the passage of water. However, such flow conditions cannot always be achieved in many low permeability concretes and are not representative of an actual working environment. Moreover, the considerable length of time required to test the concretes, and the difficulties of attaining a steady state outflow, can be regarded as disadvantages for the steady flow method.

In cases of good quality concrete, there is no flow of water through the concrete, making the steady flow method not suitable for measuring the permeability of concrete. Under such conditions, the depth of penetration method is a good choice for measuring the permeability of concrete. In the penetration method, one end of the unsaturated concrete specimen is subjected to a pressure head while the other end is free in normal atmospheric conditions. The measure of water penetration is achieved either by measuring the volume of water entering the sample or by splitting the cylinder and measuring the average depth of discoloration, due to wetting, taken as equal to the depth of penetration. Provided the flow of water is uniaxial, the water penetration depth can be approximated to the coefficient of permeability equivalent to that used in Darcy's law as developed by Valenta (1969):

$$k = \frac{x^2 v}{2hT} \tag{2.23}$$

where:

x = depth of penetration of concrete (m)
v = the fraction of the volume of concrete occupied by pores
h = hydraulic head (m)
T = time under pressure (s).

The value of v represents discrete pores, such as air voids, which are not filled with water except under pressure, and can be calculated from the increase in the mass of concrete during the penetration test. Bearing in mind that only the voids in the part of the specimen penetrated by water would be considered, we can write:

$$v = \frac{\Delta W}{\rho A x} \tag{2.24}$$

in which ΔW is the gain in weight of the specimen during the penetration test, ρ is the density of concrete, A is the area, and x is the penetration depth.

Typically the effective porosity, v, lies between 0.02 and 0.06 (Vuorinen 1985). Based on the depth of penetration method, it is possible to use the depth of penetration of water as a qualitative assessment of concrete permeability: a penetration depth of less than 50 mm classifies the concrete as "impermeable"; a depth of less than 30 mm as "impermeable under aggressive conditions" (Neville 1996). It is now acknowledged that the steady flow method suits concretes with relatively higher permeability, while the depth of penetration method is most appropriate for concrete with very low permeability. It is important to note that the scatter of permeability test results on similar concrete at the same age, and using the same equipment, is large.

For any set of tests, the value of K_1 in Eq. (2.19) depends on both the medium and the fluid and, therefore, represents the permeability of the medium at a specified temperature.

2.3.2.2 Chloride diffusion test

There are basically two methods in evaluating the chloride diffusion in concrete. One method is called the rapid permeability test (ASTM C1556-04) and another is the diffusion cell test method (Li *et al.* 1999). The latter is considered a reliable test method for the chloride diffusivity of concrete due to the difference of concentration. In this method, the specimens of $\Phi 100 \times 20$ mm ($\Phi 3.94 \times 0.79$ in.) slice are placed between two chambers and the edges are sealed with an epoxy resin (see Figure 2.7). After the epoxy resin is cured, saturated calcium hydroxide solution is poured into chambers and the specimens are immersed in the solution for five days. This procedure is to avoid an anomalous effect due to sorption rather than diffusion of chloride ions. Then a NaCl solution with a concentration of 5 Mol is added into the Chamber A to start the chloride diffusion test. The chambers are maintained at $23 \pm 2°C$ and the concentration of the chloride diffused through the specimens in Chamber B is measured periodically. The testing process is as follows:

- Prepare the solution with a prepared concentration of 0.1 Mol, 0.01 Mol and 0.001 Mol sodium chloride in saturated calcium hydroxide in advance.
- To check the chamber with high concentration of sodium chloride, use the 0.1 Mol and 0.01 Mol solution for calibration.
- After calibrating the meter, withdraw 1 ml aliquot from the chamber tested.
- Dilute it into 100 ml with saturated calcium hydroxide solution and add 2 ml of sodium nitrite (5 Mol concentration) to the solution to increase the sensitivity of the instrument.
- Put the chloride ion selective electrode into the solution and record the data when the reading of the meter becomes stable.
- Add the appropriate amount of sodium chloride to maintain 5 Mol concentration.
- To test the chamber with a low concentration of sodium chloride, use 0.01 Mol and 0.001 Mol solution for calibration.

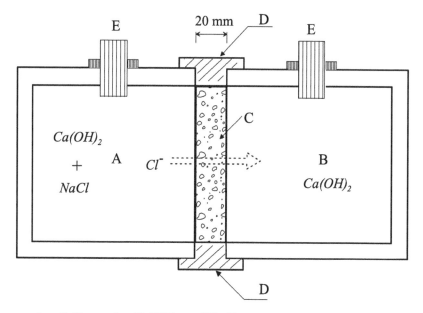

A – Cell contains Ca(OH)$_2$ and NaCl
B – Cell originally contains Ca(OH)$_2$ only
C – 20 mm thick concrete sample
D – Epoxy seal
E – Lid

Figure 2.7 Chloride diffusion test.

- Withdraw 10 ml aliquot from the solution in the chamber, and add 0.2 ml of sodium nitrite (5 Mol concentration) into the solution.
- Put the chloride ion selective electrode into the solution and record the reading when the meter becomes stable.
- Refill the chamber with 10 ml solution of appropriately prepared concentration.

A typical curve of chloride concentration in Chamber B obtained from the experiment has a strong nonlinearity between chloride concentration and time at beginning (see Figure 2.8). However, when the test time exceeds about 30 weeks, the curve becomes quite linear. This implies that the chloride diffusion reaches a steady state. The linear relation between concentration of chloride ions and time can be expressed as:

$$C = kt - A \tag{2.25}$$

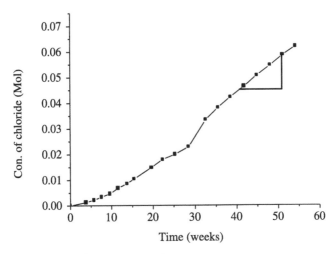

Figure 2.8 Diffusion test results.

where C is the cumulative concentration of chloride ion diffused into Chamber B at time t and k is the slope of chloride concentration-time curve in steady state. A is a constant.

On the other hand, in steady state, the flux of chloride ions through the specimens can be described by Fick's first law of diffusion. For a case of steady state, this law can be written in the following form:

$$C = \frac{DC_A}{l^2} t - A \tag{2.26}$$

where D is the diffusion coefficient in steady state; C_A is the concentration of chloride ion in Chamber A; and l is the thickness of the slice specimen.

Comparing the above two equations, the diffusion coefficient can be expressed as follows:

$$D = \frac{kl^2}{C_A} \tag{2.27}$$

It is clear that once the slope of chloride concentration–time curve (flux) in the steady state is known, the diffusion coefficient can be calculated.

In some instances, additives are introduced into concrete to accelerate the curing. There are two reasons for this, first, to achieve a quicker turn round of molds or formwork, and, second, to allow concreting to take place in cold weather. A constituent of these additives is calcium chloride ($CaCl_2$). There is still some doubt as to the effect of calcium chloride on reinforced concrete, but evidence is emerging that it can have a disastrous effect by corroding the rebars. Chloride ions in concrete are not always distributed uniformly and their concentration has resulted in severe corrosion of rebars in structural members, especially where they are exposed to the weather.

2.3.3 Deterioration due to physical action

Progress loss of mass from a concrete surface can occur due to surface wear. Concrete surfaces can be produced to have a high degree of resistance to surface wear, including abrasion, erosion and cavitation. The resistance of a concrete surface is largely dependent on achieving a hard and durable surface which is flat and free from cracks. Abrasion of a concrete surface can be defined as the worn process by dry attrition, repeated rubbing, rolling, sliding, or frictional (attrition) processes (Mehta 1980). Surface abrasion is mainly caused by dry attrition. Pavements and industry floors are most likely susceptible to dry attrition, so are hydraulic structures. Two categories of abrasion are possible for concrete roads: mechanical abrasion (such as attrition, scraping and impact) due to heavy trucking and vehicles with chains or studded tires; and polishing of the surface due to traffic action resulting in loss of skid resistance. In hydraulic structures, the abrasive action results from the cutting action of suspended solids. Concrete abrasion resistance depends mainly on: (1) concrete strength; (2) cement content; (3) water/cement ratio; (4) aggregate properties; (5) finishing methods; (6) good compaction and curing; and (7) use of topping. The abrasion resistance of concrete varies in proportion to both the compressive strength and cement contents and inverse in proportion to the water/cement ratio regardless of aggregate quality (Mailvaganam 1992).

Concrete strength has a significant influence on the abrasion resistance of concrete surface. The finishing method applied to the surface is also important, but its influence on abrasion is difficult to be quantitatively evaluated. Aggregate property also plays an important role in concrete resistance to abrasion. Concrete containing harder coarse aggregate is normally more resistant to abrasion than that containing softer aggregate. To increase resistance to surface abrasion, the finishing procedure should be done to maintain as much coarse aggregate near the top surface as possible. Good abrasion resistance concrete, such as micro silica (MS) concrete and densified with small particles (DSP) concrete, may be used for concrete structures that are most liable to be subject to dry attrition. Under conditions of severe wear, aggregates featuring high resistance to abrasion should be selected. Commonly such materials include: (1) metallic types, such as pearlitic iron turnings and crushed cast iron chilled grit; and (2) non-metallic types, such as silicon carbide grains (carborundum); fused alumina grains (corundum); and natural emery grains (alumina and magnetite) (Campbell-Allen and Roper 1991). In cases where slip resistance is very important, or where rusting of the metals would be objectionable, the non-metallic types are preferred. Fiber reinforced concrete has also been demonstrated to be much superior in wear resistance to conventional concrete (Neville 1975).

Penetrating sealing and hardening treatments can significantly increase abrasion resistance (Sadegzadeh and Kettle 1988).

Erosion refers to concrete surface wear by the abrasive action of fluids with suspended solid particles. Erosion occurs in hydraulic structures, for instance, canal linings, spillways, piers and concrete pipes for water or sewage transport. When a fluid containing suspended solid particles is in contact with concrete, the impinging, sliding, or rolling action of the particles will cause surface wear. The rate of surface erosion depends on the porosity or the strength of concrete, and on the amount, size, shape, density, hardness, and velocity of the moving particles. Better resistance to erosion is achieved by using hard aggregate; MS concrete and DSP concrete; and concrete of strength greater than 41 MPa cured at least for 7 days. Concrete surface quality has a significant influence on resistance to surface wear. The properties of the concrete surface layer are determined largely by the quality of the concreting operations and the timing of these operations, particularly the timing of finishing and curing. To improve surface quality, it is suggested to do surface finishing right before the initial set and/or to add topping with low water/cement ratio and hard aggregates.

Hydraulic structures may be intensively subjected to physical erosion which arises from particles of rock, ice or debris carried in the flowing water; from the collapse of vapor bubbles formed by pressure changes in high-velocity flows; and from the fluctuations of water pressure in and on the concrete under conditions of unsteady flow (Campbell-Allen and Roper 1991). These vapor bubbles flow downstream and, on entering a zone of higher pressure, collapse with great impact. Repeated collapse of such cavities near the surface of concrete will cause pitting. Erosion of hydraulic structures easily leads to cavitation, which relates to loss of mass by the formation of vapor bubbles and their subsequent collapse due to a sudden change of direction in rapid flowing water. Cavitation erosion is readily recognized from the nature of the holes or pits formed, which are quite different from the smoother worn surface produced by abrasion from solids. In hydraulic structures, once cavitation damage has started, the roughened surface provides a source for new cavities to form and the damage can be extended far downstream. Advanced stages of the damage show an extremely rough honeycomb texture with some holes that penetrate deep into the concrete. The erosion of concrete by sand and gravel in water can be equally as severe as that caused by cavitation.

Erosion progresses rapidly after an initial roughening of the surface occurs. The material beneath the surface is more vulnerable to attack because of the tendency of the erosion to follow the mortar matrix and undermine the aggregate. Better resistance to erosion is achieved by using high strength concrete containing the maximum amount of hard coarse aggregate and/or very strong concretes having strengths over 100 MPa with high-range water-reducing admixtures and silica fume (Holland 1983).

Better resistance to cavitation in hydraulic structures is achieved by using fiber reinforced concrete to increase the lifetime of hydraulic structures.

For hydraulic structures, damage can also be initiated by chemical attack in addition to the strictly mechanical type of damage that results from wear and erosion. In such cases, things can be even worse. Mineral water with an acidic pH can cause the leaching of lime or can etch the surface of concrete. The weakened concrete and the roughened surfaces are then more vulnerable to the intense attrition caused by the forces of erosion. This kind of acid attack produces the disintegration of the concrete which is then eroded away.

Impact by high speed water can also damage concrete. Nonlinear flow at velocities exceeding 12 m/s may cause severe damage to concrete surface through cavitation. The high velocity water jet can cause cavitation to form on concrete surface with irregularities. The impact of a very high velocity jet of water can strike the concrete surface and erode the cement paste, resulting in the boring of a hole through the concrete. Useful measures to improve concrete cavitation resistance, caused by high speed water jet, can be the removal of surface misalignments or abrupt changes of slope.

2.3.4 Degradation caused by steel corrosion

It is thought that a high percentage of the defects and deterioration in concrete structures arises from the corrosion of the reinforcing steel. Originally, in a properly designed, constructed and used structure, there should be little problem of steel corrosion within the concrete during the design life of that structure because that concrete provides a very high alkali environment with Ph value of 12.5–13.5 in which steel is well passivated. Unfortunately, this highly desirable condition is not always found during the service period of concrete structures, resulting in that corrosion of reinforcement has become probably the most frequent cause of damage to reinforced concrete structures. The processes of the degradation of concrete and the corrosion of reinforcing steel are closely related. The former provokes destruction of the concrete cover and causes microcracking that compromises its protective characteristics.

On the other hand, corrosion attack, because of the expansive action of corrosion products, generates cracking or delamination of the concrete and reduces the bonding between the reinforcement and surrounding concrete. Based on these considerations, the service life of reinforced-concrete structures can be divided into two distinct phases: the initiation of corrosion and the propagation of corrosion (Bentur *et al.* 1997; Bazant 1979). During the initial phase, aggressive substances, such as carbon dioxide and chlorides, can penetrate from the surface into the bulk of the concrete and depassivate the steel. The duration of the initiation phase depends on the cover depth and the penetration rate of the aggressive agents as well as on the concentration necessary to depassivate the steel. Once the protective layer on reinforcing steel surface has been destroyed, in the presence of water and oxygen on the surface of the reinforcement, corrosion will occur and propagate. The corrosion rate can vary considerably depending on temperature and humidity. Steel corrosion can be a very severe durability problem and can be attributed to the following factors (Bentur *et al.* 1997):

1   In many cases, concretes are made with high water/cement ratio with their specified strength achieved that leads to permeable structure.
2   In practice, there is a strong need to optimize structures, which leads to minimized structural member sizes, and subsequently leads to a tendency to reduce concrete cover.
3   The service environment can become more severe compared with those the structure was originally supposed to sustain. For example, de-icing by chloride salts is commonly used to meet the needs of the heavier winter traffic in cold climates, leading to higher chloride ingress in bridge decks and parking garages and a higher probability of corrosion of steel induced by chloride.

The signs of corrosion of reinforcing steel can be identified as rust stains and minute cracking over the concrete surface. This can be attributed to the increase in volume associated with the formation of the corrosion products and leaching of the rust. If repairs are not undertaken at the early stage when these corrosion damages occur, the corrosion of the steel will proceed further, causing severe damage through forming a longitudinal crack in parallel to the underlying reinforcement, delamination and spalling of concrete cover, as well as exposure of the steel and reduction of its cross-section that may cause a safety hazard. The damage cost of corrosion of reinforcing steel can be huge. For example, a survey by the China Academy of Engineering in 2002 reported that the annual cost due to corrosion of reinforcing steel reached 100 billion Chinese Yuan in China. So it is desirable to know clearly the mechanism of corrosion of steel and figure out the possible methods for repairing the damage caused by corrosion of reinforcing steel.

There are basically two kinds of corrosion of reinforcing steel in concrete: carbonation-induced corrosion and chloride-induced corrosion. Generally speaking, chloride-induced corrosion is more serious than carbonation-induced corrosion.

2.3.4.1 Carbonation-induced corrosion

Carbonation-induced corrosion is caused by the general breakdown of passivity by neutralization of the concrete. Carbonation occurs as a result of penetration of carbon dioxide from the atmosphere. In the presence of moisture, this forms carbonic acid which neutralizes the alkalinity of the cement matrix. The reduction in alkalinity destroys the passive environment and leaves the reinforcement in a condition where it is susceptible to corrosion.

Carbonation-induced corrosion can take place on the whole surface of steel in contact with carbonated concrete.

The rate of carbonation increases with an increase in the concentration of carbon dioxide especially in concretes with a high water/cement ratio, carbon dioxide and moisture creating the carbonic acid agent. The rate of carbonation also depends on both environmental factors (humidity, temperature, concentration of carbon dioxide) and factors related to the concrete (mainly its alkalinity and permeability) (Bertolini *et al.* 2004). The rate of carbonation varies with the humidity of the concrete. As we know, in a totally dry or wet environment, there is no carbonation. The carbonation rate may be correlated to the humidity of the environment (ibid.). The interval of relative humidity most critical for promoting carbonation is from 60–70 percent. It has also been found that the microclimate plays an essential role on real structures. The carbonation rate may vary from one part of a structure to another or passing from the outer layers of the concrete to the inner ones, thus the carbonation of concrete can be very variable even in different parts of a single structure. As the concentration of carbon dioxide increases in the environment, the carbonation rate increases. It has been suggested that, based on experimental findings, the porosity of carbonated concrete, with a high concentration of carbon dioxide, is higher than that obtained by exposure to a natural atmosphere. If the other conditions are similar, temperature has the most important influence on carbonation rate.

In common with most chemical reactions, the carbonation reaction can be accelerated by increased temperature. The quality of the concrete is regarded as the most important parameter controlling the rate of carbonation (Bentur *et al.* 1997). The quality of the concrete is a function of the composition of the binder (i.e. whether Portland cement or blended cement was used), the water/cement ratio, the water/binder ratio, and the curing conditions. The permeability of concrete has a remarkable influence on the diffusion of carbon dioxide and thus on the carbonation rate. The penetration of carbonation slows down as the decrease in the water/cement ratio due to the decrease of the capillary porosity of the hydrated cement paste. The type of cement also influences the carbonation rate. For blended cement, hydration of pozzolanic materials or GGBS leads to a lower carbon hydroxyl content in the hardened cement paste which may increase the carbonation rate. Consequently, the depth of carbonation is greater for the blended cements than for the Portland cement concrete when compared on the basis of equal water/binder ratios (ibid.). The fly-ash blended cement concretes show higher corrosion rates compared with Portland cement concrete with the same water/binder ratio (ibid.). Nevertheless, the water/cement ratio and curing are the most important factors influencing the rate of carbonation. The denser structure of concrete may slow down the diffusion of carbon dioxide. If cured properly, the lower alkalinity of cements with the addition of fly ash or blast furnace slag can be compensated for by the lower permeability of their cement pastes.

The chemistry of carbonation is that carbon dioxide in the air in the presence of moisture reacts with hydrated cement minerals, carbon dioxide and moisture creating the carbonic acid agent. That is to say, carbon dioxide molecules that penetrate into the concrete can react with solid calcium hydroxide, with C–S–H gel, and with the alkali and calcium ions in the pore solution. Calcium hydroxide is the hydrate in the cement paste that reacts most readily with carbon dioxide. Consequently, the $Ca(OH)_2$ carbonates to $CaCO_3$. The reaction can be written schematically as:

$$CO_2 + Ca(OH)_2 == CaCO_3 + H_2O \hspace{3cm} (2.28)$$

This is the reaction of main interest, especially for concrete made of Portland cement, even though the carbonation of C–S–H is also possible when calcium hydroxide becomes depleted. Other cement compounds can be decomposed, namely hydrated silica alumina with ferric oxide being produced. Carbonation may cause the concrete to shrink. In the case of the concrete obtained with Portland cement, carbonation may even result in increased strength (Bertolini *et al.* 2004). Carbonation itself is not a problem, but it is the consequences of this on reinforced concrete which has major repercussions. Those chemical reactions during carbonation result in a drastic decrease in the alkalinity of the concrete, so that the alkalinity of the concrete is reduced from average values of 12–14 down to 8 or 9 with the consumption of calcium hydroxide. It is the high pH values which protect the steel reinforcement from corrosion. The reduction in alkalinity destroys the passive environment and leaves the reinforcement in a condition where it is susceptible to corrosion. Therefore, when carbonation penetrates through the rebar cover, and oxygen and moisture are present, corrosion will take place. The depth of carbonation in reinforced concrete is an important factor in the protection of the reinforcement; the deeper the carbonation, the greater the risk of corrosion of steel. The second consequence of carbonation is that chlorides bound in the form of calcium chloroaluminate hydrates and otherwise bound to hydrated phases may be liberated, making the pore solution even more aggressive (ibid.).

The corrosion rate is usually expressed as the penetration rate of carbonation and is measured in μm per year. The depth of penetration of carbonation into concrete is proportional to the square root of time. The corrosion rate can be considered negligible if it is below 2 μm/year, low between 2 and 5 μm/year, moderate between 5 and 10 μm/year, intermediate between 10 and 50 μm/year, high between 50 and 100 μm/year and very high for values above 100 μm/year (ibid.). In high-quality concrete, the rate of carbonation-induced corrosion is negligible for relative humidity below 80 percent. It is then assumed that corrosion propagates only while concrete is wet (i.e. R.H. > 80 percent). Corrosion rate tends to decrease with time. Besides, corrosion products can reduce corrosion rate (Alonso and Andrade 1994). Page (1992) has illustrated the relationship between the corrosion rate in carbonated concrete and relative humidity of the environment, which indicates that the maximum corrosion rates, on the order of 100–200 μm per year can only be reached in very wet environment with relative humidity approaching 100 percent. For typical conditions of atmospheric exposure, i.e. R.H. = 70–80 percent, maximum corrosion rates are between 5 and 50 μm per year.

2.3.4.2 *Chloride-induced corrosion*

The chloride-induced corrosion in structural concrete is primarily caused by the presence of sufficient free chloride ions in the matrix. Chloride can get into the concrete at the time of mixing, either as an admixture component or in chloride-contaminated aggregates or mix water, or penetrate the hardened concrete later on from external sources, such as sea water, slat spray, or de-icing salt placed on concrete pavements. Hence, the chloride-induced corrosion is caused by localized breakdown of passive film on the reinforcing steel. Chloride-induced corrosion is more serious for marine structures in coastal areas. In concretes of higher water/cement ratio, i.e. more porous concretes, the chloride ion penetration rate is substantially greater than that in dense concrete. The incorporation of certain mineral admixtures, such as silica fume and slag, can lower the rate of penetration to very low values in suitably dense concretes. When sufficient chlorides are present at the time of mixing, the corrosion may start at very early service stage. In the case of chloride in the atmosphere penetrating the concrete, the corrosion will not start until it reaches a certain level of accumulation of chlorides. The penetrated chloride ions diffuse through a concrete cover to the rebar surface first. Then sufficient quantities of chloride ions have to be accumulated. Next, when the concentration of chloride ion in concrete reaches a certain level (0.6–0.9 kg per cubic meter of concrete at pH value of 12.5–13.5), it dissolves the protective oxidized film, thus a localized breakdown of the passive film on the steel is formed where oxidation occurs and a galvanic cell is created. The local active area behaves as anodes, while the remaining passive areas become cathodes where reduction takes place.

The effect of the separation of anode and cathode has significant consequences for the pattern of corrosion. In the concrete adjacent to the anodic areas, the concentration of positive ions increases, causing the pH to fall and allowing soluble iron-chloride complexes to form. These complexes can diffuse away from the rebar, permitting corrosion to continue. Some distance from the electrode, where the pH and the concentration of dissolved oxygen are higher, the complexes break down, iron hydroxides precipitate, and the chloride is free to migrate back to the anode and react further with the steel. The process thus becomes autocatalytic and proceeds with the deepening of corrosion pits rather than spreading corrosion laterally along the rebar. As the steel increases its state of oxidation, the volume of the corrosion products expands. The unit volume of Fe can be doubled if

FeO forms. The unit volume of the final corrosion product, $Fe(OH)_3 \cdot 3H_2O$, can be as large as six and half times the original Fe. This expansion creates cracking and spalling inside concrete, and finally destroys the integrity of the structural concrete and causes a failure of buildings and infrastructures.

As a result, chloride-induced corrosion is localized, with penetrating attacks of limited area surrounded by non-corroded areas. The threshold concentration of chloride ion, i.e. the critical level of chloride ion concentration in concrete at which the surface protective layer of reinforcing steel generated in high alkalinity can be broken, is a function of the pH value, i.e. the hydroxyl ion concentration. Hausman (1967) has suggested on the basis of measurements in $Ca(OH)_2$ solutions that the threshold chloride ion concentration is about 0.6 times the hydroxyl ion concentration. Gouda (1970) has suggested another relationship in which the threshold chloride ion concentration for the pore solution of a given pH region is expressed in a logarithmic form. Both Hausman's and Gouda's data were derived from experiments in solution rather than in concrete, and other effects in concrete may influence the threshold value. In fact, most specifications and guidelines specify the total content of chloride in concrete in terms of percentage by weight of the original cement used. The permitted chloride contents in many specifications and recommendations are smaller than about 0.2 percent of the cement content of the concrete (Bentur *et al.* 1997). In the range of 0.2–0.4 percent, there is risk of inducing corrosion, but not always. Sometimes, the critical level of chloride is specified in terms of weight of chlorides per unit volume of concrete, mostly in the range of 0.6–1.2 kg/m^3. This kind of specification is needed when assessing the chlorides in existing structures, where the total chloride content of concrete can be determined experimentally, but not the chloride content in cement. The present specifications for chloride content recommended by the American Concrete Institute are given in Table 2.3. ACI and many other specifications even require that no calcium chloride-containing admixtures be used in the production of prestressed concrete.

Table 2.3 Limit on chloride ions (% wt. of cement)

1. Prestressed concrete	0.06
2. Conventionally reinforced concrete in a moist environment and exposed to external sources of chloride	0.10
3. Conventionally reinforced concrete in a moist environment but not exposed to external sources of chloride	0.15
4. Above-ground building construction where the construction will stay dry. Does not include locations where the concrete will be occasionally wetted	no limit[a]

Source: After ACI.

Note: [a] No limit for corrosion control, if calcium chloride is used as admixture, a limit of 0.2% is generally recommended for reasons other than corrosion.

Under the normal atmospheric environment, the chloride-induced corrosion rate can vary from several tens of μm per year (of steel) to localized values of 1 mm per year (of steel) as the relative humidity rises from 70 to 95 percent and the chloride content increases from 1 percent by mass of cement to higher values. High corrosion rates always appear on heavy chloride-containing structures such as bridge decks, retaining walls and pillars in sea water. The corrosion rate increases when the temperature changes from a lower one to a higher one. Once the corrosion attack begins in chloride-contaminated structures, it can lead in a relatively short time to an unacceptable reduction in the cross-section of the reinforcement, even under conditions of normal atmospheric exposure. The lower limits of relative humidity near which the chloride-induced corrosion rate becomes negligible are much lower than those that make carbonation-induced corrosion negligible (Bertolini *et al.* 2004). The influence of temperature and humidity on the corrosion rate is shown in the electrochemical reactions at the steel/concrete interface and through their influence on ion transport between anodes and cathodes. It was thought that the concrete resistivity was strongly related to the corrosion rate at moderate or low temperature (Alonso *et al.* 1988; Glass *et al.* 1991). In a given set of conditions in terms of humidity and temperature and provided corrosion has initiated, the higher resistivity of blended cements results in a lower corrosion rate than that of Portland cement.

2.3.4.3 Corrosion mechanisms

Corrosion of the reinforcing steel in concrete is an electrochemical process, consisting of both oxidation and reduction reactions, in which the metallic iron is converted to the voluminous corrosion product ferric oxide. The process is associated with the presence of anodic and cathodic areas and the potential difference between the two areas arising from lack of homogeneity in the surrounding liquid medium or even in the steel itself. The differences in potential are due to the inherent variation in structure and composition (e.g. porosity and the presence of a void under the rebar or difference in alkalinity due to carbonation) of the concrete cover, and differences in exposure conditions between adjacent parts of steel (e.g. concrete that is partly submerged in sea water and partly exposed in a tidal zone). In general, corrosion cells are formed due to: (1) contact between two dissimilar materials, such as steel rebar and aluminium conduit pipes; (2) significant variations in surface characteristics, including difference in compositions, residual strain due to local cold working, applied stress, etc.; or (3) different concentrations of alkalies, chloride, oxygen, etc.

Four components must be present for corrosion to occur in a macro cell including anode, cathode, electrolyte and metallic path. The anode is the electrode at which oxidation occurs. Oxidation involves the loss of electrons and formation of metal ions. Corrosion occurs at the anode. The cathode is the electrode where reduction occurs. Reduction is the gain of electrons in a chemical reaction. The electrolyte is a chemical mixture, usually liquid, containing ions that migrate in an electric field. The free electrons travel to the cathode, where they combine with the constituents of the electrolyte, such as water and oxygen, to form hydroxyl (OH^{-1}) ions. The metallic path between anode and cathode is essential for electron movement between the anode and cathode. For the steel corrosion in concrete, the anode, cathode and metallic path are on the same steel. The electrolyte is the moisture in concrete surrounding the steel. The corrosion will stop if any of these components is removed. This is the basis for corrosion control. The process of corrosion is illustrated in Figure 2.9 with necessary notations of chemical reactions.

At the anode site, the iron dissociates to form ferrous ions and electrons:

$$Fe \rightarrow Fe^{2+} + 2e- \qquad \text{(anodic reaction)} \qquad (2.29)$$

The electrons move through the metal towards the cathodic site but the ferrous ions are dissolved in the pore solution.

At the cathodic site, electrons combine with oxygen and water to form hydroxyl ions:

$$4e- + O_2 + 2H_2O \rightarrow 4(OH)- \qquad \text{(cathode reaction)} \qquad (2.30)$$

The hydroxyl ions move to the anode through the pore solution. At the anode, we have:

$$Fe^{2+} + 2Cl- \rightarrow FeCl_2 \qquad (2.31)$$

$$FeCl_2 + 2H_2O \rightarrow Fe(OH)_2 \text{ (ferrous hydroxide)} + 2HCl$$
$$(Cl^- \text{ regenerated}) \qquad (2.32)$$

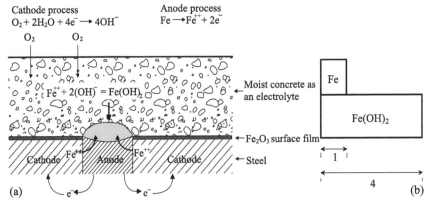

Figure 2.9 Reinforcing steel corrosion and expansion of corrosion products: (a) corrosion process; (b) volume expansion of corrosion products.

$$Fe^{2+} + 2(OH)- \rightarrow Fe(OH)_2 \text{ (ferrous hydroxide)} \tag{2.33}$$

$$Fe^{3+} + 3(OH)- \rightarrow Fe(OH)_3 \text{ (ferrous hydroxide)} \tag{2.34}$$

$$4\,Fe(OH)_2 + 2H_2O + O_2 \rightarrow 4\,Fe(OH)_3 \text{ (ferric hydroxide)} \tag{2.35}$$

It can be seen from the above equations that oxygen and water are needed for the initiation and propagation of the corrosion. There is no corrosion in a completely dry atmosphere, probably below a relative humidity (RH) of 40 percent (Mailvaganam 1992). It has been suggested that the optimum RH for corrosion is 70 to 80 percent. At higher relative humidity (RH greater than 80 percent) or under immersion conditions, the diffusion of oxygen is considerably reduced and the environmental conditions are more uniform along the steel. Consequently, there is little corrosion (Page and Treadway 1982). When corrosion occurs, ions need to travel through pores in surrounding concrete; oxygen needs to diffuse through the concrete cover; corrosion rate is therefore affected by electrical resistance, diffusivity, and cover thickness of the concrete.

The extent of steel corrosion in concrete depends upon the conductivity of the electrolyte, the difference in potential between the anodic and cathodic areas, and the rate at which oxygen reaches the cathode. This controls the velocity of the anodic reaction. For steel in concrete, the strong polarization of the anodic zones under aqueous and highly alkaline conditions raises its potential close to that of the cathode causing the surface of the steel to be passivated by the formation of an oxide layer. The passivating film prevents further reaction so that the steel remains unaltered over long periods. Another modifying effect of concrete on corrosion is that of increased electrical resistivity, which reduces the flow of electrical currents within the concrete. This is particularly true of high-density concrete (Page and Treadway 1982; Verbeck 1975). Because serious corrosion is very dangerous to concrete structures, it is necessary to detect the degree of corrosion when deciding on maintenance. The detection methods include visual inspection, half-cell potential measurement, radiography, ultrasonic, magnetic perturbation/flux, and acoustic emission technique. Some commonly used methods will be introduced in Chapter 3.

Because of the magnitude of the costs of rebar corrosion, significant efforts have been made to protect the steel in last decades. To prevent corrosion in concrete, it is essential to reduce the content of chloride to less than 0.6 kg/m³. In this case, the protection layer of rebar will remain intact and thus no corrosion will be initiated. In most instances, the corrosion control methods can be described as passive. Durability performance is obtained by proper design and control of the concrete cover. Such means are usually specified in design codes: minimum concrete cover thickness, the inherent concrete properties (in terms either of design strength or maximum water/cement ratio) and the maximum allowable crack width permitted. Obviously, concrete quality is the most important parameter that controls the rate of carbonation and chloride ingress, hence the extent of reinforcing steel corrosion. Improving the quality of concrete has thus been considered the primary protection method. Next, we have to consider the four components of a corrosion cell. If we can cut out one of them, we can stop corrosion. It is obvious that the electrolyte in concrete is made up of the moisture condition and existing air. If we make denser concrete, there will be less chance for moisture and air to get in and thus reduce the possibility of forming electrolyte and prevent corrosion. Also, we can use cathodic protection method to protect reinforcing steel. For instance, using zinc as an anode can protect the steel because the corrosion occurs at the zinc anode.

Other protection strategies include increased cover, epoxy-coated rebar, stainless steel, and corrosion inhibitors. The time-to-corrosion of embedded reinforcing steel can be significantly influenced by the amount of concrete cover over rebar. However, as the cover increases, the rebar becomes less effective and the potential for cracking due to tensile stress, shrinkage and thermal effects increases. Epoxy coating of reinforcing steel can enhance the durability performance by serving as a barrier preventing the access of aggressive species to the steel surface and providing electrical insulation. Epoxy-coating can be applied in various ways, either as a liquid or as a powder which is fused on the surface. ASTM A775/A775M-04a (Standard Specification for Epoxy-Coated Steel Reinforcing Bars) has addressed the basic requirements for epoxy-coated reinforcing steels by the electrostatic spray method. In general, the performance of epoxy-coated rebars in bridge decks and parking garages in chloride environments, where chloride de-icing salts are applied during winter, has been demonstrated to be excellent (Clifton *et al.* 1975; Satake *et al.* 1983). However, several notable problems with corrosion of epoxy-coated bars were reported in substructures off the Florida Keys in the USA (Clear 1992; Smith *et al.* 1993). The amount of damage to the epoxy coating prior to concrete casting was considered the major contributory factor to the poor performance (Smith *et al.* 1993). Thus, training is necessary to properly produce, handle and construct the coating, and repair of field damage to epoxy-coated bars.

Besides, the bond properties between concrete and the epoxy-coated rebars are not as strong as that between concrete and conventional rebars, which should be improved. There are a few studies on the use of stainless steel bars. It has been shown that the chloride threshold value for initiation of corrosion in non-welded AISI 304 (a kind of stainless steel) rebar is three to five times higher than that of conventional rebar. However, welding of the bar reduced the critical chloride level by 50 percent. In addition, use of stainless rebars is an expensive solution. Zinc coating of steel (galvanized steel) is also considered a good method of corrosion resistance. It acts both as a sacrificial and barrier-type coating. But the disadvantage is that, like other metal coatings, zinc coating corrodes over time. The rate of corrosion under the given environmental conditions will determine the loss of coating thickness and the time period during which it will be effective. Generally, there is a fairly linear relationship between the metal thickness and the duration of its effective service life for galvanized steel exposed to industrial atmosphere (Chandler and Bayliss 1985). The stability of zinc is dependent on the pH of the surrounding solution where the zinc coating is exposed. It has been found that zinc is stable at pH below about 12.5, but it tends to dissolve at an increasing rate as the pH increases above this level. The corrosion products of zinc may be deposited at the surface of the zinc coating and seal it, thus arresting the evolution of H_2 gas and leading to passivation of the zinc coating.

However, if the galvanized rebars are mixed used with ungalvanized bars, the galvanized bars will be accelerated depleted. So if the galvanized and ungalvanized bars are being used in the same structure, special care should be taken to ensure complete electrical isolation of these two (Broomfield 1997). Corrosion inhibitors are considered useful not only as preventative measure for new structures but also as preventative and restorative surface-applied admixture for existing structures. Various corrosion inhibitors (see Trabanelli 1986) can be classified into: (1) adsorption inhibitors, which act specifically on the anodic or on the cathodic partial reaction of the corrosion process or on both reactions; (2) film-forming inhibitors, which block the surface more or less completely; and (3) passivators, which favour the passivation reaction of the steel. The mechanistic action of corrosion inhibitors is thus not against uniform corrosion but localized or pitting corrosion of a passive metal due to the presence of chloride ions or a drop in pH value (Bertolini *et al.* 2004). Corrosion inhibitors admixed to the free concrete can act in two different ways: these inhibitors can extend the corrosion initiation time and/or reduce the corrosion rate after depassivation has occurred (Hartt and Rosenberg 1989).

Mixed-in inhibitors are regarded as more reliable since it is easier and more secure to add the inhibitors to the mix. Some laboratory testing has shown that certain corrosion inhibitors do not significantly affect the amount of chloride ion required to initiate corrosion, but can reduce corrosion rate. The field performance of these products lasts only a relatively short period and cannot be conclusive in determining their effectiveness. For example, the field tests with proprietary vapor-phase inhibitors in a parking garage with chloride-contaminated precast slabs did not show encouraging results (Broomfield 1997). Corrosion-rate measurements showed a reduction of 60 percent in areas with initially intense corrosion but also an increase of corrosion rate in areas with low corrosion rates.

2.3.5 Degradation caused by alkali–silica reaction

Alkali–Aggregate Reaction (AAR) is a reaction between alkalis from cement and certain forms of silica presented in aggregate, which results in excessive expansion of concrete sections and leads to severe cracking thereafter. Two general types of attacks can occur: (1) alkali-carbonate attack with dolomitic limestone aggregate (some argillaceous dolomites); and (2) alkali-silica attack with siliceous aggregates containing certain forms of amorphous or poorly crystalline silica (such as some chert, flint, opal, tridymite, cristoballite chalcedony, volcanic glasses and some limestones). The alkali content of cement depends on the materials from which it is manufactured and also to a certain extent on the details of the manufacturing process, but it is usually in the range of 0.4–1.6 percent. In concrete mixes there may be a contribution to alkali from other cementitious materials, such as pulverized fuel ash or ground granulated blast-furnace slag, which is present. It is well known that Na_2O (sodium oxide) and K_2O (potassium oxide) are present in the cement clinker in a small amount. It is thus conventional to express the results of chemical analysis of cement in terms of the oxides, Na_2O and K_2O. Furthermore, the alkali content in cement is generally expressed as an equivalent percentage of Na_2O by mass of cement. Since the molecular weights of Na_2O and K_2O are respectively 62 and 94, the equivalent percentage of Na_2O is calculated by the formula:

$$\%Na_2O_{eq} = \%Na_2O + 0.658 \cdot \%K_2O \qquad (2.36)$$

In the cement paste, Na_2O and K_2O form hydroxides and raise the pH level from 12.5 to 13.5. The concentration of these hydroxides increases as Na_2O_{eq} increases. In such highly alkaline solutions, under certain conditions, the silica can react with alkaline to form a silicate gel, which is hygroscopic. When the gel absorbs moisture, it will swell and apply pressure to the surrounding paste, causing the concrete to crack, sometimes severely, with accompanying loss of structural integrity. The degree of AAR is affected by: (1) presence of water, if there is no water, there is no expansion; (2) alkali content, if the alkali content (Na_2O and K_2O) is less than 0.6 percent, there is no reaction and the concrete contains more than 3 kg/m^3 of alkali can be considered to have a high alkali content; (3) concrete porosity, the internal stress may be relieved in concrete with high porosity. ASR can occur only in a moist environment: it has in fact been observed that in environments with a relative humidity below 80–90 percent, alkali and reactive aggregate can co-exist without causing any damage. With low effective alkali content in the concrete, i.e. the equivalent content of Na_2O in concrete is less than 3 kg/m^3, deleterious ASR can be prevented. The expansive effect, hence ASR effect, can be negligible. The extent of reaction depends on the amount of reactive silica present in the aggregate mix while the reactivity of silica minerals depends on their crystal structure and composition. The porosity, permeability and specific surface of aggregates and the presence of Fe- and Al-rich coatings can influence the kinetics of the alkali silica reaction.

It should be noted that blended cements containing pozzolana, fly ash or blast furnace slag give a resulting alkaline solution of slightly lower pH. Addition of silica fume leads to the lowest pH. Hence, use of pozzolana materials such as fly ash, GGBS can even prevent damage caused by ASR. It has also been found that by adding calcium nitrite into the concrete mix, the concrete's resistance to ASR can be significantly improved. However, the mechanism is not clear (Li *et al.* 1999; 2000). These mineral additions reduce the concentration of OH^{-1} ions in the pore solution of cement paste. This is because hydroxyl ions are consumed by the pozzolanic reaction occurring during hydration. Furthermore, the alkali transport is slowed down because of the lower permeability of pozzolanic and blast furnace cement paste, which helps reduce the ASR. Finally, the hydration products of the mineral additions bind alkali ions to a certain extent, preventing them from taking part in the reaction with the silica. The temperature also influences the alkali silica reaction. Normally the ASR increases as the temperature increases. AAR can be deleterious to concrete due to the expansion and possible cracking of the concrete associated with the reaction. What's more, development of alkali silica reaction may be very slow and its effects may show even after long periods (up to several decades). Consequently, cracking caused by alkali silica reaction usually takes many years and is often preceded by pop-outs on the concrete surface.

It is believed that AAR was first identified in 1940 in California, by Stanton (Mehta and Monteiro 2006), but only limited numbers of examples were observed in practice until fairly recently. AAR causes deleterious expansion and cracking of the concrete and reduces the tensile strength, which may have consequences for the structural capacity. The significance of AAR or ASR for concrete structures have led to a surge in research activities: (1) to determine the exact nature of the reaction (which is not fully understood as yet); (2) to define the acceptable limits of alkali content, moisture content, and reactive aggregate content; and (3) to determine methods to reduce the degree of destructive expansion. Extensive research has been made in two directions. One is the development of testing methods, which are designed to reveal whether an aggregate is potentially reactive and can cause abnormal expansion, and cracking of concrete. The other is the development of effective methods to prevent the damage induced by AAR.

Although it is possible to determine what types of aggregates have a trend of AAR, it is impossible to predict whether their use will result in excessive expansion or not. It has been found that a critical amount exists for each type of aggregate, and only this can result in serious expansion, an amount smaller or larger than this value will not cause significant swelling. Measuring the expansion of test specimens has been considered the most dependable way to evaluate aggregate reactivity, and a number of test procedures have been devised. The testing methods commonly used include standard test methods such as the mortar bar test (ASTM C227), the rock cylinder method (ASTM C586), and rapid test methods such as ASTM C289. One of the disadvantages of the standard testing methods is that most of them are very time-consuming, which is obviously incompatible with the demands of the construction industry. On the other hand, the rapid testing method,

ASTM C289, only gives the potential reactivity of the aggregate, but it does not necessarily predict expansion in a real situation.

To overcome the limitations of the above-mentioned test methods, other rapid testing methods have been developed, such as the dynamic modulus test and the gel fluorescence test. Dynamic modulus can be obtained by measuring resonant frequency and pulse velocity. It has been shown that it can provide a good indication of deterioration due to AAR. The measurement can even detect deterioration before any expansion and visible cracking occurs. It is also sensitive to the changes in environmental conditions, which activate or suppress the AAR. In the gel fluorescence test method, 5 percent solution of uranyl acetate is applied to the surface of the specimen, then the specimen is viewed under an ultraviolet (UV) light. A yellowish green fluorescent glow means that AAR is present.

Since AAR can cause significant deterioration and damage to concrete structures, many research studies have been conducted to decrease the effect of the reaction. There are numerous recommendations for minimizing the risk caused by AAR. These recommendations include:

1   Use non-deleterious aggregate and/or non-reactive aggregate when the alkali content of the cement is high.
2   Use low-alkali Portland cement or blended cements with sufficient amounts of fly ash or slag when the silica content of aggregate is high (more than 3.0 kg/m^3).
3   Keep the concrete dry (relative humidity of the concrete less than 80 percent). However, the choice of types of cements and aggregates on a construction site is usually very limited, and the environment surrounding the concrete is obviously unchangeable. In many cases, the contents of silica in aggregates and/or alkalis in cement paste cannot be reduced effectively, either. The only effective way to reduce the risk of AAR is to control moisture migration in the concrete since no AAR will occur in dry concrete even there is some silica present. Controlling moisture diffusion in concrete can be implemented in two different ways. One is to control the local moisture diffusion around the boundary of each aggregate, that is, to control the moisture exchange between the aggregate and surrounding cement paste. The other is to control the diffusion of moisture into and out of the surface of concrete members.
4   Local diffusion control coating. The control of local moisture diffusion is very important because AAR occurs right on the boundary of the aggregate. Distributions of the chemicals, Na_2O, SiO_2, K_2O, and CaO, around the aggregate boundary have shown quantitatively that the reaction rim is in the range of 300 micrometers. Local diffusion control coating is developed based on the crystallization technique. The coating product consists of powders of finely ground rapid setting Portland cement, treated silica sand (de-alkaline silicate), and proprietary chemical additives. They are mixed with water to form slurry. The slurry will

be applied to the reactive aggregate before the cast of concrete specimens. A coating is then formed around the aggregates. Application of
this technique is not aimed at completely eliminating AAR but at
reducing and slowing down the rate of AAR. As a result of a slow
reaction, the product of the reaction can be accommodated and
deposited in large capillary pores. Thus the detrimental damage due to
the expansion will be avoided.

5   Global diffusion control coating. This method applies a sunlight curing
    hydrophobic weather-resistant coating. This is a newly developed coat-
    ing, which can be applied to the surface of concrete members to reduce
    moisture diffusion. In addition to the high resistance to moisture pene-
    tration, the coating has other two useful features. One is its ability to
    reduce or to eliminate VOC (Volatile Organic Compounds) which are
    detrimental to environment. The other is that it can be cured under low
    power UV. UV curing is very important for long-term applications since
    it yields paint or coatings with excellent durability. However, con-
    ventional UV curing requires a special high power UV lamp, while this
    coating requires only natural light from the sun, that is, sunlight curing.

The alkali–silica reaction is most widespread and best understood. In the
case of alkali–silica reaction, alkali hydroxides in the hardened cement paste
liquor attach to the silica to form an unlimited swelling gel, which draws in
any free water from osmosis and thus expands, disrupting the concrete matrix. Expanding gel products exert internal stress within the concrete, causing
characteristic map cracking of unrestrained surfaces, but cracks may be
directionally oriented under conditions of restraint imposed by reinforcement, pre-stressing or loading. Cracking resulting from alkali–silica reactions
was long held not to pose any structural threat to affected members.

Consideration must always be paid to the effects of deep penetrating
cracks on the durability of reinforcement and to the self-stress induced by
the expansive reactions caused by alkali reactivity. This may be advantageous in confined sections of normally reinforced concrete members, but
could prove catastrophic in the case of pre-stressed structures. Avoidance of
the problem caused by alkali reactivity prior to construction is of greatest
importance. This could be achieved by avoiding the use of reactive aggregates, by the use of low alkali Portland cement, of slag cement or of
pozzolanic admixtures. As the swelling phenomenon is dependent on water
absorption, most control measures are taken by decreasing water ingress to
the concrete by the use of surface coatings or impregnation materials
(Campbell-Allen and Roper 1991).

2.3.6 Degradation caused by sulfate attack

Sulfate attack is one of main factors causing deterioration of concrete
durability. In most cases, the sulfates are external to the concrete, i.e. in a

solution in ground water or trade effluent and from contaminated aggregates (e.g. in the Middle East). Portland cement itself also contains sulfate as gypsum. If saline water, such as sea water, is used for mixing concrete, it definitely brings sulfate to the mixture. Solutions of sulfates of sodium, potassium, magnesium, and calcium are commonly slats that may bring severe deterioration to concrete. The total sulfate content in concrete mixture becomes a controlling factor of the degree of sulfate contact. The acceptable concentration of sulfate in concrete is about 4 percent by weight of cement. Sulfate attack of concrete is generally regarded as an expansion that can cause concrete cracking due to the reaction of sulfate with some hydration products in cement paste. When concrete cracks, its permeability increases and the aggressive water penetrates more easily into the interior, thus accelerating the process of deterioration.

Hence, the deterioration of concrete due to the sulfate attack can be considered a complicated phenomenon of physical and chemical process. Sulfate attack of hydrated cement takes place by the reaction of sulfate ions with calcium hydroxide and hydrated calcium aluminates to form gypsum and ettringite and the formation of these two products is responsible for expansion and cracking of concrete. Sulfate attack can also manifest itself as a progressive loss of strength of the cement paste due to loss of cohesion between the hydration products and loss of adhesion between the hydration products and the aggregate particles in concrete. The sulfate reacts with the tricalcium aluminate (C_3A) in the cement to form the compound ettringite, which causes the expansion of concrete. Calleja (1980) has pointed out that in all the sulfates, magnesium sulfate attack has the most damaging effect because Mg^{2+} and Ca^{2+} ions associate well, they have equal valence and similar ionic radii, which can lead to a reaction between magnesium sulfate and C–S–H gel. The deterioration of Portland cement concretes exposed to sulfates normally may be ascribed to the following reactions (Campbell-Allen and Roper 1991):

(i)   The conversion of calcium hydroxide derived from cement hydration reactions to calcium sulfate, and the crystallization of this compound with resulting expansion and disruption.

$$Ca(OH)_2 + SO_4^{2-} + 2H_2O \rightarrow CaSO_4 \cdot 2H_2O + 2OH^- \qquad (2.37)$$

(ii) The conversion of hydrated calcium aluminates and ferrites to calcium-sulfo-aluminates and sulfo-ferrites or the sulfate enrichment of the latter minerals. The products of these reactions occupy a greater volume than the original hydrates, and their formation tends to result in expansion and disruption.

$$C_3A + 3CaSO_4 + 32H_2O \rightarrow 3CaO \cdot Al_2O_3 \cdot 3CaSO_4 \cdot 32H_2O \qquad (2.38)$$

(iii) The decomposition of hydrated calcium silicates.

In the presence of calcium sulfate, only reaction (ii), can occur, but with sodium sulfate, both (i) and (ii) may proceed. With magnesium sulfate all three may occur. The severity of attack depends on the type of sulfate. Calcium sulfate undergoes expansion reaction with ettringite (arising from the calcium aluminate in cement), which gives rise to greater expansive effects than gypsum. Part of the formed ettringite is commonly located in the interface between paste and aggregate, resulting in loss of bond. Sodium sulfate (Na_2SO_4) also reacts with calcium hydroxide to form gypsum, which reduces paste strength and stiffness. Magnesium sulfate ($MgSO_4$) reacts to form gypsum and destabilizes C–S–H, the strength-governing phase in the cement paste. Severity of attack therefore increases from calcium sulfate to sodium sulfate to magnesium sulfate. As a reference, ACI 201.2R-92 (1994a) has classified the severity of exposure, as given in Table 2.4.

Sulfates in solution can combine with the tri-calcium aluminate (C_3A) in Portland cement, to form a sulfoaluminate hydrate (ettringite) in the form of needle-like crystals, causing expansion of the matrix which turns white and become soft. The ettringite occupies a greater volume within the concrete than the calcium aluminate hydrates. The expansion generates tensile stresses in the cement paste and as a result cracks develop in the concrete. Cementitious repairs in these conditions should use sulfate-resisting cement, which has a low C_3A content. Attack on dense concrete will be a surface effect so rich, well-compacted mixes must be used. The form of sulfate present in uncontaminated groundwater is normally calcium sulfate, which has limited solubility. Some other salts, such as magnesium sulfate, are much more readily soluble in water and can form stronger solutions, so they are more dangerous (Allen *et al.* 1993).

For Portland cement-based concrete, the resistance to sulfate attack depends primarily on the type and amount of cement in the concrete, the type and amount of mineral additives in the concrete, and the permeability and diffusivity of the concrete. Low permeability and diffusivity provide the best defense against sulfate attack by reducing sulfate penetration. To achieve a low permeability and diffusivity, the water/powder ratio should be as small as possible and pozzolanic additives that can reduce the calcium hydroxide content and refine the pore structure of the matrix through

Table 2.4 Classification of severity of sulfate environment according to ACI 201.2R-92

Exposure	Concentration of water-soluble sulfate expressed as SO_4	
	In soil (%)	*In water (ppm)*
Mild	<0.1	<150
Moderate	0.1 to 0.2	150 to 1500
Severe	0.2 to 2.0	1500 to 10000
Very severe	>2.0	>10000

pozzolanic reaction should be used. Reducing the water/cement ratio can make concrete less permeable and more difficult for sulfates to penetrate. It has been proved to be more effective than reducing the calcium aluminate content for improving the sulfate resistance of concrete. The severity of sulfate attack depends on the content of C_3A and, to a lesser extent, of C_4AF in the cement. Sulfate-resisting cement has a low tricalcium aluminate content and hence less potential for expansive reaction and better resistance to sulfate attack. Reducing the amount of calcium hydroxide and hydrated calcium aluminate is also helpful in improving the sulfate resistance of concrete. Blended cements with pozzolanic materials or blast furnace slag show enhanced resistance to sulfate attack (Mehta 1988a). For example, low-calcium fly ash is an effective blending material to combat the sulfate attack of concrete. Also the incorporation of silica fume into a concrete mixture greatly improves the sulfate resistance, owing to the reduced amount of gypsum formation, compared with mixtures that do not contain silica fume.

Also, the incorporation of silica fume can largely reduce the permeability of concrete, thus improving the sulfate resistance. The mixture of silica fume and fly ash should be a much better choice for production of high performance concrete. On the other hand, Mehta (1988a) has noted that it is the mineralogy rather than the chemical composition which determines the resistance to attack, and empirical guidelines based on chemical compositions of a mineral additive do not prove to be reliable. Nevertheless Mehta (1980, 1988b) has concluded that as a first approximation, a blended cement will be sulfate resistant when made with a highly siliceous natural pozzolanic or low-alumina fly ash or slag, provided that the proportion of the blending material is such that most of the free calcium hydroxide can be used up during the course of cement hydration. Under severe conditions, hydraulic cements other than Portland cement should be used, for example supersulfated cement and high-alumina cement. Supersulfated cement offers very high resistance to sulfates, especially if its Portland cement component is of the sulfate-resisting variety. It should be noted here that high-alumina cement should not be used in continuously warm, damp conditions, or in mass construction from which the heat of hydration cannot be easily dissipated. High-pressure steam curing improves the resistance of concrete to sulfate attack due to the change of C_3AH_6 into a less reactive phase, and also to the removal of $Ca(OH)_2$ by the reaction with silica.

The concrete attacked by sulfate shows a whitish appearance. Usually, damage starts at the edges/corners, followed by progressive cracking/spalling. Rate of sulfate attack also increases with concentration of sulfate and replenishment rate (e.g. sulfate attack to concrete is faster in flowing groundwater due to faster replenishment rate). Tests on sulfate resistance are normally conducted by storing specimens in a solution of sodium or magnesium sulfate, or a mixture of the two. The test may be accelerated with wetting/drying cycles that will induce salt crystallization in the pores. Effect of exposure can be estimated from: (1) change in dimension; (2) loss of strength; (3) change in dynamic modulus of elasticity; and (4) loss of weight. The test method of ASTM C1012-04 uses the immersion of well-hydrated mortar in a sulfate solution, and considers excessive expansion as a criterion failure under sulfate attack. But this method is only for mortar not concrete. Besides, this method is slow and normally takes several months. As an alternative, ASTM C 452-89 prescribes a method in which a certain amount of gypsum is included in the original mortar mix. This speeds up the reaction with C_3A but this method is not appropriate for use with blended cements. The criterion of sulfate resistance in this method is the expansion at the age of 14 days.

2.3.7 Acid attack

Acid solutions, both mineral and organic, are about the most aggressive agents for concrete since Portland cement concrete is highly alkaline, Depending on the type of acid, the attack can be mainly an acid attack, or a combination of acid followed by a salt attack. In normal concrete, the hydrated components in the cement matrix of concrete are in equilibrium with the pore solution having a high pH value, due to the presence of OH^{-1}. When concrete comes into contact with acid solutions, these compounds may dissolve at a rate depending on the permeability of the concrete, the concentration and the type of acid. Moreover, acids can also attack calcareous aggregates such as limestone in concrete. If the acid penetrates the concrete through cracks down to the steel bars, the steel bars can corrode and spall the concrete, causing deterioration of concrete. Owing to the highly basic character of Portland cement, an acid cannot penetrate dense concrete without being neutralized as it travels inwards. Therefore, it cannot cause deterioration in the interior of the specimen without the cement paste on the outer portion being completely destroyed. The rate of penetration is thus inversely proportional to the quantity of acid-neutralizing material, such as the calcium hydroxide, C–S–H gel, and limestone aggregates (Mailvaganam 1992). In summary, concrete can be attacked by liquids with a pH value below 6.5 but the attack is severe only at a pH value below 5.5. When a pH value is below 4.5, the attack is very severe.

Acid can naturally occur from rain in industrial regions, which is dilute solutions of CO_2, and SO_2, and CO_2 bearing groundwater. Water containing dissolved carbon dioxide can be a common acid source in concrete. With the presence of CO_2 in water, calcium hydroxide can be converted into: (1) calcium carbonate, which is stable but it will reduce alkalinity of concrete; (2) calcium bicarbonate, which can dissolve in water and leach out of the concrete. Once reaction occurs on the surface, CO_2 needs to travel a longer distance for further reaction. Similar to dry oxidation, its penetration rate is approximately proportional to the square root of time. So the acidic attack progresses at a rate approximately proportional to the square root of time. If these acids come into contact with concrete they can react with the alkali present in the cement hydrates to produce soluble products that can be removed by solution. In flowing water, the reaction products are carried away, exposing fresh surfaces to attack. Under static conditions, the water adjacent to the structure may become saturated. If this happens, the reaction ceases after only surface attack. The degree of acidic attack on the cement matrix depends on the solubility of the salts formed and thus on the nature of the anions involved (Bertolini 2004). For more soluble reaction products, the attack rates are higher than for insoluble products. For hydrochloric acid, soluble calcium chloride is formed, while with sulfuric acid, calcium sulfate (gypsum) is formed, which is much less soluble. Sulfuric acid with the presence of SO_2 is particularly aggressive because the sulfuric acid attack $Ca(OH)_2$ and C–S–H takes place apart from the attack in the alumina phase.

Though various physical and chemical tests on the resistance of concrete to acids have been developed, currently there are no standard procedures. In general, acid attack can be reduced by: (1) use of low water/cement ratio to reduce permeability; (2) use of low cement content to reduce C–S–H; and (3) use of pozzolanic materials, such as ground granulated blast-furnace slag, pozzolanas, and especially silica fume, to reduce the $Ca(OH)_2$ content. Where the acid attack possibility is very high, such as floors of chemical plants, the surface can be treated by sodium silicate (water glass) which reacts with $Ca(OH)_2$ and forms calcium silicate to block pores, so that reduces the acid attack.

2.3.8 Frost attack

Concrete is a porous material. As the cement and water in fresh concrete react to form a hardened paste binding the coarse and the fine aggregates together, voids are left in the originally water-filled space among the cement grains. These voids are known as capillary pores with a size range from approximately 5 nm to 1 μm and sometimes even larger. In addition to capillary pores, cement paste also contains a significant volume of smaller pores called gel pores. Water contents and capillary forces in small volumes of these pores are very important to the durability of hardened concrete, especially for those subjected to repeated cycles of freezing and thawing, which can cause disintegration of concrete surface layers.

Concrete deterioration caused by freezing and thawing is linked to the presence of water in concrete but cannot be explained simply by the expansion of water on freezing. While pure water in the open freezes at 0 °C, in concrete the water is really a solution of various salts so that its freezing point is lower. Moreover, the temperature at which water freezes is a function of the size of the pores (Figure 2.10). In other words, the temperature at which water can freeze in concrete pore decreases as the size of the pore decreases. In concrete, pore sizes have a wide range so that there is no single freezing point. As an example, for saturated Portland cement paste, free water in pores larger than 0.1 mm freezes between 0 and −10 °C; water in

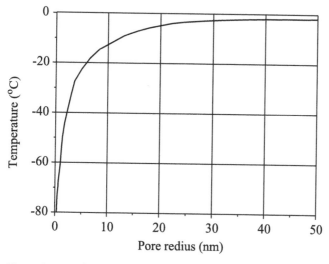

Figure 2.10 Relationship between the size of capillary pores and the temperatures at which ice formation (from pure water) is possible inside their pores.

pores between 0.1 and 0.01 mm freezes between −20 and −30 °C and gel-water (pores less than 10 nm) freezes below −35 °C (Beddoe and Setzer 1988; 1990). Freezing begins in the outer layers and in the largest pores and extends to the inner parts and to smaller pores only if the temperature drops further. Specifically, the gel pores are too small to permit the formation of ice, and most of the freezing takes place in the capillary pores. We can also note that larger voids, arising from incomplete compaction, are usually air-filled and are not appreciably subjected to the initial action of frost. During the freeze–thaw cycle attack on concrete, the presence of de-icing salts, like calcium and sodium chloride, in contact with concrete is a detrimental factor. The outer layers where these salts are present are more strongly affected by frost despite their effect in lowering the freezing point, probably due to increased water saturation caused by their hygroscopic effect, with the early appearance of scaling and detachment of cement paste which covers the aggregate.

When water freezes, there is an increase in volume of approximately 9 percent. Pockets in concrete that can fill with water, or with materials that absorb water are also source of trouble because the water in them will expand if it freezes, disrupting the surrounding concrete (Allen *et al.* 1993). As the temperature of concrete drops, freezing occurs gradually inward and puts hydraulic pressure on the unfrozen water in the capillary pores due to the volume expansion of ice. Such pressure, if not relieved, can result in internal tensile stresses that may cause local failure of the concrete. On subsequent thawing, the expansion caused by ice is maintained so that there is now new space available for additional water which may be subsequently imbibed. During re-freezing, further expansion occurs. Thus repeated cycles of freezing and thawing have a cumulative effect. It is the repeated freezing and thawing, rather than a single occurrence of frost that causes damage. Frost action is an important factor causing concrete degradation in cold region. When the solutions contain de-icing chemicals, though the de-icing salts can lower the temperature of ice formation, which may be viewed as a positive effect, they may also bring the following negative effects: (1) an increase in the degree of saturation of concrete due to the hygroscopic character of the salts; (2) an increase in the disruptive effect; (3) the development of differential stresses as a result of layer-by-layer freezing of concrete due to salt concentration gradients; (4) temperature shocks; and (5) salt crystallization in supersaturated solutions in pores (Mehta and Monteiro 2006). Overall, the negative effects far outweigh the positive effect.

There are two other processes that can increase hydraulic pressure of the unfrozen water in the capillaries. First, since there is a thermodynamic imbalance between the gel water and the ice, diffusion of gel water into capillaries can lead to a growth in the ice body and thus to an increase of hydraulic pressure. Second, the hydraulic pressure is increased by the pressure of osmosis caused by local increases in solute concentration due to the removal of frozen (pure) water from the original solution. The extent of damage caused by repeated cycles of freezing and thawing varies from surface scaling to complete disintegration as layers of ice are formed, starting at the exposed surface of the concrete and progressing through its depth. In general, concrete members that remain wet for long periods are more vulnerable to frost than any other concrete. It is clear that the hydraulic-pressure mechanism of frost damage has more severe consequences in a system of fully saturated pores, because in that case the pressure can only be released if the microstructure expands, which may quickly result in cracking.

In general, the loss of mass or the decrease of dynamic modulus is used as the index of degradation. The general influence of saturation of concrete, in the deterioration mechanism, is related to a value known as the critical saturation, below which the concrete is quite resistant. The critical saturation depends on the size of the body, on its homogeneity, and on the rate of freezing. Frost resistance is determined by the number of freeze–thaw cycles that a particular concrete can withstand before reaching a given level of degradation (Bertolini 2004). Another important parameter to frost resistance is the water/cement ratio, which largely determines the porosity of the cement matrix. In order to prevent a concrete from damage caused by repeated cycles of freezing and thawing, air-entraining agent can be used. An air-entraining agent is a chemical admixture that can deliberately entrain the air into cement paste with a closed space (<0.3 mm).

When mixed with water, air-entraining admixtures produce discrete bubble cavities which are incorporated in the cement paste. The essential constituent of the air-entraining admixture is a surface-active agent which lowers the surface tension of water to facilitate the formation of the bubbles, and subsequently ensures that they are stabilized. The surface-active agents concentrate at the air/water interfaces and have hydrophobic (water-repelling) and hydrophilic (water-attracting) properties which are responsible for the dispersion and stabilization of the air bubbles. The bubbles are separated from the capillary pore system in the cement paste and they never are filled with the products of hydration of cement as gel can form only in water. The main types of air-entraining agents are: (1) animal and vegetable fats and oils and their fatty acids; (2) natural wood resins, which react with lime in the cement to form a soluble resinate. The resin may be pre-neutralized with NaOH so that a water-soluble soap of a resin acid is obtained; and (3) wetting agents such as alkali salts of sulfated and sulfonated organic compounds.

The performance of an air-entraining admixture should be checked by trial mixes in term of the requirements of ASTM C 260-77 or BS 5075: Part 2 (1982). The essential requirement of an air-entraining admixture is that it rapidly produces a system of finely divided and stable foam, the individual bubbles of which resist coalescence; also, the foam must have no harmful chemical effect on the cement. The beneficial effect of air entrainment on concrete subjected to freezing and thawing cycles is to create space for the movement of water under hydraulic pressure. However, there are some further effects on the properties of concrete, some beneficial, others not. One of the most important is the influence of voids on the strength of concrete at all ages.

2.3.9 *Deterioration caused by the marine environment*

The deterioration of concrete structures in the marine environment is a combination of mechanisms, both chemical and physical in nature as shown in Figure 2.11. Attack of concrete by sea water can be of various types: superficial erosion by waves or tides, abrasion by sand in suspension and by ice, repeated freeze–thaw and wet–dry attacks, swelling caused by crystallization of salts, chemical attack by salts dissolved in the water, alkali-aggregate reaction, chloride-induced corrosion of reinforcement steel, and sulfate attack. The reaction between sulfate ions and both C_3A and C–S–H takes place in concrete exposed in marine environment, resulting in the formation of ettringite which, as well as gypsum, fortunately, are soluble in the presence of chlorides and can be leached out of the concrete by the sea water (Lea 1970). This happens with no expansion but gradual material loss, which is quite different from sulfate attack in other environments.

In marine structures, deterioration is most severe at the tidal and splash zones, due to combination of: (1) physical erosion; (2) salt crystallization pressure; (3) leaching of sulfate attack products; (4) freezing/thawing in cold regions; and (5) chloride penetration. The tidal zone is the range between

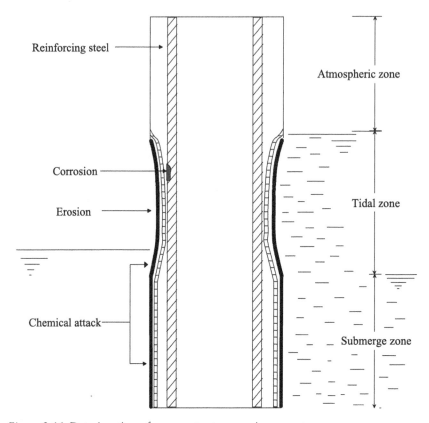

Figure 2.11 Deterioration of a concrete structure in sea water.

mean high and mean low water levels. This zone is periodically immersed, generally on a daily basis. Corrosion of steel and spalling of concrete easily occur in this zone. Reinforcement corrosion is the most common mechanism in the deterioration of concrete structures. In a marine environment, corrosion of the reinforcement is primarily due to chloride penetration of the concrete, which results in expansion of the reinforcing steel surface ultimately spalling the concrete surface.

Additionally, abrasion of structural elements from ice and debris occur within this zone. The alternating motion of waves and tides also contributes to the deterioration of the concrete. The high velocity wave, depressions and irregularities of the surface layer can cause cavitation to form in marine concrete structures. If the absolute pressure at points of surface irregularities approaches the vapor pressure of the water, minute bubbles will form and quickly collapse. The collapse of these bubbles can produce minute water jets having extremely high velocities and result in an intense impact wave splash concrete surface. This effect is very destructive to high strength concrete. The deterioration caused by cavitation can take the form of tearing out of large pieces of concrete. Cyclic drying and capillary suction occur in the concrete just above sea level, and water carries the dissolved salts into the concrete. Subsequent evaporation causes these salts to crystallize in the pores, producing stresses that can cause micro-cracking. Below the low tide line is an area where continual immersion occurs. In general, this is an area of less severe attack to steel and concrete structures compared with those of exposed at tidal or splash-zones. Popovics (1987) has found that sulfate attack from sea water is less than what would be expected from its ion concentration. Thus, Popovics has proposed that chlorides, reacting with the tri-calcium aluminates, inhibit the expansive phenomena associated with sulfate reactions.

Keys to improved durability against deterioration caused by the marine environment include: (1) using concrete with low permeability; (2) limiting the C_3A content of the cement; or (3) using pozzolans for partial cement replacement. Mehta has indicated that permeability of concrete is the most important factor influencing concrete performance exposed to marine environment (Mehta 1980; 1988b). Low permeability can reduce penetration of salt, sulfate and water; besides low permeability concrete normally has high strength and good erosion resistance to the marine environment. The rate of deterioration caused by seawater depends on the quantity of seawater absorbed by the concrete, so that all factors that contribute to obtaining a lower permeability will improve concrete structures' resistance to attack by seawater and marine environments. Mehta (ibid.) has noted that concretes containing even high tri-calcium aluminate cements have excellent service lives in a marine environment if the permeability is sufficient low.

Low permeability concrete can be achieved by the use of a low water/ cement ratio, an appropriate choice of cementitious materials, good compaction, absence of cracking and well curing. Since salt in water can contribute to the corrosion of reinforcing steel in concrete structures, normally 50–75 mm of dense concrete cover is required for reinforced concrete structures in the marine environment. In order to avoid the alkali silica reaction, Mehta (1988b) has recommended using cement content not below 400 kg/ m^3; a content of tricalcium aluminate below 12 percent and preferably between 6–10 percent; and the use of good quality aggregate. For ocean structures, it is preferable to use blast furnace slag cement, fly ash cement or pozzolanic cement. Due to pozzolanic reaction of slag, fly ash and pozzolans, a good amount of calcium hydroxide will be consumed and a large amount of secondary C–S–H generated. As a result, concrete produced by these cements has a finer pore structure, which largely reduces the transport rate of both sulfate and chloride ions.

2.4 Degradation caused by disasters

2.4.1 Degradation caused by fire

The development of natural fires can be divided into three phases: (1) pre-flashover (or growth period); (2) post-flashover (or fully developed fire); and (3) the decay period. In the pre-flashover period, any combustion is restricted to small areas of the compartment; therefore, only localized rises in temperature can occur which may still be substantial. The overall or average rise in temperature within the bounded fire compartment will be very small and indeed at this stage there may be no obvious signs of a fire. A large number of incident fires never get beyond the pre-flashover stage since there may be neither sufficient fire load nor air supply to allow the fire to grow. Flashover occurs when the fire ceases to be a local phenomenon within the compartment and spreads to all the available fuel within the compartment. In post-flashover period, the rate of temperature rise throughout the compartment is high and 200 °C per minute can be reached. In compartment fires, maximum temperatures of over 1000 °C are possible. The rate of temperature rise continues until the rate of fuel combustion decreases. It is during the post-flashover period that structural elements are exposed to the strongest effects of the fire, during which loss of integrity is likely. Once the temperature reaches a peak, the fire proceeds into its decay phase. In this stage, the temperature starts to decrease. However, owing to thermal inertia, the temperature inside the structure will continue to increase for a short while in the decay period, i.e. there will be a time lag before the structure starts to cool.

The effect of fire on concrete depends on the fire environment and the properties of the concrete itself. The environment caused by fire includes the highest temperature reached and the length of time that the highest temperature is maintained, and the rate of temperature rise, the total length of the fire. The resistance of a concrete to fire depends mainly on cement type, water/cement ratio, aggregate type, microstructure, cement content, and the thickness of the cover to reinforcement. In general, concrete is relatively good at resisting the effects of damage by fire and has a good service record with regard to fireproofing. Concrete is a poor conductor of heat and, in fact, the thermal conductivity is reduced as the temperature increases. Unlike wood and plastics, concrete is incombustible and does not emit toxic fumes on exposure to high temperature. Unlike steel, which loses its 90 percent strength at 700 °C, when subjected to temperature of the order of 700–800 °C, concrete is able to retain strength for reasonably long periods, thus permitting rescue operations by reducing the risk of structural collapse. For example, in 1972 when a 31-storey reinforced concrete building in São Paulo, Brazil, was exposed to a high-intensity fire for over 4 hours, more than 500 people were rescued because the building maintained its structural integrity during the fire.

The effect of increasing temperature on cement paste depends on the degree of hydration and moisture state. If the cement paste is seriously weakened, the concrete is likely to disintegrate. A well-hydrated cement paste consists of mainly C–S–H, CH, and ettringite. By the time the temperature reaches about 100 °C, the free water and absorbed water have evaporated, combined with loss of the ettringite in the cement paste. This desiccation process is accompanied by a shrinkage cracking. The degree of desiccation is dependent on the rate of temperature rise and the length of exposure to high temperature. It may be noted that due to the considerable heat vaporization needed for conversion of water into steam, the temperature of concrete does not rise until all the evaporated water has been removed. By the time the temperature reaches about 300 °C, a small amount of chemically combined water has been lost. The actual water loss depends on time and temperature. With this loss of chemically combined water there is a drop in the strength of the concrete. During this stage, there is a more pronounced chemical change in the cement paste. Many concretes show a distinct color change when exposed to temperatures in this range, usually taking on a noticeable pinkish tinge. Further dehydration of cement paste due to decomposition of CH begins at 500 °C, but a temperature of 900 °C is required for complete decomposition of the C–S–H. In summary, concretes differ considerably in their individual capabilities of fire resistance, which is usually governed by three factors (Malhotra 1956): (1) the amount of chemically combined water that is lost; (2) chemical changes that destroy the bond between the cement paste and the aggregate; (3) the gradual degradation of the hardened cement paste. When the concrete cools, the calcium oxide absorbs moisture, converting to calcium hydroxide and disintegration of the affected concrete occurs.

The porosity and mineralogy of aggregate have an important influence on the behavior of concrete exposed to fire. Porous aggregates, depending on the rate of heating and aggregate size, permeability, and moisture content, may themselves be susceptible to disruptive expansions leading to pop-outs. Low-porosity aggregates should, however, be free of problems related to moisture movement. Siliceous aggregates containing quartz, such as granite, and sandstone, can cause distress in concrete at about 573 °C because the transition of quartz in microstructure is associated with a sudden expansion. They may undergo rapid changes in volume at certain temperatures and this sudden increase in volume may cause disruption of the concrete. In the case of carbonate rocks, a similar distress can begin above 700 °C. The experiment results show that concrete made of carbonate and lightweight aggregates has a superior performance at a high temperature compared to concrete made of siliceous aggregates. At 650 °C, the former retained 75 percent of original strength but the latter only 25 percent. Except for the reasons presented earlier, less difference in the coefficient of thermal expansion between the matrix and aggregate is another reason for achieving the better performance during a fire. Aggregates normally expand progressively when heated, while hardened cement paste expands only to a certain point and then begins to shrink. The combined effects of expanding aggregates and shrinking matrix result in a weak concrete and can lead to a crack (Mailvaganam 1992). Igneous rocks are reasonably stable up to about 1000 °C, but there may be some spalling with a rapid rise and fall in temperature. Some artificial lightweight aggregates, such as expanded clay and shale, sintered pulverized fuel ash, and blast furnace slag, are manufactured at temperatures above 1000 °C, and, hence, very stable at temperature below this level. In general, concrete with high coefficient of thermal expansion are less resistant to temperature changes than concretes with lower coefficients.

It is interesting to note that the specimens tested hot with pre-load in compression demonstrate a better performance than the cases without pre-load. This is almost certainly due to the closing effect of the pre-load to the cracks formed due to any thermal incompatibility between the aggregate and the matrix. Compared to its compressive strength, the flexural strength and elastic modulus of a concrete drop more rapidly under same high temperature. This can be because flexural strength and elastic modulus of a concrete are more sensitive to micro-cracking in the transition zone formed during fire than compressive strength of the concrete is. Besides, the loss of strength is considerably lower when the aggregate does not contain silica (Malhotra 1956). Mixes also show some influence on the strength loss of concrete when subjected to fire. Leaner mixes appear to suffer a relatively lower loss of strength than rich ones. Normal concrete made with siliceous or limestone aggregates show a color change with temperature. This change is not permanent and reflects the maximum temperatures reached during the fire. Thus, the residual strength can be approximately judged; generally, concrete whose color has changed beyond pink is suspect and concrete past the grey state is probably friable and porous (Mailvaganam 1992). As far as creep is concerned, it is much greater at elevated temperatures than at ambient conditions.

The damage caused by fire in concrete include expansive spalling due to steam and water vapor (especially for high performance concrete), cracking caused by different thermal expansion between cement paste and aggregates at above 500 °C, and breakdown of cement paste due to complete loss of chemically combined water when the temperature approaches 950 °C. Depending on the moisture content, explosive spalling may occur in the early stages of exposure to fire. There is a series of violent disruptions, each of which removes a shallow layer of concrete from the surface over a localized area. Internal cracks develop almost parallel to the faces and eventually a plate of concrete becomes detached. The detachment often occurs at the position of the outside face of the reinforcement. There will be stresses built up in the concrete structural elements by the temperature rise to the maximum reached during the fire and the return to ambient temperature when the fire is extinguished. Concrete damaged by fire exhibits significant strength loss on cooling. The thermal expansion and contraction stresses developed in concrete structures depend on a number of factors including the restraint imposed on the elements by other members in the concrete structures and the thermal gradient through the members.

2.4.2 Degradation caused by earthquakes

Earthquakes are considered one of the most devastating natural disasters. Before discussing renovation methods for earthquake-damaged structures, the structure of the earth and some related fundamental knowledge on seismology should be understood. The earth is roughly spherical with an average radius of 6400 km, consisting of three spherical shells with quite different physical properties in its inner structure. The outer shell is a thin crust of thickness varying from a few kilometers to a few tens of kilometers; the middle shell is the mantle, about 2900 km thick; and the inner shell contains the core with a radius of approximately 3500 km. The core can be further distinguished into inner core and outer core. The crust consists of various types of rock, differing in thickness depending on its position in the ocean or on a continent. The crust in the continental part has a thickness about 30–40 km, but reaching 70 km under high mountains. The crust under the oceans has a thickness of only about 5 km. The mantle is mainly made of uniform ultra-basic olivine rock. Its outer 40–70 km shell together with the crust is usually referred to as the lithosphere, directly under which is a layer of soft visco-elastic asthenosphere a few hundreds of kilometers thick. The wave velocity in the asthenosphere is obviously lower than those in its neighboring rocks. The lithosphere and asthenosphere consist of the upper mantle, below which the lower mantle extends to a further layer with around 1900 km in thickness. The core is composed of outer and inner cores. It is generally regarded that the outer core is in a liquid state because it has been found that transverse waves cannot spread through the outer core.

Earthquakes are naturally occurring, broad-banded vibratory ground motions, resulting from a number of causes including tectonic ground motions, volcanoes, landslides, rockbursts, and man-made explosions (Chen and Scawthorn 2003). They occur as a result of a sudden rupture of the rock plate which constitutes the earth. The vibrations are generated by the occurrence of sudden rupture and are termed earthquake motion or earthquake. Among various causes, tectonic-related earthquakes are the largest and most important. According to the theory of plate tectonics, the fracture and sliding of rock along faults within the earth's crust induce strong vibrations and energy release, thus earthquakes (ibid.). A fault is a zone of the earth's crust within which its two sides have moved. Deformation due to plate tectonics is a very slow but persistent process. In a very long time period, deformation accumulates elastic strain in the crust. Strain energy is then accumulated in the crust and it may crack suddenly when the strength of rock is overcome and the accumulated energy is transferred into earthquake waves. An earthquake thus occurs. Earthquakes initiate a number of phenomena or agents, termed seismic hazards, which can cause significant damage to the built environment. For most earthquakes, shaking is the dominant and most widespread agent of damage. Shaking near the actual earthquake rupture lasts only during the time when the fault ruptures, a process that takes seconds or at most a few minutes. The seismic wave generated by the rupture then propagates long after the movement on the fault has stopped.

The rupture of rocks causing an earthquake extends over a quite some distance. The point at which the rupture initiated is called the focus (or hypocenter). Its depth, distance from focus to site, is called focal (hypocentral) depth. The point on the earth's surface straight above the focus (hypocenter) is called epicenter, which is the projection on the surface of the earth directly above the hypocenter. The distance from focus or epicenter to any point is called hypocentral distance or epicentral distance (see Figure 2.12).

Seismic waves originating at the hypocenter spread in all directions and reach the earth's surface following different paths. Earthquake energy is radiated over a broad spectrum of frequencies through the earth, in body wave and surface wave. The original body seismic waves can be distinguished as P and S waves. Physically a P wave is compressive in nature and vibrates parallel to the direction of propagation, so P wave is also called longitudinal wave. An S wave oscillates in a direction perpendicular to the direction of propagation. Its motion is primarily transverse. Hence an S wave is also called a transverse wave. An S wave is further classified as one having apparently only a horizontal component, which is called an SH wave, and one having only a vertical component, called an SV wave.

The propagation velocity of P and S waves is determined by the density

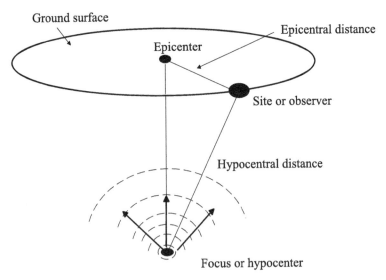

Figure 2.12 Earthquake location notation.

and elastic properties of the medium in which the wave travels. The velocities of the P wave can be expressed as:

$$C_L^2 = \frac{E}{\rho} \frac{1 - v}{(1 + v)(1 - 2v)} = \frac{\lambda + 2\mu}{\rho} \tag{2.39}$$

for the plain strain case; and

$$C_L^2 = \frac{E}{\rho} \tag{2.40}$$

for a plain stress case.

The velocity of the S-wave can be expressed as

$$C_T^2 = \frac{\mu}{\rho} \tag{2.41}$$

Thus, for the plane strain case, we have:

$$\frac{C_L^2}{C_T^2} = \frac{2(1 - v)}{1 - 2v} = 2 + \frac{2v}{1 - 2v} \tag{2.42}$$

For the plane stress case, we have:

$$\frac{C_L^2}{C_T^2} = 2(1 + v) \tag{2.43}$$

Thus it can be seen that the P wave travels faster than S-wave does and will reach the earth's surface first. In general, the velocities of the P-wave are 3–8 km and those of the S wave are 2–5 km. The fault in the earth can be represented by 9 dipole forces. The displacement field can be expressed as:

$$U_n(x, t) = M_{ij} * G_{ni, j}$$

$$= \frac{15 v_n v_i v_j - 3 v_n \delta_{ij} - 3 v_i \delta_{nj} - 3 v_j \delta_{ni}}{4\pi\rho} \frac{1}{r^4} \int_{r/C_1}^{r/C_2} \tau M_{ij}(t - \tau) d\tau$$

$$+ \frac{v_n v_i v_j}{4\pi\rho C_1^3 r} M_{ij}\left(t - \frac{r}{C_1}\right) \tag{2.44}$$

$$+ \frac{6 v_n v_i v_j - v_n \delta_{ij} - v_i \delta_{nj} - v_j \delta_{ni}}{4\pi\rho C_1^2} \frac{1}{r^2} M_{ij}\left(t - \frac{r}{C_1}\right)$$

$$-\frac{6v_nv_iv_j - v_n\delta_{ij} - v_i\delta_{nj} - 2v_j\delta_{ni}}{4\pi\rho C_2^2}\frac{1}{r^2}M_{ij}\left(t - \frac{r}{C_2}\right)$$

$$-\frac{v_nv_i - \delta_{nj}}{4\pi\rho C_1^3 r}v_jM_{ij}\left(t - \frac{r}{C_1}\right)$$

The near field terms in this displacement field for a dislocation are proportional to r^{-4}, and far field terms are proportional to r^{-1} as well as particle velocity. The intermediate-field terms are proportional to r^{-2}. This is, however, a slightly misleading name, because there is no intermediate range of distance in which these terms dominate. In practice, they are found to be small in far field and are often comparable to the near-field displacements at distances where the latter are appreciated. The distance between the seismograph and the focus of the earthquake can be expressed as:

$$d = \frac{\Delta t_{p-s}}{1/C_T - 1/C_L} \tag{2.45}$$

where Δt_{p-s} is the difference in arrival time for the first P wave and the S wave.

The P wave and the S wave are sometimes called body waves since they are generated in the earth and can spread in all directions at whatever depth. Since there are a number of discontinuities in the earth's crust, these body waves are subject to complex phenomena, such as reflection, refraction, diffraction, and attenuation. Reflection of the P and S waves at these discontinuities follows Snell's law. The modulus of elasticity of the medium usually increases with depth in the earth. Accordingly, the angle of incidence of body waves originating at great depth comes closer to 90° near the surface, following Snell's law. As such, the seismic wave is frequently assumed to be incident normally at the earth's surface. The reflection/refraction of a body wave caused by discontinuity in the propagation medium generates other waves. In particular, the waves spread along the earth's surface. These waves are thus called surface waves. It should be noted that it is easy to generate surface waves in the process of wave propagation since the near-surface part of the crust is made of layers of materials of different mechanical properties. Surface waves are classified into a Rayleigh wave and a Love wave.

Rayleigh waves, produced by the interaction of P waves and SV waves with the earth's surface, involve both vertical and horizontal particle motion. A Rayleigh wave can also be generated at the free surface of a half space. In this case, their particle motion is in the vertical plane only, with a trace of reverse ellipse having a ratio of vertical to horizontal amplitudes about 3:2. The amplitude attenuates very fast along the vertical axis, reducing to about one-fifth after one wavelength. Rayleigh waves exist near the surface of a homogeneous elastic half-space. On the other hand, if the half-space is overlain by a layer of material with lower body wave velocity, Love waves can develop. Love waves were discovered first in real earthquake records and their existence was proved theoretically by Love (Kramer 1996). A Love wave results from the interaction of an *SH* wave with a soft surficial layer and has no vertical component of particle motion. Thus, a Love wave has only the horizontal displacement component which is normal to the direction of propagation, while a Raylaigh wave has both horizontal and vertical displacement components.

Body waves and surface waves can be recorded during an earthquake by the corresponding transducers. The main component causing maximum acceleration is generally to be an *S* wave but some authors have also attributed it to a surface wave. Very often, no distinction is made between body and surface wave in strong motion records. This is because a body wave is distorted due to irregularities in the topography or medium, and a surface wave is subject to secondary excitation. An *S* wave has been considered significant in earthquake engineering to date and is assumed to be incident normally at the earth's surface.

Earthquakes are complex multidimensional phenomena, the scientific analysis of which requires measurement. Earthquake strength is defined in two ways: first, the strength of shaking at any given place (called the intensity) and second, the total strength (or size) of the event itself (called magnitude). Prior to the invention of modern scientific instruments, earthquakes were qualitatively measured by their effect or intensity. With the deployment of seismometers, an instrumental quantification of the entire earthquake event, i.e. the unique magnitude of the earthquake, becomes possible. The amplitude of ground motion does increase with earthquake size for a measuring point at the same epicentral distance and having the same soil properties. Hence if we take into account the differences in epicentral distance and soil properties, it is possible to estimate the magnitude of the ground motion.

Quantitative instrumental measures of magnitude include engineering parameters such as peak ground acceleration, peak ground velocity, the Housner spectral intensity, and response spectra. An individual earthquake is a unique release of strain energy – quantification of this energy has formed the basis for measuring the earthquake event. Magnitude is a quantitative measure of the size of an earthquake, related indirectly to the energy release and independent of the place of observation (Dowrick 2003). Earthquake magnitude is calculated from amplitude measurements on seismograms, and is on a logarithmic scale expressed in ordinary numbers and decimals. Richer was the first to determine earthquake magnitude. He proposed a formula (1935):

$$M_L = \log_{10} A + k \log_{10} \Delta + C \qquad (2.46)$$

where M_L is the local magnitude, A is the maximum amplitude, of ground motion, in microns recorded on a standard Wood-Anderson short-period torsion seismometer, at a site 100 km from the epicenter; Δ is epicentral distance; k and C are constants. The equation is valid for the epicentral distance, Δ, less than 600 km, and hence Richter magnitude is now more precisely called local magnitude to distinguish it from magnitude measured in the same way but from recordings on long-period instruments, which are suitable for more distant events. Subsequently, a number of other magnitudes have been defined, the most important of which are surface wave magnitude M_S, body wave magnitude M_B, and moment magnitude M_W. Gutenberg and Richter (1956) proposed a surface wave magnitude scale, which is based on the amplitude of Rayleigh wave with a period of about 20 sec.

$$M_S = \log_{10} A + 1.66 \log_{10} \Delta + 2.0 \tag{2.47}$$

where A is the maximum ground displacement in micrometers and Δ is the epicentral distance of the seismometer measured in degrees (360° corresponding to the circumference of the earth). M_S is most commonly used to describe the size of shallow (less than about 70 km focal depth), moderate to large earthquakes.

Due to the observation that deep-focus earthquakes commonly do not register measurable surface waves with periods near 20 sec, for those deep-focus earthquakes, the body wave magnitude should be used because the surface is so small for such a case (Gutenberg and Richter 1956).

$$M_b = \log_{10} A - \log_{10} T + 0.01\Delta + 5.9 \tag{2.48}$$

where A is the p-wave magnitude in micrometers and T is the period of the p-wave (usually about 1 second). Body wave magnitudes are more commonly used in eastern North America, due to the deeper earthquakes there. All the three scales M_L, M_b and M_S suffer from saturation at higher values.

For strong earthquakes, the measured ground-shaking characteristics become less sensitive to the size of the earthquake than for small earthquakes. This phenomenon is referred as saturation. To describe a strong earthquake, the moment magnitude should be used (Kanamori 1977; Hanks and Kanamori 1979):

$$M_w = \frac{\log_{10} M_0}{1.5} - 10.7 \tag{2.49}$$

where the seismic moment, M_0, is defined as (Wyss and Brune 1968):

$$M_0 = \mu AD \tag{2.50}$$

where μ is the shear modulus of the material along the fault, A is the area of the dislocation or fault surface, and D the average amount of displacement or slip on that surface. The moment magnitude M_w is regarded as the most reliable and generally preferred magnitude (Dowrick 2003) and it overcomes the saturation problem of other magnitude scales by incorporating seismic moment into its definition. Seismic moment is a modern alternative to magnitude, which avoids the shortcomings of the latter but is not so readily determined. Comparatively, M_w and M_S are almost identical up to magnitude 7.5 (Chen and Scawthorn 2003).

Magnitude, which is defined on the basis of the amplitude of ground displacements, can be related to the total energy in the expanding wave front generated by an earthquake. Hence, the total energy released during an earthquake is often estimated from the relationship between the energy and magnitude by Richter:

$$\log_{10} E = 11.8 + \log_{10} M \tag{2.51}$$

where E is the energy expressed in ergs (1 erg = 10^{-7} N-m). M can be either surface or moment magnitude. It implies that a unit change in magnitude corresponds to a $10^{1.5}$ or 32-fold increase in seismic energy release, two magnitude units increase is equivalent to 1000 times more energy. A magnitude 5 earthquake therefore would release only about 0.001 times the energy of a magnitude 7 earthquake.

Another measurement for the size of earthquake is seismic intensity scale or earthquake intensity. The term "earthquake intensity" has been used for more than 170 years, before the term "earthquake magnitude" and is still widely used. Earthquake intensity is an index of the local destructive effect of an earthquake in a small area (Hu *et al.* 1996). This term was introduced to be a physical quantity, but through qualitative or fuzzy definitions. Earthquake intensity is measured through four categories of macroseismic phenomena: human feeling; damage to artificial structures; response of objects; and change in natural conditions. This term is used as a simple measure of earthquake damage; as a simple macroseismic scale for the strength of earthquake motion for seismologists; and as a rough but convenient index to sum up experience on earthquake engineering construction and to depict zones of seismic hazard (Hu *et al.* 1996). Since earthquake intensity depends on macroseismic phenomena, the intensity of ground motion varies considerably from place to place even for the same earthquake. There are a number of different scales for earthquake intensity and, sometimes, they cause confusion. In general, seismic intensity is a metric of the effect, or the strength, of an earthquake hazard at a specific location. The seismic intensity scale is decided by human judgment based on various structures.

This macroseismic idea of earthquake intensity was a natural product when no instruments were available to measure any physical quantity of ground motion. The seismic intensity is usually applied for qualitative measures of location-specific earthquake effects. There is no one-to-one correspondence with physical quantities of earthquake motion such as acceleration, velocity and displacement. It is also difficult to find a reliable relationship between magnitude, which is a description of the earthquake's total energy level, and intensity, which is a subjective description of the level of shaking of the earthquake at specific sites, because shaking severity can vary with building type, design and construction practice, soil type, and distance from the event (Chen and Scawthorn 2003). However, the seismic-intensity scale bears some correlation to maximum acceleration over some range.

Numerous intensity scales had been developed in pre-instrumental times. The most popular intensity scale is called MMI (Mercalli Modified Intensity, initially, after the Italian Mercalli, then modified by Richter) (Wood and Neumann 1931). MMI is a scale defining the level of shaking at specific sites on a scale of I–XII. The details of the scale are as follows:

I   Not felt except by a very few under especially favorable circumstances.

II   Felt only a few persons at rest, especially on upper floors of buildings; delicately suspended objects may swing.

III   Felt quite noticeably indoors, especially on upper floors of buildings, but many people do not recognize it as an earthquake; standing motor cars may rock slightly; vibration like passing of truck.

IV   During the day felt indoors by many, outdoors by few; at night some awakened; dishes, windows, doors disturbed; walls make cracking sound; sensation like heavy truck striking building; standing motor cars rocked noticeably.

V   Felt by nearly everyone, many awakened; some dishes, windows broken; a few instances of cracked plaster; unstable objects overturned; pendulum clocks may stop.

VI   Felt by all, many frightened and run outdoors; some heavy furniture moved.

VII   Everybody runs outdoors; some chimneys broken.

VIII   Damage slight in specially designed structures; considerable in ordinary buildings; with partial collapse, great in poorly built structures; panel walls thrown out of frame structures; fall of chimneys, factory stacks, columns, walls; heavy furniture overturned; persons driving motor car disturbed.

IX   Damage considerable in specially designed structures; well designed frames thrown out of plumb; great in substantial buildings, with partial collapse; building shifted off foundations; ground sand and mud; underground pipes broken.

X   Some well-built wooden structure destroyed; most masonry and frame structure destroyed; ground badly cracked; rails bent; water splash over banks.

XI Few structures remain standing; bridges destroyed; underground pipe-line completely out of service.

XII Damage total; practically all the works of construction are damaged greatly or destroyed; wave seen on ground surface; lines of sight and level are distorted; objects thrown into the air.

Generally speaking, there are two strategies in designing earthquake-proof buildings and infrastructures. One is to strengthen the infrastructure by providing additional reinforcement at critical sections. The other is to pro-vide a base isolation system. Strengthening is the conventional approach and has been used for many years. The limitation of this strategy is that the stiffer the structure, the larger the force generated by earthquakes due to increased stiffness. This approach always results in high floor accelerations for stiff buildings, or large inter-story drifts for flexible buildings. Because of this, the building contents and non-structural components may suffer sig-nificant damage during a major earthquake. As an alternative, in order to minimize inter-story drifts, in addition to reducing floor accelerations, the concept of base isolation is increasingly being adopted. The base isolation has also been referred to as passive control. Base isolation system is promis-ing in its ability to decouple the upper structure from the potentially dam-aging earthquake vibrations induced by ground motion. This decoupling is achieved by increasing the flexibility of the system, together with providing appropriate damping, through installation of certain devices between the building and the supporting foundation, so as to separate or isolate the motion of the building from that of the ground. The base isolator tries to prevent or reduce the energy transmitted into the structure, to absorb or consume the energy in the structure by specially designed energy dissipation devices, and/or to modify the structural property to minimize the structural vibration with or without energy input.

The applicability of the concept of base isolation can also be applied to the isolation of sensitive equipment installed inside a building from unwanted floor vibrations through installation of an isolation system between the equipment base and the supporting floor. Moreover, hospitals, police and fire stations, and telecommunication centers always contain valuable equipment and these infrastructures should remain operational immediately after an earthquake. Protection of valuable equipment in these facilities through, for example, the base isolation is thus desirable. There are generally two common methods to create base isolation. One is achieved by providing a pure sliding system and the other is achieved by using seismic isolators. The first method provides a sliding or friction surface between the foundation and the base of the structure to increase flexibility in a structure (Li *et al.* 1989). The shear force transmitted to the superstructure across the isolation interface is limited by the static friction force (see Figure 2.13). The coefficient of friction is usually kept as low as is practical. Sliding isolation systems have been successfully used for nuclear power plants,

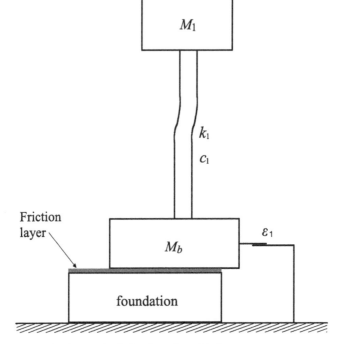

Figure 2.13 Base isolation based on friction.

emergency fire water tanks, large chemical storage tanks, and other important structures.

In the second method, some bearings of relatively low horizontal stiffness,
but high vertical stiffness are installed between the upper structure and its
foundation. With such devices, the natural period of the structure will be
significantly lengthened and shifted away from the dominant high frequency
range of the earthquakes. The rubber bearing is the most popular type of
base isolator and is easy to manufacture (see Figure 2.14). A typical elastomeric bearing is composed of alternating layers of steel and hard rubber and,
therefore, is known as a laminated rubber bearing (Chen and Scawthorn
2003). This type of bearing can be made very stiff in the vertical direction to
sustain vertical loads and sufficiently flexible in the horizontal direction to
isolate the horizontal vibration. The ability to deform horizontally enables
the bearing to significantly reduce the shear forces induced by the earthquake. This type of base isolation is very good at reducing the high acceleration, or the high frequency motion. In reality, the reduction in seismic loads
transmitted to the superstructure through installation of laminated rubber
bearings is achieved at the expense of large relative displacements across the
bearings. Normally substantial damping is introduced into the bearings or
the isolation system to alleviate this large displacement problem.

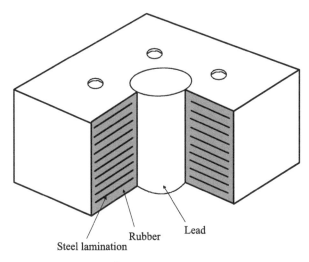

Steel lamination
Rubber
Lead

Figure 2.14 Base isolator.

In summary, base isolation is very attractive and its main advantages are (Hu *et al.* 1996):

1   very small relative deformation in the superstructure which avoids damage to structural and nonstructural elements;
2   economical and easy design of the superstructure;
3   possibility of building structures for the very worse conditions such as very high seismicity;
4   safety of important structures or systems during unexpected strong earthquakes.

Strong motion observation and test results all show that base isolation, when properly designed, is reliable and effective, reducing high-frequency motion to a half or one-fifth (Hisano *et al.* 1990). Numerous researchers have studied the dynamic behaviors of base-isolated structures under earthquakes using different devices for seismic isolation. Kelly (1993) gave a detailed procedure for the analysis of rubber isolation systems mounted on building structures. To overcome the very large design displacements for the isolators, Kelly (1999) suggested some strategies, instead of introducing supplemental dampers alongside the isolators (Hall 1999; Hall and Ryan 2000), such as the adoption of a gradual increasing curvature for the disk of the friction pendulum system, and the use of increased stiffness and increased damping for elastomeric isolators.

3   Inspection and evaluation

Inspection and evaluation are the beginning of any rehabilitation and renovation works. Inspection and evaluation of infrastructure condition are typically conducted by a multidisciplinary team that includes architectural,
structural, mechanical, and electrical engineers. During the inspection and
evaluation process, the participants verify whether the infrastructure is generally built up in conformance with the original construction documents,
sketch out some of the observed structural elements and note their condition. Inspection of existing structures is characterized as the process by
which any deterioration in the structure is observed and recorded while
evaluation normally refers to all works to determine and assess the condition of the structure based on the results obtained during inspection. It is
intended to evaluate the general condition of the structure and identify any
areas of deficiency. The followed repair work is largely based on the conclusions of the previous inspection and evaluation. According to the FIP *Guide
to Good Practice of Inspection and Maintenance of Reinforced and
Pre-stressed Concrete Structures* (Bilger *et al.* 1986), the main reasons for
inspection and maintenance of structures are:

1   to control the functional requirements and provide assurance that the
    structure is safe and fit for its designated use;
2   to identify actual and potential sources of trouble and misuse at the
    earliest possible stage, and to prevent serious deterioration and failure,
    consequently increasing the service life;
3   to monitor the influence of the environment since there is a relation
    between the aggressiveness of the environment and the durability of
    concrete structures; the increasing content of aggressive agents in the
    atmosphere, acid rain and the use of de-icing salts considerably diminish
    the life expectancy of concrete structures;
4   to provide feedback of information for designers, constructors and
    owners on the factors governing maintenance problems, to which neces-
    sary attention should be best paid during the design and construction
    stages;
5   to provide information on which decisions concerning preventive

measures and work can be made; it is cheaper to repair minor damage at an early stage than to replace major components or the total structure in the event of failure.

A regular inspection can thus: (1) avoid unexpected damages that would lead to failure and casualties; (2) ensure continued operation of a facility; and (3) protect capital investment. Whether a structure is inspected for (2) and (3) depends on the owner and on economic concerns. For example, if a bridge is owned by a city, (2) and (3) are important concerns. On the other hand, for a privately owned factory building, the owner may have been more tolerant of damage and usually does not inspect their buildings until signs of serious degradation appear. Under normal conditions, frequent and well-organized inspections (and repair at an early stage of deterioration) are probably the most effective method of reducing maintenance costs over the lifetime of the structure.

Inspection intervals depend on engineering judgment and have to be decided according to many factors such as:

1 the type and importance of the structure;
2 the loading conditions and the severity of loading;
3 the consequences of failure;
4 the presence of an aggressive environment that influences the speed of deterioration;
5 the history of the structures;
6 the location of the structures;
7 economics.

The FIP *Guide to Good Practice* has listed three structure classes as follows:

Class 1: where possible failure would have catastrophic consequences and/or where the serviceability of the structure is of vital importance to the community;
Class 2: where possible failure might cost lives and/or where the serviceability of the structure is of considerable importance;
Class 3: where it is unlikely that possible failure would lead to fatal consequences and/or where a period with the structure out of service could be tolerated.

Also, the guide has listed three environmental and loading conditions as:

1 Very Severe: the environment is aggressive and there is a cyclic or fatigue loading;
2 Severe: the environment is aggressive, with static loading, or the environment is normal, with cyclic or fatigue loading;
3 Normal: the environment is normal, with static loading.

The proposed inspection intervals (in years), for three classes of structure and for three environmental and loading conditions, according to the FIP Guide to Good Practice (Bilger *et al.* 1986), are summarized in Table 3.1. These intervals should be regarded as absolute maxima and, in most cases, more frequent inspections should be planned (Campbell-Allen and Roper 1991). It should be noted that regular inspection is only effective when combined with consistent and orderly reporting. The results of all inspections should be accurately and fully recorded, so that a complete service history of the structure is readily available at any time. Special forms for recording observations during inspection and evaluation have to be developed and levels of deterioration well defined to eliminate subjective judgments of the inspector as much as possible. Certainly, design of such forms needs experience and careful thought.

A structure can be investigated in a variety of ways, depending on the type of structure, the apparent condition of structure, and whether the original design drawings are available. The American Society of Civil Engineers has published SEI/ASCE 11-99, *Guideline for Structural Condition Assessment of Existing Buildings* (ASCE 2000). This standard recommends a multilevel approach to structural assessment of buildings. The inspection generally includes a preliminary investigation (desk study and condition survey), detailed inspection, and assessment of the conditions. Some of the information about the deterioration can be obtained through a systematic review of service records and the original design and construction details of the structure, i.e. a desk study. If the records are incomplete or unavailable, intelligent observation and sound judgment should be applied in planning a field investigation program. In fact, a detailed field investigation is normally required following a desk information collection and review. This detailed investigation program may include visual examination, non-destructive testing, and collection of specimens for laboratory testing. The results of the field investigation and laboratory tests will have a significant influence on the selection of materials and methods of repair. By such inspections, it is

Table 3.1 Proposed inspection intervals by FIP

Environmental and loading conditions	Classes of structure					
	1		2		3	
	Routine	Extended	Routine	Extended	Routine	Extended
Very Severe	2*	2	6*	6	10*	10
Severe	6*	6	10*	10	10	–
Normal	10*	10	10	–	Only superficial inspections	

Note: *Midway between Extended Inspections.

then possible to build up a database relating to a particular structure, or to a class of structures. The development of defects and deterioration can then be traced and decisions can be made on the point in time when repair becomes an essential and economical measure.

3.1 Preliminary investigation

As one part of a systematic inspection and evaluation of structures, preliminary investigation helps in understanding the history and current condition of structures, predicting the extent of possible defects so that the following detailed investigation could have a definite object in view. It also provides a sound basis for determining the materials and structural tests in detailed investigations, which tests are necessary and which ones are not. Hence, it can minimize the investigation costs by eliminating unnecessary material and structural test. Sometimes, a preliminary investigation is the only action needed or the results need to be presented to the building owners to let them to decide whether a thorough checking of conditions of structures is needed or not. Generally speaking, according to Chandler (1991), the purpose of preliminary investigations is:

1   to obtain an elementary description of the structure from a study of its documented history and a close look at the structure and fabric;
2   to get an overview of the state of the concrete structure with regard to its present safety;
3   to advise the designing engineers and the building proprietor on the significance of performing a detailed registration and evaluation of the state;
4   to plan the necessary sampling and site investigations of the concrete structure which are needed to determine the causes and extent of damage.

The preliminary investigations generally include a desk study and a condition survey.

3.1.1 The desk study

The desk study is the first task of inspection and evaluation, and it is also called gathering information (Allen *et al.* 1993). The purpose of desk study is to obtain a preliminary description of history and current condition of the structure through studying its documented records. The information obtained during desk study can help when estimating the possible deterioration the structure may most likely be subject to, making preliminary judgment of condition of structures and planning the next step of the survey. The investigation of reasons for the structural damage during the desk study can be conducted by observation, studying records and asking questions, supplemented if necessary by a certain amount of testing, and then interpreting the information obtained (Allen *et al.* 1993). All the sources of information concerning the design, construction and service life of the structure should be studied to learn as much as possible about the past and the current condition of the structure. The information can be used to determine the amount of investigatory work that is necessary. Attempts should also be made to clearly reconstruct the original design assumptions and theories since the design codes may have changed after the structure was built. It is always good practice to gain familiarity with the history of the structure as much as possible and to obtain any information or special circumstances such as ground subsidence or settlement in the locality during the desk study. A list of information that can be useful during desk study may include:

1   the original project plans and specifications, and permit and license applications;
2   the original design – drawings, structural calculations, modifications, and records;
3   construction materials information – reports, tests, and shop drawings;
4   construction records – correspondence, inspection reports, contractor's reports, site photographs, test results, "as built" records, survey notes, building inspection reports as well as public accounts of construction;
5   design and construction personnel – individuals involved may be interviewed to gain inside information regarding problem encountered;
6   service history of building – records of current and former owners, maintenance records, repair records, alternation records, settlement or unusually severe exposure, weather records, and seismic activity record;
7   any previous reports, surveys, and investigations.

Much of this documentation is generally unavailable for older structures, with increasing available documentation with younger structures. In general, these documents may be found in:

1   the owner's files;
2   city archives;
3   original design firms;
4   original contractor companies;
5   testing agencies;
6   building management firms;
7   large subcontracting companies;
8   equipment installers;
9   local record office and fire authority;
10 insurance companies;
11 engineering publications or the general press.

For historic buildings, design documents may also be found in:

1 university and local libraries;
2 historic societies;
3 state preservation offices.

For historical buildings, the desk study should also check what code requirements were applicable at the time of design. These should be compared to presently applicable codes and standards. Alteration plans and change orders should also be examined.

3.1.2 The condition survey

The recorded details obtained during desk study should always be subjected to checks on site, thus the condition survey is necessary since there may be discrepancies between what is shown on drawings and what was actually constructed. A properly planned condition survey is an important key to any renovation and rehabilitation project. A condition survey is aimed at carrying out a close and more intensive examination of all elements of the structure. It enables the engineer to confirm the correctness of the existing design information and to assess the current condition of the structure. A condition survey may also be required for other reasons such as routine maintenance, rehabilitation, modification of the service conditions, investigation of structural stability, and study of the performance of a material under a specific exposure condition (Mailvaganam 1992). The condition survey should examine variations from the original design and note the evidence of structural modification, deterioration of materials, weakness in structural members or connections, settlement or foundation problems, or unusual structural features. As a summary, a condition survey should evaluate the existing conditions of the structure, the level of deterioration, and the mechanism which leads to the existing conditions.

The condition survey could range from a cursory visual assessment to a complex operation involving time-intensive planning, execution and in-situ testing. In the case of a small structure, the whole job can be completed within hours, while for a large structure, comprehensive planning and a rehearsal of the full inspection are needed. However, the prime objectives of the condition survey are to identify the possible causes of any visible distress and to establish the structural integrity and satisfactory performance of the structure. Items required to carry out the condition inspection include: (1) safety facilities for the inspector; (2) tools for making notes of observations – papers, tape recorder, and so on; (3) removal of everything that prevents good visual access; (4) a tower or scaffold should be available for close inspections; and (5) a camera and video camera for recording deterioration conditions of the structure to keep in archive or to examine in detailed investigation.

Visual inspection, with some basic tests followed by a report on the findings, is probably the single most reliable and informative source of information for the condition survey. The purpose of the visual inspection is to make initial assessments of the structural form and detect all the symptoms of deterioration and defects. The visual inspection provides part of the primary evidence and produces a short list of potential causes, which can be used to plan the detailed investigation. Certainly the reliability of visual inspections depends on the perceptiveness and judgment of the inspector. Typical basic tests that may be conducted during the visual inspection include a measurement of geometry, deflection, displacement, identification of cracks spalling, erosion of concrete and other damages. During visual inspection, normally the following should be addressed:

1   verification of information gathered during the desk study;
2   the appearance of the concrete surface – texture, color, spalling and erosion;
3   presence of cracks – location, width, direction, and pattern;
4   exposed reinforcement;
5   deformation of the structure;
6   presence of water leaks;
7   damage to structural elements and to finishes;
8   any further defects and/or deterioration of the structure;
9   examination of the apparent defects reported and investigation of cause of the defects.

Once all the pertinent data has been collected during the desk study, a detailed review can then be carried out and a checklist for further assessing the condition of the structure should be prepared. An outline of such a checklist based on American Concrete Institute's recommendation (1984) is shown below. Certainly, the items in the list may vary depending on the type of structure being investigated and the purposes of investigation. The answers to items in the checklist should be put in files for the expert committee to make judgments and also as historical information for future investigation:

1   description of structure;
2   present condition;
3   nature of loading and detrimental elements;
4   original condition;
5   materials of construction;
6   construction practices;
7   initial physical properties of concrete;
8   special items for massive structures, i.e. dams, nuclear containment structures;
9   special items for pavements and bridges;

10 review of requirements of regulatory authorities;
11 preconstruction data, i.e. soil investigation, seismic information;
12 requirements for in-service testing;
13 acceptance criteria.

Part of such a condition survey involves finding out about the materials and construction methods originally used and often this information is not available if the inspection is carried out at some time after construction has been completed. The data obtained during condition survey should be presented graphically as detailed as possible. Normally, several possible outcomes of a preliminary investigation may include:

1 The structural system is adequate for the intended use.
2 The structural system is adequate for the existing condition but may not be adequate for intended future use.
3 The analysis is inconclusive, and further investigation is needed.
4 Not desirable to proceed with a further detailed investigation. This may be due to (a) excessive damage where the structural integrity cannot be economically restored; or (b) the owner's objectives cannot be satisfactorily met.

After preliminary investigation, an engineering judgment from experts with experience in conjunction with simplified analysis techniques can be made to determine rational structural demands and capacities for the critical structural members. Based on the preliminary investigation, an interim report should be prepared. The following items should be addressed in this interim report: (1) description of loading and performance criteria considered; (2) description of the structure; (3) description of preliminary evaluation process; (4) discussion of preliminary findings; and (5) recommendations regarding particular actions such as conducting a detailed assessment. Depending on the simplicity of the structure and the experience of the engineer, as well as the assistance of a material testing engineer, it is sometimes possible to make recommendations for rehabilitation without further investigation or a detailed study after preliminary investigations.

3.2 Detailed investigation and evaluation

Once the preliminary investigation has been completed and the search for existing records has been closed, the detailed investigation can be planned with more confidence. The detailed investigation is also normally required when the information provided in the preliminary investigation is not conclusive and further examination is required. The detailed investigation may also be performed directly by mandatory actions or by the client. The extent and scope of detailed investigations will depend on many factors, including the structural form, the quantity and quality of existing records, the degree of anticipated change of use and alteration of the structure, and the extent of deterioration (Kay 1992). The individual activities depend on whether an assessment of structural capacity or condition is being undertaken. To this extent, the purpose of making a detailed investigation of the state of the structure will be:

1   to identify the cause or causes of the defects and damage found;
2   to determine the size and extent of defects and damage;
3   to evaluate whether the defects and damage may be expected to spread further in the concrete structure to areas where no damage has yet been found (this applies especially to progress corrosion);
4   to estimate the significance of the damage in respect to the normative safety of the concrete structure;
5   to identify the areas of the concrete structure where rehabilitation (i.e. protection, repair or strengthening) is considered necessary;
6   to identify deficiencies of the structure and recommend alternatives for rehabilitation if it is determined to be carried out in the assessment of preliminary investigation.

The use of various testing methods may become necessary in the stage of detailed investigation though it may cost a lot. These methods may include some destructive tests, some non-invasive tests, some non-destructive tests, measurements of structural components, mapping of cracks, and installation of instrumentation to monitor movements of cracks. Testing techniques and equipment should be determined based on the extent and type of deterioration or damage and to the importance of the structure (FIP 1986). As far as is possible, non-destructive test methods should be used. Destructive testing refers to the process of observing, inspecting, and/or measuring the properties of materials, components, or system in a manner which may change, damage, or destroy the properties or affect the service life of the test specimen (ASCE 2000). Destructive testing of portions of a structure may be carried out on the structure in situ or on samples obtained on site and tested in the laboratory. Common destructive tests for concrete structures may include:

1   coring of existing concrete structures;
2   pull-out test;
3   reinforcing steel tensile test;
4   concrete core compression test;
5   cement content test of concrete;
6   petrographic analysis;
7   rapid soluble-chloride test;
8   beak out.

The non-invasive tests may include:

1   concrete cover depth measurement;
2   concrete surface hardness, such as rebound hammer test;
3   delamination determination, such as chain drag test;
4   dust sampling.

Non-destructive test can be defined as the measurement, inspection, or analysis of materials, existing structures, and processes of manufacturing without destroying the integrity of materials and structures. The non-destructive tests may include:

1   reinforcement location and size;
2   carbonation front;
3   chloride content;
4   sulfate content;
5   extent of cracking and discontinuities;
6   areas of poor consolidating;
7   presence of voids and honeycombs;
8   extent of deterioration;
9   level of moisture;
10   thickness;
11   moisture;
12   dynamic modulus.

The detailed specifications for each test should be provided by the engineer or the testing agency. The laboratory testing report should include everything that is noticed during sample preparation and testing, which should be combined with the records at condition survey, such as the sample locations in the structure and methods for obtaining them.

Once all the available information and reports are ready, they should be put together and an evaluation process can thus be executed. If the deteriorations can be eliminated in a cost-effective way, the succeeding renovation job can then be done after evaluation. If the cause cannot be eliminated in a cost-effective manner, or it has been concluded that the structure has already degraded to an unacceptable extent, correction measures may be required. In such situations, if the structure, as originally designed (but not necessary as built) is adequate for future use, it can be rehabilitated to reach its as-designed condition. If the load-carrying capacity is not enough, strengthening may be a good way to correct it. The detailed evaluation should then result in recommendations of the appropriate course of action, which is to be addressed in the report. The possible action plans can be the following (ASCE 2000):

1   accept the structure as is;
2   rehabilitation of the structure to correct deficiencies identified;
3   change the use of the structure;

4   phase the structure out of service.

The information gathered during investigation and evaluation can now be organized into a report that is submitted to the owner of the structures. Based on the SEI/ASCE 11-99 recommendations, the following format is suggested:

1   an optional executive summary;
2   introduction;
3   description of the existing structure;
4   the desired design loading and other performance criteria;
5   description of the investigation and evaluation processes;
6   conclusions and findings;
7   recommendations;
8   the proposed renovation steps.

3.3 Non-destructive tests

Non-destructive tests can be defined as the measurement, inspection, or analysis of materials, existing structures, and processes of manufacturing without destroying the integrity of materials and structures. The common terms used in non-destructive inspection and evaluation are as follows:

NDT   Non-destructive test
NDI   Non-destructive inspection
NDE   Non-destructive evaluation
QNDE   Quantitative non-destructive evaluation.

NDT techniques can be characterized as active or passive, and surface, near surface or volumetric methods. The active techniques are those where energy in some form is introduced into or onto the specimen, and an observable change in the input energy is expected if an anomaly is present. On the other hand, passive techniques are those that monitor or observe the item in question in either the "as-is" state, under the influence of a typical load environment or a proof cycle, or with a visual enhancing liquid covering the surface (Bray and Stanley 1997). For passive techniques, the presence of a defect is determined by some response or reaction from the specimen being tested. Surface techniques are those where only surface flaws can be detected. If the surface techniques are improved to be able to inspect depths below the surface, they are classified into near surface or volumetric techniques. Non-destructive testing is normally economical through reduced failures, more efficient design, better performance, etc. Non-destructive testing is also beneficial in reducing the frequency of unscheduled maintenance. What's more, with the proper use of NDT, scheduled maintenance periods may be lengthened. The application fields of NDT include: (1) quality control such as processing of rolling steel, ceramics, curing of polymer, wielding and/or porosity in a finished (manufactured) products; (2) in-service inspection such as fatigue; cracks; corrosions; debonding; and impact damage; (3) allowable tolerant philosophy, such as allowable flaw size; detection limit; crack growth rules; inspection interval; and (4) probability of detection (POD). The advantages of NDT include: (a) testing can be performed on the actual structure rather than on a compaction sample of it; (b) tests can be conducted at many locations; and (c) tests can be conducted at multiple times or even continuously, that is, real-time testing (Taylor 1992).

Non-destructive testing is an important practice in the inspection of metals, but its application in concrete is relatively new due to its in-homogeneity, complex composition and non-uniformity among its individual products. Large-scale experiment work of NDT in civil engineering only started in 1980s. However, in the past few decades, there has been great progress in the development of non-destructive methods in concrete, and several methods have been standardized by ASTM (2003b) and ACI (1998), the Canadian Standard Association (CSA), the International Standards Organization (ISO), the British Standards Institute (BSI) and others. In civil engineering, NDT-CE stands for non-destructive test in civil engineering. Estimating the strength of concrete is only one of the functions of NDT. The other major areas of NDT involve determining the extent of cracking and discontinuities, location and size of crack in concrete members, presence of reinforcement corrosion, areas of poor consolidation, presence of voids and honeycombs, condition of hydration process, integrity of concrete structures, locations and size of reinforcing steels, level of moisture, fatigue, debonding of fibers or FRP, detection of impact damage, and extent of deterioration that may have occurred due to service or damage from unusual loading conditions. NDT is becoming more and more popular because of its low cost, rapid evaluation and minimal damage to concrete structure.

3.3.1 Testing objects and problems of NDT for inspection

NDT for building and infrastructure inspection is a relatively young area and its activities for infrastructure inspection can be viewed in different categories. One point of view is testing objects, or problems, the other is the various testing methods. However, these categories interweave. Every single non-destructive technique has its own limitations, and, normally, a single technique should not be used on its own to make a final decision of importance. A real problem may require several different NDT techniques to be utilized together to ensure a correct conclusion. Or a technique can be used for different testing objects.

(A) TESTING OBJECTS

1   Buildings: masonry, reinforced concrete, pre-stressed concrete;
2   Bridges: steel, reinforced concrete, pre-stressed reinforced;
3   Highway and airport runways: asphalt concrete, Portland cement concrete; fiber reinforced concrete;
4   Other infrastructures: plate forms, towers, chimney, poles, and tunnels.

(B) TESTING PROBLEMS

1   Strength: concrete compressive strength gain or loss, residual strength after fire or earthquake, steel yield stress;
2   Cracks and fractures: surface cracks, internal cracks, crack length and width, fracture toughness;
3   Thickness: concrete cover for reinforcement, layers in pavement, restraining wall;
4   Moisture: distribution, saturation;
5   Corrosion: occurrence, rate, section area reduction, position;
6   Debonding: ceramic tile system; FRP strengthening layers, concrete covers;
7   Volume stability: shrinkage, creep, chemical reaction caused expansion;
8   Hydration process: setting time, early strength gain, de-mounding time;
9   Quality control: welding quality, grouting quality, compaction quality, epoxy filling quality; and
10   Structural integrity: internal damage, stiffness degradation, connection condition, reinforcement condition.

3.3.2 Techniques of non-destructive testing

An outline of the various techniques of non-destructive testing (NDT) of concrete and concrete structural components is described as follows. The theoretical background and general procedures of each method, as well as the equipment, are introduced. Typical applications of each method in testing of concrete and concrete structural components are also addressed. The reliability of the results, as well as the factors influencing the results, is evaluated and commented on.

3.3.2.1 Mechanical wave techniques (MWT)

These techniques take the mechanical waves as the working agency. The principle of these techniques is the generation, propagation, reflection and transmission laws regarding mechanical waves. Ultrasonic waves, acoustic waves, sub-acoustic waves are mechanical waves. A special feature of these waves is that they are directly related to the mechanical properties of the media they propagates, hence this is their remarkable advantage used to determine the mechanical performance of the materials. The mechanical wave NDT techniques are widely used in concrete structures. Mechanical waves NDT technique can test a very wide range as a tool of measurement, a detector, and a monitor. They can meet most of the requirements for NDE in civil engineering. They are accurate in determining shape, size and depth of the defective areas, with high sensitivity, deep penetration, low cost, easy and fast to operate, and convenient for in-situ use. Mechanical waves are not harmful to the human body, which is the obvious important advantage over radiation.

Mechanical wave techniques can be active or passive. In the active ones, the testing apparatus produces mechanical waves. In the passive techniques, the mechanical waves stem from the tested object itself. The main active mechanical wave NDT techniques are Ultrasound Testing techniques and Impact Echo method, and main passive one is Acoustic Emission.

(A) ULTRASONIC TESTING TECHNIQUES (PULSE VELOCITY, PULSE-ECHO)

The use of ultrasonic pulses to test concrete was developed in Canada and England in the mid-1940s (Mailvaganam 1992). Ultrasonic testing of concrete is based largely on pulse velocity measurements by using through-transmission techniques. This method is mainly used to study surface and sub-surface deterioration, cracking, and location of usually large voids in a hardened concrete mass, to determine the general condition of concrete structures, to assess the strength of the concrete, and to estimate the thickness of layers where the layers have differing sound-propagating properties. It is particularly useful in detecting the existence and extent of internal cracks and the depth of visible surface cracks, and in estimating the approximate depth of damage due to causes such as frost, sulfate attack and fire.

The ultrasonic waves are mechanical waves (in contrast to light or X-rays, which are electromagnetic waves) that consist of oscillations or vibrations of the atomic or molecular particles of a substances about the equilibrium position of these particles. Ultrasonic waves have a frequency higher than the hearing range of the normal human ear, which is typically considered to be 20 kHz. Ultrasonic wave propagation requires the presence of a medium such as a fluid or solid. Generally low-frequency ultrasonic wave can travel a long distance in many solid and liquid materials, while high-frequency ultrasonic waves tend to attenuate rapidly. Ultrasonic stress waves are generally used in pulse velocity measurement and pulse–echo techniques. As we know, the wave velocity depends upon the elastic properties and mass of the medium, and hence, if the mass and velocity of wave propagation are known, it is possible to assess the elastic properties of the medium. For an infinite, homogeneous, isotropic elastic medium, the compression wave velocity is given by:

$$C_L = \sqrt{\frac{K \cdot E}{\rho}} \qquad (3.1)$$

where C_L = compression wave velocity (km/s):

$$K = \frac{(1-v)}{(1+v)(1-2v)}; \qquad (3.2)$$

E = dynamic modulus of elasticity (N/m^2);
ρ = density (kg/m^3);
v = dynamic Poisson's ratio.

In this expression, the value of K is relatively insensitive to variations of the dynamic Poisson's ratio v, and hence, provided that a reasonable estimate of this value and the density can be made, it is possible to compute E using a measured value of wave velocity C_L. For a particular concrete mix, the elastic modulus varies with the square root of strength if other factors remain the same. Hence, measurement of the velocity of sound in Eq. 3.1 provides a means of estimating strength.

Ultrasonic waves can be generated by many means such as applying small, high-frequency displacements to surfaces or solid materials. In real applications, normally high frequency sound waves are introduced into materials. Most ultrasonic inspections are done at frequencies between 0.1 and 25 MHz (Newman 2001). Most commercial equipment uses pulses of ultrasound and the basic layout is shown in Figure 3.1. This test equipment provides a means of generating a pulse, transmitting this to the materials, receiving and amplifying the pulse and measuring and displaying the time taken. During transmission, the waves are reflected at interfaces. The electronic timing device measures the interval between the onset and reception of the pulse. The reflected ultrasonic wave or first arriving time can be displayed, either on an oscilloscope or as a digital readout, and then can be analyzed to determine the presence and location of flaws or dislocations. In general, the equipment must be able to measure the transit time to an accuracy of ±1 percent.

To test concrete, an ultrasonic wave is sent through concrete. The time taken by the pulse to travel through a known length is measured as well as the amplitude of the wave after transmission. In real applications, a pulse generator causes repeated voltage pulses to be transmitted to a transducer which is coupled to the concrete surface. Similar pulses are sent to a timing circuit. The transmitting transducer causes compression waves to be generated in the concrete. These waves are picked up by a similar receiving transducer, which converts the mechanical energy of the compression waves back into pulses of electrical voltage (Kay 1992). These converted pulses are then sent through an amplifier to the timing circuit. If the distance between the

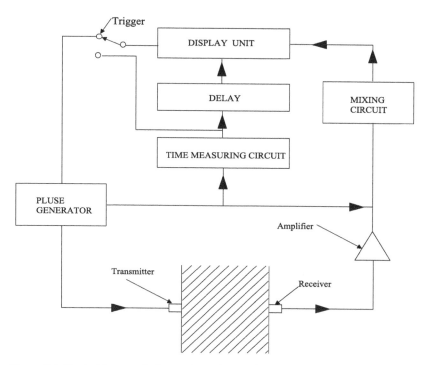

Figure 3.1 Typical layout of ultrasonic testing equipment.

transducers is known, the pulse velocity can be calculated. When ultrasonic waves are transmitted across a void in a solid body, their amplitude reduces significantly, and most of the ultrasonic waves are reflected at a discontinuity created by the void or a crack. However, the ultrasonic pulses can travel around voids or localized discontinuities in a concrete member and the measure of the time of travel provides a means of detecting the presence and approximate locations of such voids or discontinuities. Therefore, both amplitude and velocity measurements through the body of a concrete mass are required to study the interior condition of the concrete mass.

There are three basic ways in which the transducers may be placed on the test section of a concrete member, as shown in Figure 3.2. These are known as direct, semi-direct and indirect transducer configurations. In the direct configuration, the transducers are placed on opposite faces of the members under inspection and the ultrasonic pulse travels directly through the member. Since the maximum pulse energy is transmitted at right angles to the face of the transmitter, the direct configuration is the most reliable from the point of view of transit time measurement. It gives the most accurate measurement of pulse velocity and should be used whenever possible. In the semi-direct configuration, the transducers are placed on adjacent faces of the member or on opposite faces, but with the transducers not directly opposite one

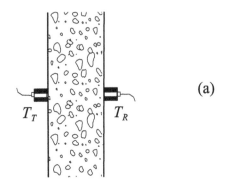

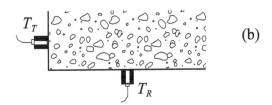

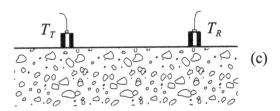

Figure 3.2 Sensor configurations in ultrasonic testing: (a) direct transmission;
(b) semi-direct transmission; (c) indirect transmission.

another. The pulse also travels through the body of the concrete. The semi-direct configuration can sometimes be used satisfactorily when the angle between the transducer is not large and the path is not too long. In the indirect configuration, the transducers are placed on the same surface of a member. The pulse travels through the concrete in a region just below the surface. The signal is dependent on the scattering of the pulse due to discontinuities and is subject to error. The received signal amplitude may be less than 3 percent of that for a comparable direct transmission (Bungey and Millard 1996). Besides, velocity of the pulses is significantly influenced by the surface which may not be representative of the interior of the concrete, and the exact path length is uncertain. The indirect method is therefore relatively insensitive and is definitely the least satisfactory. When using the indirect configuration, a special procedure is necessary to account for the lack of precision of path length, requiring a series of results taken at each location with the transducers at different distances apart. A graph is then plotted showing the relationship between transmission time and the distance between the transducers. The slope of the straight line graph drawn through the points is taken as the mean pulse velocity along the line of measurement on the concrete surface. If there is a discontinuity in this plot it is likely that either surface cracking or an inferior surface layer is present (ibid.).

In recent years, several test devices have been developed to measure the time of travel of pulses through concrete and most of them have been commercialized. Normally the test equipment is composed of a wave generator, an ultrasonic sensor which receives the wave traveling through concrete, a time measuring circuit and a display unit such as an oscilloscope or other display device. Nowadays, equipment for automatic recording of inspection results is available and the inspection costs are relatively low. Operation is relatively straightforward but requires great care to obtain reliable results. One essential point for an accurate application is good acoustical coupling between the concrete surface and the face of the transducer. When the concrete surface is relatively smooth, this can be achieved by a medium such as petroleum jelly, liquid soap or grease. Only a thin layer is required and the transducer should be pressed firmly against the surface to avoid air pockets. When the surface is very rough and uneven, a stiffer and thicker layer of grease may be necessary to provide a smooth surface for transducer application. The use of ultrasonic pulse velocity measuring equipment on concrete is described in ASTM C597 (ASTM 2003a).

Ultrasonic techniques have found wide applications in both laboratory and in-situ (Bungey and Millard 1996; Malhotra and Carino 2004). The main uses described are: (1) assessing uniformity of concrete in the field; (2) indicating changes in the characteristics of concrete with time; and (3) assessing the degree of deterioration and/or cracking when surveying concrete structures.

These techniques are excellent for investigating the uniformity of concrete. However, there are several factors may affect the accuracy of the measured results. These factors can be divided into two categories: (1) factors resulting directly from the concrete properties; and (2) other factors related to the environment and the testing process (Malhotra and Carino 2004). The first category of factors includes (a) aggregate; (b) cement type; (c) water/cement ratio; (d) admixtures; and (e) age of concrete. The second category of factors is (f) the operating temperature; (g) transducer contact; (h) moisture conditions of concrete; (i) path length; (j) stress history; (k) size and shape of the specimen; and (l) presence of the reinforcement. The influence of these factors has been addressed by Bungey and Millard (1996) and Malhotra and Carino (2004). The ultrasonic pulse velocity gives the least reliable application for strength estimation of concrete, hence, the use of pulse velocity method to estimate the compressive and/or flexural strengths of concrete is not recommended (Malhotra and Carino 2004). Besides, it has been found that ultrasonic inspection may not produce good results with objects having complex shapes or rough surfaces, and is difficult to detect and measure the length of small, tight cracks and crack-like discontinuities. Also, the flaw image during ultrasonic inspection is also complex. This technique is also highly dependent on user expertise, and requires careful calibration of equipment (Newman 2001).

Recently, a new embedded ultrasonic measurement system has been developed and applied in hydration process monitoring of a cement-based material (Zhang *et al.* 2008). It uses embedded piezoelectric composites as the ultrasonic transducers as shown in Figure 3.3. The test result shows that the embedded ultrasonic system can successfully measure the development of wave velocity in the cement-based material (see Figure 3.4). Compared to

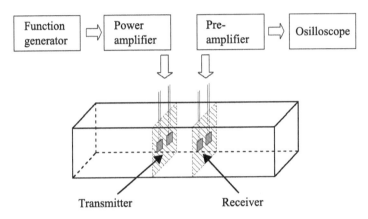

Figure 3.3 Embedded piezoelectric composites as ultrasonic transducers.

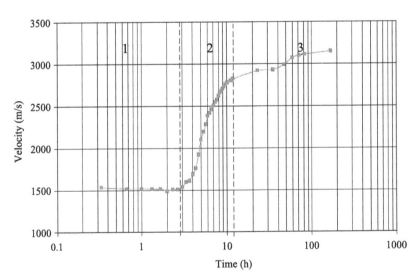

Figure 3.4 Development of wave velocity in cement-based material.

traditional ultrasonic non-destructive methods, which have conventional commercial ultrasonic transducers fixed on the surfaces, the new method eliminates the coupling and surface contacting problems and is inexpensive and effective for any scale structures. For construction of a concrete structure, early-age performance monitoring can be performed by the new system to provide guidance for construction control. After the concrete is cured and the strength is fully developed, the system can be used to conduct structural health monitoring to detect the damage accumulation or even disaster development. It meets all the requirements of concrete structure health monitoring from fresh stage to hardened stage.

(B) IMPACT–ECHO METHOD

The principle of the impact–echo method can briefly be illustrated as follows. A transient stress pulse is introduced into a test object by a mechanical point contact impactor on the surface. The generated stress pulse propagates into the specimen along spherical wavefronts as dilatational, compression, or *P* wave in a direction normal to the wavefronts and as distortional, shear, or *S* wave in a direction perpendicular to the direction of the *P* wave propagation. In addition, surface waves (*R* waves) travel on the surface away from impact point and their amplitude decreases exponentially with depth. The *P* and *S* waves travel through the material and are partly reflected back by interfaces, i.e. from defects or external boundaries. The arrival of these reflected waves at the surface where the initial impact was generated produces surface particle displacements, which are converted into electrical voltage by a receiving transducer. The data, mainly the arrival time differences among various waves, are then analyzed either in time domain or frequency domain to derive properties of the medium. Impact–echo is actually a very old non-destructive testing method. In ancient times, people already used this method to evaluate the quality of stone by striking the stone and listening to the ringing sound with the human ear.

However, modern impact–echo methods are relatively new techniques for detecting concrete defects based on the use of transient stress waves for non-destructive testing, although the idea of using impact to generate a stress pulse is an old idea. It has the great advantage of eliminating the need for a bulky transmitting transducer. The principle of the impact–echo technique is illustrated in Figure 3.5. In this method, transient stress waves are introduced by mechanical impact on the concrete surface. These waves are reflected by internal discontinuities or defects (such as delaminations, or honeycomb sections), propagate back to the surface, and are recorded by a receiver placed near the point of impact and mounted on the concrete surface. The picked signals are then transformed into the frequency domain. The frequency content provides information about the condition of the concrete, working on essentially the same principle as the rebound hammer test. In practice, these frequency contents of the digitally recorded waveform are

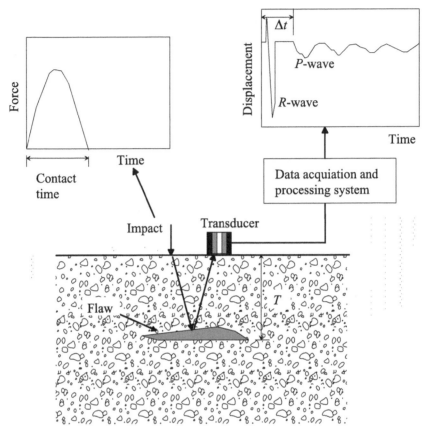

Figure 3.5 Principle of the impact–echo method.

obtained by using the Fast Fourier Transform (FFT) technique, which gives the relative amplitude of the component frequencies in the waveform.

An impact–echo test system is normally composed of three components: an impact source; a receiving transducer; and a digital processing oscilloscope or waveform analyzer which is used to capture the transient output of the transducer, store the digitized waveforms, and perform signal analysis. The force–time history of an impact may be approximately a half-sine curve, and the duration of the impact is the "contact time", which determines the frequency content of the stress pulse generated by the impact, thus, the shorter the contact time, the higher the range of frequencies contained in the pulse. Thus, the contact time determines the size of the defect which can be detected by impact–echo testing. As the contact time decreases and the pulse contains higher frequency components, smaller defects can be detected. In using the impact–echo method to determine the locations of flaws within an object, tests are performed at regularly spaced points along "scan" lines marked on the surface. Examination of the amplitude spectra from these scans reveals the depth and approximate size of defects which may be present. Internal discontinuities can thus be located by this technique. However, interpretation of test data requires in-depth understanding of the technique (Mailvaganam 1992). In addition, shorter-duration impacts are needed to accurately locate shallow defects. The stress pulse must have frequency components greater than the frequency corresponding to the flaw depth. The selection of the impact source is a critical aspect of a successful impact-echo test system.

Hammers are used in evaluation of piles (Steinbach and Vey 1975; Brendenberg 1980; Olson and Wright 1990). Hammers produce energetic impacts with long contact times which are acceptable for testing long, slender structures but are not suitable for detecting flaws within thin structures such as slabs and walls (Malhotra and Carino 2004). Impact sources with shorter duration impacts, such as small steel spheres and spring-loaded spherically-tipped impactors, have been used to detect flaws within slab and wall structures (Carino *et al.* 1986; Sansalone and Carino 1988; 1989). The use of small ball bearings as the impact source was regarded as one of the several key breakthroughs in impact-echo researches in the mid-1980s (Sansalone 1997). Such a source produces a well-defined and mathematically simple input which in turn generates waves with characteristics that facilitate signal interpretation. Ball bearing impactors are easy to use and the frequency content of the resulting stress waves can be made appropriate to the size of the structure and the sizes and locations of flaws that are to be detected. The data acquisition system should have a sampling frequency of at least 500 kHz. The optimal sampling frequency depends on the thickness of the test object, but, for testing relatively thin members, a high sampling rate is more effective. Nowadays, the impact-echo test equipment can be portable and has a wide range of applications in inspection of concrete both under laboratory and field conditions. The impact-echo method is more precise than the rebound hammer method. But the geometry and mass of the test object may influence the results.

Since the early 1970s, the impact–echo method has been widely used for the evaluation of concrete piles (Steinbach and Vey 1975; Brendenberg 1980; Olson and Wright 1990). In these studies, the partial or complete discontinuities of piles, such as voids, abrupt changes in cross-section, weak concrete, soil intrusions and the location of these irregularities can be detected by impact–echo technique. Carino and Sansalone initiated experimental and theoretical studies to develop an impact–echo method for testing structures other than piles (Carino *et al.* 1986; Sansalone and Carino 1988; 1989). They used the impact–echo technique to detect interfaces and defects in concrete slab and wall structures, including cracks and voids in concrete, the depth of surface-opening cracks, voids in pre-stressing tendon ducts, the thickness of slabs and overlays and delamination in slabs. The impact–echo technique has also been used in layered-plate structures, including concrete pavements with asphalt overlays (Sansalone and Carino 1989; 1990). Lin and Sansalone (1996; 1997) have developed a new method for determining *P* wave velocity in concrete on the basis of the impact–echo technique.

The development of a standard test method for flaw detection using impact–echo is difficult because of the many variables and conditions that may be encountered in field-testing, in which the type of defects and shapes of structures are two variables frequently met. ASTM C 1383 (2002) has proposed a standard test method based on the use of impact–echo method to measure the thickness of plate-like concrete members. This method includes two procedures. Procedure A is used to measure the *P* wave velocity in concrete based on the travel time of the *P* wave between two transducers at a known distance. In Procedure B, the plate thickness is determined using the measured *P* wave velocity measured in Procedure A. The data analysis procedure in ASTM C 1383 considers the systematic errors associated with the digital nature of the data in Procedures A and B.

(C) ACOUSTIC EMISSION (AE)

The acoustic emission (AE) method is a passive NDT method and it simply listens to "sounds" generated within materials. AE is defined as the class of phenomena whereby transient elastic waves are generated by the rapid release of energy from localized source within a material. The energy release causes stress waves to propagate through the material, so AE is sometimes called stress wave emission. The AE device relies on detection of stress waves generated by the energy release due to micro-crack formation and propagation, dislocation, and includes movement within materials. In concrete, these stress waves may also be generated by local stress distributions associated with the chemical action such as alkali–aggregate reaction and corrosion of the steel inside concrete. In composites, the stress waves are generated by a variety of actions involving cracking and separation in the fibers and the matrix. The AE technique can be used to monitor the change of material conditions in real time and to determine the location of these emission events as well. It cannot detect a crack that already exists but has not propagated. Since AE is a result of local stress redistribution, any micro-mechanisms that will cause stress release can be the source of the AE. A general listing of typical AE sources, classified by material, is as follows (Bray and Stanley 1997):

Concretes
  Micro-crack and macro-crack initiation and propagation
  Separation of reinforcing members
  Mechanical rubbing of separated surfaces
  Corrosion of reinforce steel inside concrete
Composites
  Matrix cracking and fiber debonding at low strain level in composite

Fiber fracture at medium strain level
Delamination in composite materials at high strain level
Fiber pullout.

There are a number of different ways in which AE signals may be evaluated including event counting, rise time, spectrum analysis, AE source location (detect location), energy analysis, signal processing, signal duration, and so on. When an AE event occurs at a source within the material, the stress wave travels directly from the source to the receiver in the form of body waves. Surface waves may then arise from mode conversion. When the stress waves arrive at the receiver, the transducer responds to the surface motions that occur. AE signals cover a wide range of energy levels and frequencies. Modern instrumentation can record an AE signal in the range of 50 kHz to about 1 MHz from concrete. At lower frequencies, background noises from the test equipment become a problem. At very high frequencies, the attenuation of the signals is severe. The AE signals are usually considered in two basic types: the burst type and the continuous type. The former corresponds to individual emission events while the latter is an apparently sustained signal level from rapidly occurring emission events (Bray and Stanley 1997). Most important for a data analysis method in AE technique is to acquire an integrated and meaningful AE waveform.

The main elements of a modern AE instrumentation system are schematically shown in Figure 3.6. Details of requirement for an AE measurement system can be referred to Beattie (1983). The AE transducers are mainly

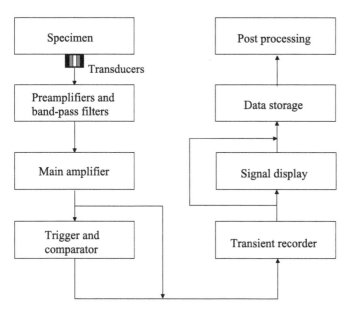

Figure 3.6 The main elements of a typical AE instrumentation system.

made of piezoelectric composites, preferably wide-band with a linear phase response and small in size. These transducers are used to convert the surface displacements into electric signals. The voltage output from the transducer is directly proportional to the strain in the PZT, which depends, in turn, on the amplitude of the surface waves. Preamplifiers are normally necessary because the output from the transducers is usually low. Bandpass filters are used to block the acoustic emission signals which lie outside the frequency range of interest. A trigger is useful and the triggering is initiated when the output voltage of the transducer exceeds a preset reference voltage (threshold), such that the lower level background noises can be rejected. A transient recorder is used to digitize a signal in real time and store the digitized data in a memory. The radiation patterns of acoustic emission sources are formed according to the same principle described before for ultrasonic transducers. In industrial applications, the AE sensors usually are located a relatively longer distance away from the source, i.e. several times the width of the crack.

The fundamental difference between AE and the ultrasonic is that AE is generated by the material itself under stress, while in ultrasonic the acoustic waves are generated by an external source and introduced into the material (Mailvaganam 1992). Therefore, the AE technique is very useful for monitoring highly stressed structures in service or undergoing some mechanical test procedure. The first representative investigation of AE from metals was carried out by Kaiser, who established the so-called Kaiser effect: "the absence of detectable acoustic emissions at fixed sensitivity level, until previously applied stress levels are exceeded". Concrete and fiber reinforced concrete are multi-phase and flaw-rich composite materials. When it is subjected to a load, micro-cracks tend to generate along flaws as the load increases, which makes it very suitable for AE monitoring. Unlike other NDE techniques, this technique indicates only active flaws and cannot determine the presence of other kinds of flaws. It responds to changes in flaw size instead of the total size. Therefore, continuous monitoring is required to detect a flaw extension whenever it occurs.

AE from concrete has been studied for the past 30 years. Almost all the applications of AE technique to concrete are focused on the fracture process zone. An extensive series of investigations was carried out by Maji and Shah (1988), Li and Shah (1994), Maji *et al.* (1990), Ouyang *et al.* (1991), and Landis *et al.* (1993). In these tests, concrete, mortar, and low-volume-fraction FRC specimens are loaded in either direct tension or bending, to study the damage initiation and propagation within the materials. Both notched (Maji and Shah 1989) and un-notched (Li and Shah 1994) specimens were investigated. They found that in the initial stages of loading, AE events generated randomly at various positions of the specimen. When the load reached about 80 percent peak load, most of the AE events came from a narrow band, which was referred to as the micro-cracking localization. Subsequently, a major crack was formed in the localization region and further

AE events were concentrated in a zone around the crack. AE events occurred not only at the crack tip, but also behind the crack tip, indicating ligament connections. Deconvolution techniques were used by Maji *et al.* (1990), Ouyang *et al.* (1991), Li (1996), and Suaris and van Mier (1995) to study the orientation and the mode of micro-crack. The relative amplitudes of AE signals at different transducer positions were used to distinguish between tension and shear micro-cracks in mortar and aggregate-matrix interfaces.

For fiber reinforced concrete, Li and Shah (1994) attempted to use the AE technique to study the tensile fracture of short steel fiber reinforced concrete. They carried out a series of uni-axial tension tests on both plain concrete and fiber reinforced concrete specimens to evaluate the relationship between micro-cracking and macroscopic deformation. They developed an innovative AE source location mechanism to detect micro-crack nucleation while macroscopic deformation was measured by LVDTs. They found that the degree of internal damage could be indicated by the rate of occurrence of AE events. The AE source location and characteristics of the micro-cracks were successfully deduced based on the AE events obtained during tests. Li *et al.* (1998) improved this AE measuring system for un-notched fiber-reinforced concrete specimens with a relative high volume fraction of fibers. An adaptive AE trigger signal identifier was developed for succeeding an adaptive trigger AE measurement system. An automated P wave arrival time determination method was developed to pinpoint the location of thousands of AE events by using an adaptive low-pass filter (Li *et al.* 2000).

The AE technique has also been explored to detect reinforcing steel corrosion in concrete. When corrosion is formed on a rebar, the corroding areas swell and apply pressure to the surrounding concrete. Micro-cracks will be formed and stress waves will be generated during the expansion process when the pressure is high enough to break the interface layer. The growth of the micro-cracks is directly related to the amount of corrosion in a corroding rebar. Thus, by detecting the AE event rate and their amplitude, the degree of the corrosion can be interpreted. Li *et al.* (1998) examined the correlation between the characteristics of the AE event and the behavior of the rebar corrosion in HCl solution and the possibility of corrosion detection of rebar inside concrete through an accelerated corrosion experimental method. The theoretical prediction and experimental results have shown that the AE technique does have the ability to detect rebar corrosion in an early corrosion stage. The AE technique has also been used to detect cracks in concrete, fracture process zone location, thermal cracking and debonding of reinforcement.

The major disadvantage is that to use the AE technique, skills and experience are needed to operate the measurement system and to interpret the information gathered. The detected data depend largely on the type of the measurement system. Also, AE technique requires a very high speed sampling device and sensitive transducers. As a result, the AE equipment is expensive and the technique is only largely used in laboratory researches.

There are a number of factors that may affect the AE from concrete. Some of the most important factors are background noises, signal attenuation, specimen geometry, type of aggregate, the interface between the sensor and the concrete surface, the setting of the threshold, the sampling rate of the measurement system and the characteristics of the transducers.

3.3.2.2 Electrical and electro-chemical methods

Many non-destructive methods are based on measuring the changes in electrical properties, including electrical resistance, dielectric constant, and polarization resistance, of concrete and these methods can be classified into electrical and electro-chemical non-destructive method. The electrical resistance of an electrolyte is directly proportional to the length and inversely proportional to the cross-sectional area and is expressed by:

$$R = \rho \frac{L}{A} \tag{3.3}$$

where:

   R = resistance in ohms;
   ρ = resistivity in ohm-m;
   L = length in m;
   A = cross-sectional area in m^2.

The changes in electrical properties of concrete are closely related to the evaporable water content in concrete, which varies with water/cement ratio, degree of hydration, and degree of saturation. The ion concentration in water varies with time, too. The conduction of electricity by moist concrete is essentially electrolytic and can be used to interpreter properties of concrete, despite the complex relationship between the concrete's moisture content and its dielectric constant. The electrical resistance depends on the size of concrete specimen while the resistivity is essentially a material property, so that electrical resistivity is more widely used to characterize concrete property in electrical method. Resistivity measurement could provide a rapid non-destructive assessment of concrete surface areas, crack size, reinforcing steel corrosion and cement hydration. To detect steel corrosion, the ability of the corrosion current to flow through the concrete is assessed in terms of the electrolytic resistivity of the material. An in-situ resistivity measurement set-up in conjunction with half-cell measurement is shown in Figure 3.7. In this test technique, normally low frequency alternating current is applied and current flowing between the outer probes and voltage between the inner probes are measured. Since the ability of corrosion current to flow through concrete increases with decreasing resistivity, the measured resistivity ρ can be used together with potential measurements to assess

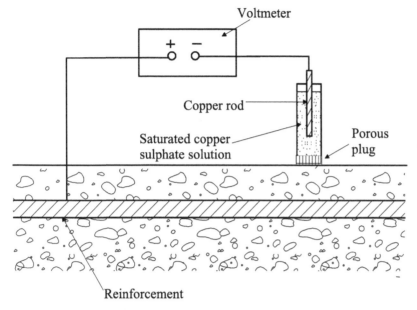

Figure 3.7 Reinforcement potential measurement by half-cell method.

the likelihood of corrosion. Classification of the likelihood of corrosion can be obtained following the values in Table 3.2 when half-cell potential measurements show that corrosion is possible. When the test results are interpreted, highly negative corrosion potential or high gradient in potential map imply that corrosion is thermodynamically favorable. If resistivity is also low, there is a high chance of having significant corrosion actually occurring in the reinforcing steel.

The resistivity can also be used as a measure of the degree of hydration of

Table 3.2 General guides to interpretation of electrical test results

Half-cell potential (mV) relative to copper/ copper sulphate reference electrode	*Percentage change of active corrosion*
<−350	90%
−200 to −350	50%
>−200	10%
Resistivity (ohm-cm)	Likelihood of significant corrosion (non-saturated concrete when steel activated)
<5,000	Very high
5,000 ~ 10,000	High
10,000 ~ 20,000	Low/Moderate
>20,000	Low

Source: Based on ASTM C 876 and Langford and Broomfield (1987).

cement-based materials. In the past, such measurement always meant using contact electrical resistivity apparatus which has many drawbacks. Recently, Li and Li (2002) and Li *et al.* (2003) have invented a non-contacting electrical resistivity apparatus to measure the resistivity of cement-based materials. The measuring system of this non-destructive testing method is shown in Figure 3.8, which adopts the principle of a transformer. The electrical circuit contains a primary coil with wound wires and a ring-type cementitious specimen acting as the secondary coil of the transformer. The primary coil is the input coil of the transformer and the secondary coil is the output coil. When an AC is applied to the primary coil, mutual induction causes the current to be induced in the secondary coil. With the measurement of the induction current in cement-based specimen, the curve of resistivity versus time can be obtained. By interpreting the behavior of resistivity curve, the hydration characteristics of cement-based materials can

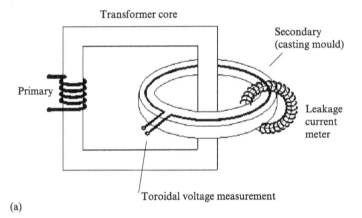

(a)

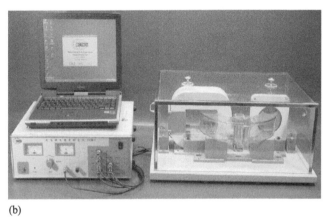

(b)

Figure 3.8 A non-contact resistivity test apparatus for cement-based materials: (a) illustration of measurement theory; (b) the whole package testing apparatus.

be obtained. The characteristics include hydration stages, setting times, and strength development. It has been found that the non-contact electrical resistivity measurement provides a good non-destructive way to determine and assess the setting time and mechanical properties of the cement-based materials during the entire setting process. In general, fresh concrete behaves essentially as an electrolyte with a resistivity in the order of 1 ohm-m, a value in the range of semiconductors, while hardened concrete has a resistivity in the order of 10 ohm-m, a reasonable good insulator.

The Wenner four-probe technique has been used for soil testing for many years and it has recently been adopted in application to in-situ concrete. In this method, four electrodes are placed in a straight line on, or just below, the concrete surface at equal spacing as shown in Figure 3.9. A low frequency alternating electrical current is passed between the two outer electrodes while the voltage drop between the inner electrodes is measured. The apparent resistivity is calculated as:

$$\rho = \frac{2\pi s V}{I} \tag{3.4}$$

where s is the electrode spacing, V the voltage drop and I the current. This method can also be found in applications of detecting pavement thickness. Since concrete and subgrade pavement material have different electrical characteristics, the change in slope of the resistivity vs. pavement depth

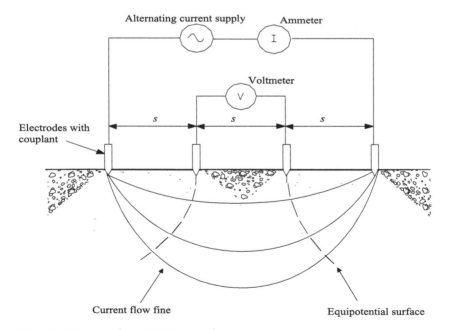

Figure 3.9 Four-probe resistivity test of concrete.

curve will indicate pavement thickness (Vassie 1978). When testing a concrete pavement, the electrode system may be adjusted at a 25 mm or 50 mm spacing for the initial readings and then expanded with 25 mm increments for successive readings extending to a spacing equal to the pavement depth plus 75 ~ 150 mm (Malhotra and Carino 2004).

The corrosion of steel in concrete is an electrochemical process that requires a flow of electrical current for the chemical corrosion reactions to proceed. The electrochemical method can be used to detect signs of rebar corrosion based on an electrochemical process through measuring the electric potential of reinforcing bars (Newman 2001). Some tests use the half-cell method of measurement described in Chapter 2, in which a copper/copper sulfate half-cell is used. In this technique, an electrical connection is made to the reinforcement at a convenient position. The electric potential difference between the anode and the cathode is measured by a voltmeter; if the values are more negative than −350 mV, the probability of corrosion is in excess of 90 percent. In practice, a series of measurements can be taken at grid points to map the probable corrosion activity. This technique has been standardized by ASTM C876. According to this, the difference between two half-cell readings taken at the same location with the same cell should not exceed 10 mV when the cell is disconnected and reconnected. Besides, the difference between two half-cell readings taken at the same location with two different cells should not exceed 20 mV. This method is readily utilized in the field, but trained operators are required. What's more, the actual rate of corrosion (such as percentage loss of section) is not provided by this method. The interpretations of these potentials vary with investigator and agency.

Other electrical methods for detecting reinforcement corrosion are available. One such technique is based on electrical resistance measurements on a thin section of in-situ reinforcement (Vassie 1978). The electrical resistance of reinforcement bar is inversely proportional to its thickness, so, as the thin slice is gradually consumed by corrosion, it becomes thinner with a corresponding increase in resistance. In this technique, to facilitate measurement, the probe is normally incorporated into a Wheatstone bridge network (Figure 3.10). One of the probes is protected from corrosion while the other arm is the in-place portion of the reinforcement. The measured resistance ratio can be used to monitor the corrosion rate. The significant disadvantages are the need to position the exposed arm of the probes during construction and the concerns of associated sampling techniques required to locate the probes in large structures subject to localized corrosion.

3.3.2.3 The magnetic technique

The magnetic technique uses magnetic fields as an essential tool to detect or interpret materials' properties. As far as applications in concrete structures are concerned, currently there are three different kinds of magnetic field

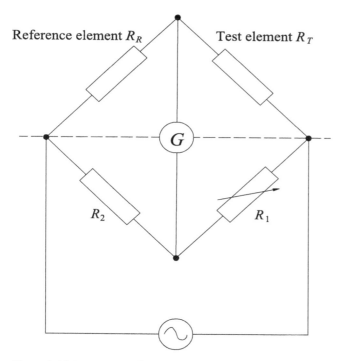

Figure 3.10 Basic circuit for electrical resistance probe technique.

phenomena: (1) alternating current excitation of conducting materials and their magnetic inductance; (2) direct current excitation resulting in magnetic flux leakage fields around defects in ferromagnetic materials; and (3) nuclear magnetic resonance (Malhotra and Carino 2004). The magnetic induction method is only applicable to ferromagnetic materials, in which the test equipment circuitry resembles a simple transformer and the test object acts as a core (Figure 3.11). With a piece of metal close to the coil of transformer, the inductance increases and the change in induced current depends on the magnetization characteristics, location, and geometry of the metal. The inductance of the coil can be used to measure coil-to-place distance if the relationship between mutual inductance and the coil-to-place distance is known. The magnetic induction theory has resulted in the development of equipment to determine the location, sizes, and depth of reinforcement or depth of concrete cover, for instance, the cover meter to detect the location of embedded steel.

Magnetic flux leakage (MFL) non-destructive testing consists of magnetizing a test part, generally a ferromagnetic material, and scanning its surface with some form of flux-sensitive sensor for the leakage field (Bray and Stanley 1997). The fundamental theory of MFL has been explained elsewhere (Bray and McBride 1992; Bray and Stanley 1997). When

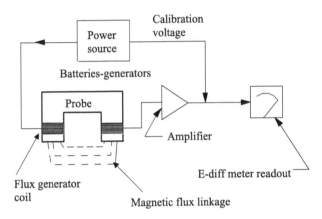

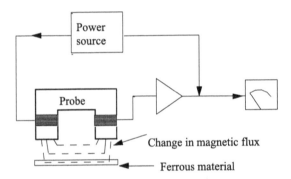

Figure 3.11 Principle of operation of induction meter used to locate reinforcement.

ferromagnetic materials are magnetized, magnetic lines of forces (or flux) flow through the material and complete a magnetic path between the poles, which increases from zero, at the center of the specimens, to increased density and strength toward the outer surface. When cracks or defects exist in materials, they can lead to perturbations of the magnetic flux. Its magnetic permeability is drastically changed and leakage flux provides a basis for non-destructive identification of such discontinuities. The amount of leakage flux produced also depends on defect geometry. Broad, shallow defects will not produce a large outward component of leakage flux. What's more, a defect whose long axis is parallel to the lines of flux will not produce leakage flux, too. By sensing the magnetic flux along a testing member, damage location can be deduced. MFL techniques are suitable for the detection of surface or near-surface anomalies in ferromagnetic materials, such as detection of area reduction in reinforcements and breakage of tendon in pre-stressing cables. Internal defects in thick parts may not be detected because the magnetic lines of flux nearly bypass the defect with little leakage. These techniques generally do not require mechanical contact with the testing object and are very amenable to automatic signal recognition schemes, both of which are of great benefit in automated, high-speed inspection.

Nuclear magnetic resonance (NMR) is based on the interaction between nuclear magnetic dipole moments and a magnetic field. Magnetic resonance occurs in electrons, atoms, molecules, or nuclei in response to excitation by certain discrete radiation frequencies as a result of a space quantization in a magnetic field. The NMR technique can be used as a basis for determining the amount of moisture content in concrete by detection of a signal from the hydrogen nuclei in water molecules. Careful surface preparation is required and the object for testing must be clean and originally demagnetized. The test requires a source of high-current electric power to magnetize the object. The operation is relatively messy and good operators' skills are needed to interpret results.

Magnetic particle testing is another kind of magnetic non-destructive method, mainly for steel structures. In this method, the object is magnetized and covered with magnetic powder. Surface and/or near-surface discontinuities in the magnetized materials may create leakage in the magnetic field, which, consequently, may affect the orientation of the particles above those areas. A variation of this test involves using wet fluorescent particles visible in back light through a borescope. The magnetic particle technique can be used for locating surface cracks, laps, voids, seams and other irregularities. Some subsurface defects can also be detected to a depth of about 0.635 mm (0.25 in) (Newman 2001). This method is relatively fast, simple to administer, and inexpensive. But there is a limited depth penetration for these methods. Magnetic particle testing has a long history and has been implemented into many standards and specifications. This method is a most effective way to non-destructively detect surface and near-surface discontinuities in ferromagnetic material.

3.3.2.4 *The electromagnetic wave technique*

Electromagnetic wave covers a very wide range of frequency in the scientific meaning. According to the different frequency range, from the lower to higher, they have different names, such as radio wave, high frequency wave (RF), very high frequency wave (VHF), ultra-high frequency wave (UHF), microwave (L-Band, S-Band, etc.), infrared lights (IR), X-rays, etc. In NDT engineering, they are usually referred as different techniques according to the kind of electromagnetic wave used as the working agency. The electromagnetic wave techniques are usually related to relative lower frequency, namely, the frequency from RF to S-band, which usually are called microwaves. Microwave testing in the reflection mode often is called the radar technique, whereas infrared lights, lights, X-rays techniques etc. are usually regarded as other independent NDT techniques. For most cases, the electrical techniques are well developed in this band.

Electromagnetic waves propagate in a medium, the wave velocity v is:

$$V = c/n \qquad (3.5)$$

where c is the velocity of light; n is the index of refraction. For a homogeneous medium,

$$n = \sqrt{\varepsilon\mu} \qquad (3.6)$$

where ε and μ are the relative permitivity and relative magnetic permeability of the material respectively. The relative magnetic permeability almost equals to unity for most of materials. For a mixture consisting of j constituents with volume V_j, the relative permitivity could be regarded as the linear combination of its constituents:

$$\varepsilon = \sum_{i=1}^{j} V_i \varepsilon_i / V \qquad (3.7)$$

where V is the total volume of the mixture. Hence, one could deduce the volume component from the measured velocity provide the materials of the constituents are known.

From the point of view of quality control of concrete, the water content is one of the most important parameters. Fortunately, the relative permitivity of water is about 81 whereas the one of cement and other aggregates is about 5, of air is 1. Hence, the measurement is very sensitive to the constants of water. The accuracy is about 3–5 percent for normal concrete mixture.

The time of flight could be used to measure the thickness by the formula:

$$\text{Thickness} = 150t/\sqrt{\varepsilon} \qquad (3.8)$$

The scattered wave usually called an echo. The amplitude of the echo is related to several material parameters. By means of some inverse procedure, one can deduce the characteristic of a layered structure (Attoh-Okine 1995). By scanning, the echoes information, the amplitudes, the phases, the TOF data of echoes, could be used to form an image or can be as a raw data for tomography.

As the concrete is a mixture, the scattered wave from the aggregate grains will form a cluster. The scattering at an interface is strongly dependent on the difference of complex permitivity of the two materials. As the difference of permitivity between aggregate and the matrix is small (4–5), the scattered waves from the aggregate are usually weaker than in the ultrasonic waves case. This is the advantage of microwaves to detect water or air defects, which have large difference of permitivity. The test results are not sensitive to the aggregate size.

(A) RADAR

A basic radar system consists of a control unit, antennas (one is used for transmitting and one receiving pulses), an oscillographic recorder, and a power converter for DC operation. In the inspection of concrete, it is desirable to use a radar antenna with relatively high resolution so that the small layer of a concrete member can be detected.

Radar can be employed in the rapid investigation of concrete structures, such as for measuring the thickness of structural members, for determining the spacing and depth of reinforcement, and for detecting the position and extent of voids and other types of defects in bare or overlaid reinforced concrete decks. In testing of concrete, normally short-pulse radar is used, which is the electromagnetic analog of sonic and ultrasonic pulse-echo methods. In this method, the equipment generates electromagnetic pulses which are transmitted to the member under investigation by an antenna close to its surface. The pulse travels through the member and it propagation velocity is determined by the electrical permittivity. The relative permittivity of the concrete is determined predominantly by the moisture content of the concrete. Typical relative permittivity values for concrete range between 5 (oven-dry concrete) and 12 (wet concrete) (Bungey and Millard 1996).

During propagation, the pulses are partially reflected and refracted at any interfaces where there are distinct changes in dielectric characteristics, such as the location of an internal void. The reflection is also influenced by moisture content and degree of reinforcement. Reflected pulses are received by a second antenna and processed by the equipment, which provide an evaluation of the properties and geometry of testing objects. There are three fundamentally different approaches for using radar to investigate concrete structures (Bungey and Millard 1996):

1   Frequency modulation: in which the frequency of the transmitted radar signal is continuously swept between pre-defined limits. This system has only some limited use to data on relatively thin walls;
2   Synthetic pulse radar: in which the frequency of the transmitted radar signal is varied over a series of discontinuous steps. This method has been used to some extent in field, as well as in laboratory, transmission lines to determine the electrical properties of concrete at different radar frequencies;
3   Impulse radar: in which a series of discrete sinusoidal pulses within a specified broad of frequency band are transmitted into concrete. This method is the most widely used in field tests of concrete structures and most commercial radar systems are of this type.

The Ground Penetrate Radar (GPR) is a type of impulse radar and very powerful. A GPR can image the profile of bridge pile in bridge scour investigation. A maximum frequency should be recommended in a GPR survey design for clutter reduction. It is suggested that the wavelength λ be tan times large than the characteristic dimension for geological materials (Davidson *et al.* 1995). For most cases, the frequency range is from 90 MHz to 1 GHz and pulse repetition frequency is 25 Hz. Following typical results could be reached: a scour depth of 1.2 m relative to the general riverbed. The filling may be a sandy layer 0.5 m deep overlying gravel 0.8 m thick at the center of the hole. The water depth is about 1 m (Davidson *et al.* 1995). The GPR measurement is fast enough to be suitable to detect a large area such as 10,000 sq. meters pavement of airports (Weil 1995).

It should be indicated that the radar signal can be attenuated as it passes through the member under investigation. The amount of attenuation is dependent on the conductivity of the material and the frequency of the signal used. Attenuation increases in conductive materials and therefore it may be difficult to penetrate saturated or salt-contaminated concrete. An increase in signal attenuation will result in a decrease in the penetration depth by radar pulse. In general, low frequency radiation penetrates further than high frequency radiation. However, low frequency signals result in a loss of resolution, or ability to pick out small objects, because of the increase in wavelength. The results may also be skewed by moisture content and steel reinforcement. The images are provided in strips trailing the radar's movement instead of a complete picture. Their interpretation depends on the experience of the personnel conducting the test (Newman 2001).

(B) INFRARED THERMOGRAPHY

Infrared is a waveband in electromagnetic spectrum that lies just beyond red and is invisible to the naked eye. All objects emit infrared radiation. The amount of radiation increases with the rise of temperature. Infrared thermography is a remote sensing technique, which is based on the principle that sub-surface anomalies in a material affect heat flow through that material. These changes in heat flow cause localized differences in surface temperature. By measuring surface temperature under conditions of heat flow, the location of the subsurface anomalies can be detected. Combined with the other test methods, voids and discontinuities altering the flow of the emission can be detected by the device.

The radiance received by an infrared camera is expressed by

$$N_{cam} = \tau_{atm}\varepsilon N_{obj} + \tau_{atm}\rho N_{env} + (1 - \tau_{atm})N_{atm} \tag{3.9}$$

where τ_{atm} = transmission coefficient of the atmosphere, ε = object emissivity, ρ = object reflectivity, N_{obj} = radiance from the surface of the object; and N_{env} = radiance of the surrounding environment considered as a black body, and N_{atm} = radiance of the atmosphere. If the transmission coefficient of the atmosphere is considered to be close to unity, Eq. 3.9 can be simplified as follows:

$$N_{cam} = \varepsilon N_{obj} + \rho N_{env} \tag{3.10}$$

If the emissivity is very high (more than 0.9), according to Kirchhoff laws, $\varepsilon = 1 - \rho$, Eq. 3.10 can be further reduced to:

$$N_{cam} = N_{obj} \tag{3.11}$$

Consequently,

$$f(N_{obj}) = I_{cam} \tag{3.12}$$

$$S(T_{obj}) = I_{cam} \tag{3.13}$$

where I_{cam} corresponds to the radiometric signal obtained on the camera calibration curve; $f(\ldots)$ and $S(\ldots)$ are the relationship which allow camera signals to be converted to radiance N_{obj} and to temperature T_{obj} values, respectively. In the case of an ideal instrument, a direct relationship can be derived between the radiometric signal and the temperature of the object.

Infrared thermography technique is a very sensitive and accurate inspection method and it can handle complex shapes. It is repeatable and economical in saving time, labor, equipment, and traffic control. This technique can quickly examine a very large area and it is a real kind of non-destructive test technique. But it needs qualified technicians and specialized equipment. Some of the scanning system is very sensitive and can identify temperature differences as small as 0.05 °Celsius. The portable display monitor portrays the image in color. Generally, the hotter the area, the brighter the image, the cooler, the darker the image. This portrayed image can be video-recorded in real time, thus this system provides an instant thermal picture. The displayed image can also be stored in a computer, where further analysis of the captured data can be carried out.

The infrared thermography method can be used to detect hidden cracks, voids, porosity, changes in compositions, and delaminations in concrete structures such as bridge decks, dams, highway pavements, garage floors, parking lot pavements, and building walls.

Li *et al.* (2000) applied the infrared thermography technique to inspect the debonded ceramic tiles on a building finish. In their test, a thermographer measures the infrared radiation emission from the building surface under certain ambient temperature and displays it in the form of a visual heat image to determine the debonding of tile finishes of tall buildings. The difference in temperature of a tile surface is related to the thermal properties of an overall structure and could thus reveal the degree of damage of the tile system (Figure 3.12). The results presented prove that infrared thermography technique could be used as a diagnostic tool to identify severe defects beneath the surface of building finishes. The infrared thermography

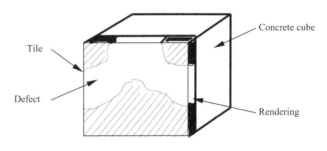

(a)

(b)   (c)

Figure 3.12 Surface temperature distribution of a debonded tile sample: (a) with upper half dry and lower half filled with water; (b) uniform heating of the inspected face; (c) after cooling for one half hour.

technique has also been utilized for assessment of concrete moisture conditions (Godfrey 1984).

(C) RADIOMETRY AND RADIOGRAPHY

The techniques of radiometry and radiography are based on radioactive sources, such as X-rays, γ-rays or neutrons. Both X-rays and γ-rays have very small wave length. The X-rays have a wavelength range from 3 nm to 0.03 nm while the γ-rays' wavelength is much smaller. As the wavelengths of light decrease, they increase energy. Thus X-rays and γ-rays are both at the high energy end of electro-magnetic spectrum and can penetrate matter with some attenuation. The attenuation of radiation passing through matter is exponential and may be expressed as:

$$I = I_0 \exp(-\mu X) \tag{3.14}$$

where:

   I = energy intensity of the beam at a particular location;
   I_0 = incident energy intensity;
   μ = linear absorption coefficient;
   X = distance from the member surface.

The absorption coefficient depends on the composition of the material. Thus, the measured intensity (with detector or a radiation sensitive film) can provide information of the material.

   In radiographic methods, a radiation source and photographic film are placed on opposite sides of a testing object. The result, after exposure of the film, is a photographic image of the member's interior. Defects inside the member can thus be identified. These techniques are generally fast and reliable, and can provide information not available by other means. However, they involve complex technology, high initial costs, and specific training and licensing requirements. X-rays can go through about 1 m of concrete. Applications include determining distribution of aggregate particles, three-dimensional configuration of air voids inside concrete, segregation, and the presence of cracks. But X-ray equipment is very expensive and operates using a high voltage. As the use of γ-rays does not require electricity, γ-ray equipment has gained a considerable market. However, γ-ray equipment has to be properly shielded. Besides, additional safety checks are required to prevent exposure to radioactive materials. The γ-ray technique is especially valuable for determining the position and condition of reinforcement, voids in the concrete or grouting quality of post-tensioned structures.

   In the radiometry methods, a radiation source and a detector are placed on the same or opposite sites of a concrete member. The number of electric pulses produced at the detector is a measure of the dimensions or physical characteristics (e.g. density or composition of the concrete member). The γ-ray is most commonly used in radiometry systems for concrete although neutron radiometry has been used for asphalt concrete and soil. For detection of γ-rays, the Gieger-Muller tube is most commonly used. When high-energy radiation passes through concrete, some energy is absorbed, some energy passes through, and a significant amount is scattered by collisions with electrons in the concrete. So when employing γ-ray examining concrete, there are basically two modes of transmission, the direct transmission mode and the backscatter mode. For the former, depending on the source, γ-rays can go through 50–300 mm of concrete. For the latter, it essentially measures concrete within 100 mm of the surface. Some examples of applications include: (1) measurement in transmission mode with internally embedded probes in fresh concrete to monitor the density and hence the degree of consolidation; and (2) non-contact backscatter measurement allows the monitoring of density for a relatively thin pavement.

3.3.2.5 Building dynamics

Building dynamic techniques are widely used in structural engineering, some simple methods are useful for assessing localized integrity, such as delamination, while more complex methods appear in pile integrity testing, determination of member thickness, and examination of the change of stiffness of members affected by cracking or other deterioration. In general, there are three categories of dynamic response methods for NDT purposes: modal analysis, resonant and damping techniques. The testing equipment for building dynamic techniques is roughly composed of two parts: one generating mechanical vibrations and the other sensing these vibrations. Pulse–echo method is one simple and easy way to apply building dynamic techniques which involves measuring the reflected shock waves caused by a surface hammer and analyzing them in time and/or frequency domains. Dynamic response testing of large structures may similarly involve hammer impacts or the application of vibrating loads. The vibration response is recorded by carefully located accelerometers. Through measuring the natural frequencies and/or the rate of attenuation (or damping) of vibrations of the building structure, the dynamic property, defects and damages of the building structure, even individual member stiffness, can be obtained.

Resonant frequency technique is also widely used to derive the dynamic property of a building structure. Since every elastic object has many resonant frequencies and these frequencies are related to its stiffness and mass distribution, many physical characteristics of the object may be determined from the characteristics of the induced vibration. Typical vibration mode shapes include flexural, longitudinal, and torsional as well as fundamental and higher order. Usually, the fundamental flexural and extensional modes are most easily excited and are important in the NDT inspection of building structures. When the testing object is made to vibrate in one of its natural or resonant modes by an applied external force, its mode shape can reveal the configuration and composition of the testing object. This method is mainly used to determine the dynamic modulus of elasticity, stiffness, and Poisson's ratio of concrete. Several factors may influence the results of the resonant frequency method in inspection of concrete structures, including: mix proportions and properties of aggregates (Jones 1962); specimen-size (Obert and Duvall 1941; Kesler and Higuchi 1954); and curing conditions (Obert and Duvall 1941; Kesler and Higuchi 1953).

Damping testing method is another kind of building dynamic method. Damping is closely related to the dynamic motion of an object. When a solid object is subjected to dynamic forces, the amplitude of its free vibration will decrease with time after the exciting forces are removed. This is because some of the internal energy of the vibrating object is converted into heat and this phenomenon is termed as damping. What's more, solid objects exhibit a hysteresis loop, i.e. the downward stress–strain curve due to unloading does not exactly retrace its upward path. In addition, engineering materials always exhibit mechanical relaxation by an asymptotic increase in strain resulting from the sudden application of a fixed stress, and, conversely, by an asymptotic relaxation in stress whenever they are suddenly strained (Bray and McBride 1992). This mechanical relaxation has an associated relaxation time, the direct result of which is the significant attenuation of vibrations whenever the imposed frequency has a period that approximates the relaxation time. Normally the damping effect is characterized by the specific damping capacity, Y, which is given by

$$Y = \frac{\Delta W}{W} \qquad (3.15)$$

where ΔW = energy dissipated in one cycle; and W = total energy of the cycle. Damping is a relaxation process, which is governed by a characteristic time that corresponds to the peak frequency and is referred to as the relaxation time. The specific damping capacity and associated dynamic response of the material are characterized by the damping coefficient, which can be expressed by

$$a = \frac{1}{N} \ln\left(\frac{A_0}{A_n}\right) \qquad (3.16)$$

where a = damping ratio; A_0 = vibration amplitude of the reference cycle; and A_n = vibration amplitude after N cycles. The specific damping capacity ($\Delta W/W$) for materials is calculated as:

$$\frac{\Delta W}{W} = 1 - e^{-2a} \qquad (3.17)$$

The damping testing methods require an input vibration pulse and an associated output signal. The test body is first caused to vibrate in one of its natural vibration mode, the input signal is then interrupted, and the vibration decay of the testing object is measured. The specific damping capacity ($\Delta W/W$) is thus determined from the resultant decay curve of the output signal. Several aspects are normally involved in investigating the damping phenomenon, including: (1) determination of amplitude decay in free vibrations; (2) determination of the hysteresis loop in the stress–strain curve during forced vibrations; (3) determination of the resonance curve during forced vibration; (4) determination of energy absorption during forced vibration; (5) determination of mechanical impedance during forced vibrations; and (6) determination of sound-wave propagation constants (Bray and McBride 1992).

3.4 Reflected and transmitted waves

The major types of body waves are longitudinal (P) and transverse (S). They are normally named according to relationships of particle motions relative to the direction of propagation. For longitudinal waves, the propagation and particle motion directions are the same. Longitudinal waves can propagate in solids, liquids, and gases and are the most widely utilized wave mode for non-destructive testing of materials and structures. On the other hand, shear waves have particle motion transverse to the direction of propagation of the wave. Share waves cannot pass through the liquid and thus are limited to solid inspection only. The reflection and transmission behavior of the wave front at an interface plays an important role in ultrasonic investigations. The wavefront defines the leading edge of a stress wave as it propagates through a medium. The reflection and transmission describe a behavior of a wavefront at the interface of two different materials. The acoustic impedance ratio of two materials plays an important role in determining the reflection and transmission parameters. The acoustic impedance is defined as a product of density and wave velocity of a material.

The reflection and transmission behavior of the wave front at an interface plays an important role in ultrasonic investigations and provides the basis of determining the presence of a flaw and other anomalies. Here, let us consider a one-dimensional wave propagation case as shown in Figure 3.13. Figure 3.13 shows that two mediums have different material properties with a boundary, in which ρ and C_L are density and longitudinal wave velocity for medium I; and ρ^A and C_L^A are those for medium II. A plane wave traveling in media I approaches the boundary from the left. This wavefront is parallel to the boundary and is designed as the incident wave and indicated by the symbol i. Upon striking the boundary, part of the energy is reflected back to media I and part of it is transmitted into media II through the boundary. Their portion of energy transmitted and reflected is a function of the properties of mediiums I and II. The displacements for incident, reflected, and transmitted wave are as follows:

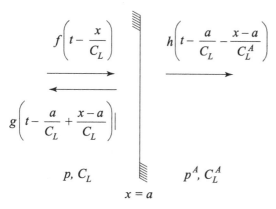

Figure 3.13 One-dimensional wave propagation at an interface between two mediums.

$$U^i = f\left(t - \frac{x}{C_L}\right)$$

$$U^r = g\left(t - \frac{a}{C_L} + \frac{x-a}{C_L}\right) \qquad (3.18)$$

$$U^t = h\left(t - \frac{a}{C_L} - \frac{x-a}{C_L^A}\right)$$

where U^i is the displacement of incident wave, U^r is the displacement of reflected wave, U^t is the displacement of transmitted wave. C_L is longitudinal wave velocity for medium I; and C_L^A is that for medium II.

At the interface $x = a$, stress continuity and displacement compatibility have to be satisfied. From these conditions, we can obtain the relationship of

$$A_R = \frac{1 - \dfrac{\rho^A C_L^A}{\rho C_L}}{\dfrac{\rho^A C_L^A}{\rho C_L} + 1} A_I = \frac{1 - z}{z + 1} A_I \qquad (3.19)$$

$$A_T = \frac{2\rho C_L}{\rho^A C_L^A + \rho C_L} A_I = \frac{2}{z + 1} A_I$$

where A_R is the displacement amplitude of reflected wave, A_I is the displacement amplitude of incident wave and A_T is the displacement amplitude of transmitted wave; and

$$\sigma_R = \frac{\dfrac{\rho^A C_L^A}{\rho C_L} - 1}{\dfrac{\rho^A C_L^A}{\rho C_L} + 1} \, \sigma_I = \frac{z - 1}{z + 1} \, \sigma_I = R\sigma_I \tag{3.20}$$

$$\sigma_T = \frac{2\rho^A C_L^A}{\rho^A C_L^A + \rho C_L} \, \sigma_I = \frac{2z}{z + 1} \, \sigma_I = T\sigma_I$$

The R and T are called reflection and transmission parameters. These expressions show that the ratio of the acoustic impedances completely determines the nature of the reflection and transmission at the interface. They can be plotted as a function of the ratio of $\rho^A C_L^A/\rho C_L$ (Figure 3.14). Let us look at a few extreme cases.

For $\rho^A C_L^A/\rho C_L = 0$, $T = 0$ and $R = -1$, which means that the incident wave is reaching a free surface. No stress can be transmitted. To satisfy the zero stress boundary condition, the displacement must be twice the displacement of the incident wave. The reflected wave has the same amplitude as incident wave but opposite polarity. It implies that a free end will reflect a compression wave to a tension wave with identical amplitude and shape and verse visa.

For $\rho^A C_L^A/\rho C_L = 1$ which means the same materials, $T = 1$ and $R = 0$. The wave is completely transmitted.

For $\rho^A C_L^A/\rho C_L$ goes to infinity, which means that an incident wave is

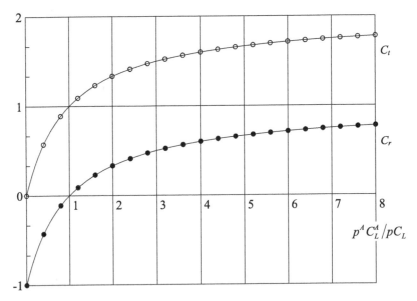

Figure 3.14 The reflection and transmission coefficients vs. acoustic impedance.

approaching a fixed end. No displacement can occur ($U_T = 0$). The stress at the boundary is twice that of the incident wave and the reflected wave has the same amplitude and polarity as the incident wave.

When the wavefronts propagate upon the boundary between two materials with different properties with an angle not normal to the interface, the reflection and transmission will depend on the angle. Let θ be the wave incidence angle. The reflection angle of the wave is also θ. However, the angle of transmission, β, is a function of the angle of incidence, θ, and the ratio of wave velocities in the two media, and is given by Snell's law:

$$\sin \beta = \frac{C_L^A}{C_L} \sin \theta \qquad (3.21)$$

Stress waves can change their mode of propagation when striking a boundary at an oblique angle. Depending on the angle of incidence, a P wave can be partially reflected as both P and S waves and can be transmitted as both P and S waves. An S wave reflects and transmits at angles, determined using Snell's law, that are less than the angles of reflection and transmission for a P wave.

3.5 Concrete strength assessment

It is generally recognized that the concrete in a structure cannot be expected to have the same strength as the concrete in the test cubes or cylinders which are made for control purposes at the time of construction. Standard tests of concrete strength using cubic or cylindrical specimens can only be possible in concrete manufacturing plants. Other strength assessment methods should be developed for in-situ concrete or concrete without mix proportion and service records. The residual strength of concrete can be evaluated by using the following methods.

3.5.1 Compressive tests on the cored specimen

The compressive strength of concrete can be determined by performing the compressive test on the specimens cored out of the inspected structures. This is considered the most direct method. A core is a cylinder of concrete cut from hardened concrete through a hollow drill barrel of a rotary cutting tool. The equipment is usually portable, but heavy (Figure 3.15). Hand-held equipment is available for cores up to 75 mm diameter. In both cases, a skilled operator is needed as poor operation will cause damage to the specimen.

Normally, 100 mm or 150 mm diameter cores are employed for strength tests while lengths of the prepared samples are usually between one and two

Figure 3.15 A core-cutting drill.

times the diameter. British and American standards require the diameter to be at least three times the aggregate size and British standards further require a minimum diameter of 100 mm. British standards also suggest that the length of the core (L) should be kept relatively short (between L/d = 1.0 to 1.2) for reasons of drilling costs, damage, and variation along length. On the other hand, ASTM recommends the use of longer lengths (L/d = 2.0) if possible, so that results are more comparable to those from standard cylinders. During drilling, the equipment must be firmly supported and braced against the concrete to prevent relative movement which may result in a distorted or broken core, and a water supply is also necessary to lubricate the cutter. After the core is broken off from the structure, it should be examined immediately. If there is insufficient length for testing, excessive reinforcement or voids, an extra core should be drilled from adjacent locations. The hole left by the core should be filled with low-shrinkage concrete, or by a cast concrete of suitable size together with cement grout or epoxy resin. Smooth-sided, straight cores are necessary for strength testing. It is preferable that the cores do not contain reinforcement and a cover meter can be used to locate the bars.

The cored specimen needs to be trimmed by a diamond saw to make the core suitable for testing. The ends of the specimen should be ground or be capped with high alumina cement mortar or a sulfur–sand mixture before testing strength to make sure parallel end surfaces normal to the axis of the

core. Many standards recommend procedures for cutting, testing and interpretation of results for compressive tests on the cored specimen. These
available standards include BS 1881: Part 120, ASTM C42, and ACI 318.
A compression test is carried out at a rate within the range 12–24
N/(mm^2–min). In the UK, cores are tested under saturated condition. To
saturate a core, it should be immersed for at least two days after capping. In
the USA, dry testing is used if the in-situ concrete is in a dry state. Splitting
failure with cracks all around the core should be observed during compression test. Diagonal shear cracks are also acceptable except in short cores or
where there is reinforcement or honeycombing.

There are many factors that influence the measured core compressive
strength, which may be divided into two basic categories: concrete characteristics and testing variables. The factors in the first category include: moisture content, curing condition and voids in the cored specimen. A saturated
specimen is 10–15 percent lower in strength than dry specimen. It is thus
very important that the relative moisture conditions of core and in-situ concrete are taken into account in determining actual in-situ concrete strengths.
It has also been found that moisture gradients within a core specimen also
influence measured strength (Bartlett and MacGregor 1994). Curing conditions of the cored and its parent concrete are different after the core is
removed from the structure. This effect is negligible for an old structure but
has to be considered for concrete less than 28 days old. Excessive voidage
can affect core strength (Bungey and Millard 1996). The factors belong to
the second category are: (1) length–diameter ratio of the core specimen; (2)
core diameter; (3) drilling direction; and (4) reinforcement. The measured
core strength decreases with core length–diameter ratio due to the effect of
end constrains from loading platens of testing machine. BSI (BS1881 1983)
has published correction factors to an equivalent length-diameter ratio of
2.0 (Bungey and Millard 1996). Vertically drilled specimens have been
shown to be stronger than horizontally drilled specimens by about 8 percent
(Concrete Society 1987). Reinforcement in the core specimen should be
avoided as much as possible. If it cannot be avoided, the corrections should
be performed. For a core with a bar perpendicular to its axis, a correction
relationship has been proposed by BS 1881: Part 120:

$$\text{Corrected strength} = \text{measured strength} \times \left[1.0 + 1.5\left(\frac{\phi_r}{\phi_c}\frac{h}{l}\right)\right] \qquad (3.22)$$

Where ϕ_r = bar diameter; ϕ_c = core diameter; h = distance for bar axis
from nearer end of core; l = core length (uncapped). If the correction is
greater than 10 percent, the result should be disregarded. If there are multiple bars within a core, the correction should be conducted according to:

$$\text{Corrected strength} = \text{measured strength} \times \left[1.0 + 1.5\frac{\Sigma(\phi_r h)}{\phi_c l}\right] \qquad (3.23)$$

It should be noted that if the spacing of two bars is smaller than the diameter of the larger bar, only the bar with higher should be considered. There are, however, some limitations of this core testing method.

The core testing technique may not be appropriate if strength needs to be determined at a large number of locations. In this case, non-destructive tests are needed to measure other related properties in place on the structure in order to interpret the concrete strength (Kay 1992).

3.5.2 Rebound hammer measurement

Rebound hammer is one of the oldest non-invasive methods of accessing variation of concrete strength within a structure. The equipment was developed by Ernst Schmidt in Switzerland in the 1940s. This is why the rebound hammer is also called as Schmidt hammer. It offers a simple means of measuring the in-situ hardness of a localized area of the concrete surface through measuring the rebound value of a spring-driven hammer mass after its impact with concrete. The rebound hammer measurement is principally a surface hardness test method with little apparent theoretical relationship between the strength of concrete and the rebound number of the hammer. However, some empirical correlations have been established between strength properties and the rebound number.

A Schmidt rebound hammer weighs about 1.8 kg and is suitable for both a laboratory and a field test. A schematic cutaway of the rebound hammer is shown in Figure 3.16. The rebound hammer is essentially made of a spring-controlled hammer mass that slides on a plunger. A cylindrical casing houses the system. Other features include a latching mechanism that locks the hammer mass to the plunger rod and a sliding rider to measure the rebound of the hammer mass. The rebound distance is recorded as "rebound number" corresponding to the position of the rider on the scale. When the plunger is pressed against the surface of a concrete member to be tested, the spring-loaded mass accumulates a certain amount of potential energy due to the spring being stretched a certain distance. The spring is then automatically released at that position, releasing the stored energy and causing the mass to rebound, carrying a rider with it on a guide scale. The rider is usually held in position by pressing a button and this helps in recording the reading. The rebound hammer measurement can be conducted horizontally, vertically upward or downward, or at any intermediate angle. Due to different effects of gravity on the rebound as the test position is changed, the rebound number will be different for the same concrete and will require separate calibration or correction charts. The use of the rebound hammer is covered by ASTM C805, which also states that the rebound method is not intended as an alternative to strength determinations, but it may be used to access uniformity of concrete, to determine areas of poor quality or deteriorated concrete, and to indicate changes in characteristics with time (Kay 1992).

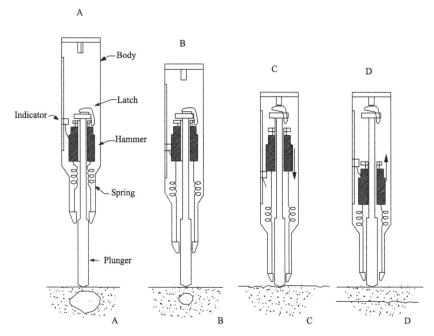

Figure 3.16 A cutaway schematic view of the Schmidt rebound hammer.

Principally, the measurement of rebound hammer is directly related to the hardness of concrete surface, not its strength. However, due to the lack of handy testing method in determining concrete strength on site and some relationship between surface hardness and compression strength, the rebound hammer has been utilized as a common tool to access concrete strength through a calibration curve between rebound number (surface hardness) and compressive strength. To make the curve meaningful in a statistical sense, a large number of experiments need to be conducted for each type of concrete.

Different rebound hammers are available calibrated for different types of concretes, while the selection of the correct hammer requires advance knowledge of the answer one is attempting to obtain. Although the rebound hammer provides a quick and inexpensive means of checking the uniformity of concrete, the accuracy of its results may not be satisfory. It can be influenced by the following factors:

1 formation and smoothness of the concrete surface;
2 age of the concrete;
3 type of cement used in the concrete mix;
4 moisture conditions of the concrete;
5 type of coarse aggregate used in concrete;
6 carbonation degree of concrete.

The type of formwork can affect the rebound number and toweled surfaces generally give higher results than formed surfaces (Kay 1992). It has been shown that toweled surfaces or surfaces made against metal forms yield rebound numbers 5–25 percent higher than surfaces made against wooden forms (Kolek 1958; Greene 1954). Besides, toweled surfaces will give a higher scatter of individual results and a lower confidence in estimated strength. Therefore, when toweled surfaces are to be used, a special correlation curve must be developed. Kolek (1958) has indicated that the rate of gain of surface hardness of concrete is rapid up to the age of 7 days and after which there is little or no gain in the surface hardness. When testing old concrete, direct correlations are necessary between the rebound numbers taken on the structure and the compressive strength of cores taken from the structure. Kolek (1969) has found that the type of cement significantly affects the rebound number readings. High alumina cement concrete may show higher rebound numbers than ordinary Portland cement concrete. The super-sulfated cement concrete can have 50 percent lower strength than that obtained from the ordinary Portland cement concrete correlation curves. The degree of saturation of the concrete, as well as the presence of surface moisture, also has a decisive effect on the evaluation of rebound hammer test results (Zoldners 1957). It has been demonstrated that well-cured, air-dried specimens, when soaked in water and tested in the saturated surface-dried condition, show rebound readings 5 points lower than when tested dry. Klieger *et al.* (1954) have shown that, for a three-year-old concrete, differences up to 10 to 12 points in rebound numbers existed between specimens stored in a wet condition and laboratory-dry samples, which represents approximately 14 MPa difference in compressive strength.

The rebound number is also affected by the type of aggregate used. Klieger *et al.* (1954) found that, for equal compressive strength, concrete made with crushed limestone coarse aggregate show rebound numbers approximately 7 points lower than those for concretes made with gravel coarse aggregate, representing approximately 7 MPa difference in compressive strength. The surface may be carbonated, which tends to significantly increase hardness or it may also be less well cured because of rapid loss of moisture from this zone. In a similar way, the hardness may vary across a structure because of changing exposure conditions. The rebound test is also sensitive to local variation in the concrete quality, for instance, the presence of a large piece of aggregate under the plunger would result in an unusually high number; conversely, the presence of a void would show a very low result (Nasser and Al-Manaseer 1987). For this reason, in most structures, it has been suggested that at least 16 readings be taken in a square of foot area (Mailvaganam 1992).

Finally, it should be indicated that the skill of an operator of rebound hammer may make a significant difference in the accuracy of a rebound hammer test. Generally speaking, an experienced operator would provide more reliable results with a rebound hammer.

3.5.3 *The surface wave measurement method*

When the surface of a solid elastic medium is disturbed by a dynamic or vibratory load, three types of stress waves are created, which are: (1) compression waves (also called longitudinal or *P* waves); (2) shear waves (also called transverse or *S* waves); and (3) surface waves (also called Rayleigh waves). A surface wave spreads along the surface of a solid with its characteristic velocity and the particle motion is retrograde elliptical.

The surface wave velocity is given by the following approximate formula (Viktorov 1967):

$$C_R = \frac{0.87 + 1.12v}{1 + v} \sqrt{\frac{\mu}{\rho}} \tag{3.24}$$

where μ is the shear modulus of elasticity and equal to $E/2(1 - v)$.

For concrete, with a typical Poisson's ratio of 0.2, the Rayleigh wave velocity is 92 percent of the shear wave, or 56 percent of the compression wave under plane strain condition. Although wave velocity is directly related to the modulus of elasticity and Poisson's ratio only, people have found some relationship exists between wave velocity and compressive strength of concrete. Thus, through a careful calibration, measurement of the velocity of the wave has been used to estimate a compressive strength of concrete.

The equipment of surface wave measurement setup, used for strength estimation of concrete, usually consists of a controlled impact-based stress wave generator, two receiving accelerometers (receivers), a digital oscilloscope, and a personal computer (Popovics *et al.* 1998). The two receivers are located on the surface of the test specimen along a line extended from the impact site. Transient stress waves generated by the impact, propagate along the surface of the specimen, first passing receiver 1 and then receiver 2 (Figure 3.17). The arrival time of the surface (*R*) wave can be determined from the signal. By utilizing the difference in *R* wave arrival times between the two receivers, the *R* wave velocity can easily be determined provided that the spacing between the two receivers is known. There is no unique relationship between *R* wave velocity and strength of concrete. The relationship depends on the type of aggregate and cement used in production of the concrete and the mix proportions. If the surface wave measurement technique is used to estimate the strength of concrete in a structure, it is necessary to establish the correlation between wave velocity and compressive strength for the particular concrete to be tested.

3.6 Surface cracking measurement

To detect the finer cracks on a concrete surface that cannot easily be observed by the naked eye, one-sided, self-calibrating surface wave

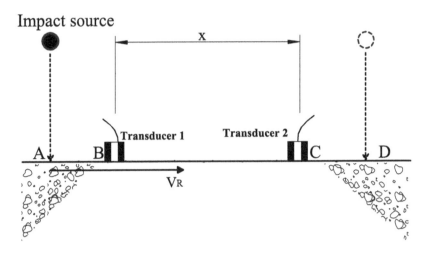

Figure 3.17 Surface wave velocity measurement for concrete strength assessment.

transmission technique can be used. The set-up of surface cracking measurement is similar to that of surface wave measurement for concrete
strength. It consists of one impact source and two receivers. The theoretical
basis for self-calibrating is summarized as follows (Achenbach *et al.* 1992).
In the frequency domain, we can represent a stress wave signal generated by
an impact (wave source) at location *A* and received by the near accelerometer at location *B* as a simple product form:

$$V_{AB} = S_A d_{AB} R_B \tag{3.25}$$

where V_{AB} is the FFT (Fast Fourier Transform) form of the captured time
domain signal, S_A the source characteristics term, R_B the receiver response
term, and d_{AB} the signal transmission function between locations *A* and *B*.
Similarly, the wave signal sent by the same impact event at location *A* and
received by the far accelerometer at location *C* is given by

$$V_{AC} = S_A d_{AB} d_{BC} R_C \tag{3.26}$$

where V_{AC} is the FFT (Fast Fourier Transform) form of the captured time
domain signal, S_A the source characteristics term, R_C the receiver response
term, d_{AB} the signal transmission function between locations *A* and *B*, and
d_{BC} the signal transmission function between locations *B* and *C*.

The S_i and R_i terms are usually difficult to be determined with a satisfied
accuracy because they contain uncertain variability caused by vibration of
impact events, and accelerometer coupling. Thus, it is desirable to eliminate
these parameters to get a reliable transmission function, d_{ij}, between the two
transducers. Moreover, the d_{ij} is more useful in interpreting the materials
properties. To eliminate the R_i and S_i terms, another complementary set of

experiments from other side of the receiving accelerometer pair along the same line can be conducted. The only difference from the previous set of tests is that the wave source is now at point D instead of point A. Thus, V_{DB}, and V_{DC} can be obtained as:

$$V_{DC} = S_D d_{DC} R_C \qquad (3.27)$$

and

$$V_{DB} = S_D d_{DC} d_{CB} R_B \qquad (3.28)$$

It should be noted that d_{BC} is same with d_{CB}. Thus, by simple manipulation of four signals: V_{AB}, V_{AC}, V_{DB}, and V_{DC}, an expression for the transmission between locations B and C can be obtained as

$$d_{BC} = \sqrt{\frac{V_{AB} V_{DB}}{V_{AB} V_{DC}}} \qquad (3.29)$$

The d_{BC} is a function of frequency and can be visualized as the ratio of the amplitude of the signal from the far accelerometer to that of the near accelerometer. Thus, a transmission value of one indicates no amplitude loss (completely transmission) as the wave propagates from B to C whereas a value of zero indicates complete signal amplitude loss (no transmission). In the case of a point source d_{BC} values should be less than 1, even for perfect transmission materials due to some loss from attenuation.

Usually, an impact event will generate waves propagated in all directions. Some wave components propagate along the surface while others reflect at free boundaries. By setting an appropriate window, surface wave or the first P-wave reflection from the opposing side of the specimen can be distinguished. The d_{BC} measurement can be used to detect the surface crack in concrete. Clearly, if there is a crack between two transducers, the d_{BC} values will suffer a severe reduction in almost all the frequency range.

3.7 Assessment of fire damage to concrete

In general, concrete structures are non-combustible, do not emit toxic fumes during a fire, and can stand in fire safely for a quite long period. However, long time exposure to fire or high temperature will cause concrete deterioration and reinforcing steel degradation. Such deterioration and degradation may lead to loss of service ability or collapse of the structure. Hence, after a fire, the concrete structure should be evaluated and the deterioration caused by fire should be assessed before any repair work is taken.

3.7.1 Determination of the maximum temperature

(A) EXAMINE THE DEBRIS FOUND AT THE SCENE

To assess the severity of a fire incident, it has been customary to examine the debris found at the scene, such as pieces of softened or melted plastics, melted metal, softened glass, or charred wood. They can be used to estimate the maximum temperature reached by the fire. For example, polystyrene melts at 150–180 °C; cellulose darken at 200–300 °C; lead melts at 300–350 °C; and silver melts at 950 °C. Table 3.3 lists the survey on the

Table 3.3 Survey on behavior of commonly used building materials at high temperature

Material	Typical examples	Conditions	Appr. temp (°C)
Polystyrene	Thin wall food containers, foam, light shades, handlers, curtain hooks, radio casings	Collapse Softens Melts and flows	120 120–140 150–180
PVC	Cables, pipes, ducts, linings, profiles, handles, knobs, house-ware, toys, bottles	Degrades Fumes Browns Charring	100 150 200 400–500
Cellulose	Wood, paper, cotton	Darkens	200–300
Wood	Doors, windows, floors	Ignites	250
Lead	Plumbing, sanitary installations, toys	Melts, sharp edges rounded Drop formation	300–350 300–350
Zinc	Sanitary installations, gutter, down pipes	Drop formation Melts	400 420
Aluminum and alloys	Fixtures, casings brackets, small mech. parts	Softens, melts Drop formation	400 650
Glass	Glazing, bottles	Softens, sharp edges, rounded Flowing easily, viscous	500–600 800
Silver	Jewellery, spoons, cutlery etc.	Melts Drop formation	950 950
Brass	Locks, taps, door handles, clasps	Melts (particularly at edges) Drop formation	900–1000 900–1000
Copper	Wiring, cables ornaments	Melts	1000–1100
Cast iron	Radiators, pipes	Melts Drop formation	1100–1200 1100–1200
Bronze	Windows, fittings, door bells, ornamentation	Edges rounded Drop formation	900 900–1000

behavior of some commonly used building materials under critical temperature and can be used as useful reference, Concretes made with siliceous or limestone aggregate show a change in color with temperature, depending on the presence of certain compound of iron, so that the maximum temperature during a fire can be estimated from the change in concrete color (Neville 1996). The color of concrete changes on heating from gray to light pink at about 300 °C. The pink darkens with further rise of temperature, attains maximum intensity around 600 °C, then begins to lighten and turn to whitish gray by 800 °C. From this point of view, visual observation of color change and spalling, possibly aided by surface tapping, may provide considerably good information with confidence of assessing fire damage of concrete. However, the change of concrete color might be influenced by type of aggregates and the exposure time of concrete in fire.

(B) THEORETICAL CALCULATION

A more accurate method to determine the maximum temperature is thermo-luminescence testing. Thermoluminescence is a visible light signal that reflects the influence of temperature. It is based on the fact that the curve of light output vs. temperature for a given sample depends upon its thermal and radiation history. Placido (1980) has proposed that the thermolumin-escence of sand extracted from concrete can form the basis of a test for fire-damaged concrete as a measure of the actual thermal exposure experienced by the concrete. However, the thermoluminescence output is affected by the length of exposure to the high temperature, which may mean that the reduc-tion in the strength of concrete exposed to fire for a prolonged period can be significantly underestimated (Chew 1993).

3.7.2 Penetration of heat

The thermal gradient, i.e. penetration of heat in concrete members, is another important factor in assessing fire damage. The penetration of heat can be conveniently studied on core samples of 60–80 mm long, taken from the walls and ceilings of fire-exposed compartments, through investigating color changes of concrete in the sample. The representative color of concrete at different temperatures has been described earlier and the thermal gradient can thus be determined. More reliable conclusions on the heat penetration can be drawn by thermogravimetric analysis. The thermogravimetric test is based on the fact that the cement paste undergoes a continuous series of dehydration in a temperature range of 100–850 °C. Thus, a specimen of fire-exposed concrete in the test will exhibit different loss of mass up to the maximum different temperature. By testing several small samples (about 500 mg each) taken from various sections of the cores, the distribution of maximum temperature reached at various depths in the compartment's boundary elements during the fire can be determined.

Since the presence of inert aggregates in the test sample tends to obscure the clarity of conclusion, it is advisable to remove most aggregates from sample. Furthermore, since the dehydrations are not all irreversible, the core samples should be taken as soon following the fire incident as possible, and the samples be kept in a desiccator until tested. Decision on the concrete layers to be removed should be based on a careful examination of the condition of entire fire-exposed surface. Sometimes it may be sufficient to make an impact on the damaged sections, using a hammer. The ringing of sound concrete and the dull thud of the weak material can easily be distinguished.

3.7.3 Fire-damage factors

The repair starts by assigning fire-damage factors to concrete and to the steel components. The whitish gray and pink concrete is considered as zero strength, and the concrete that has attained temperatures between 100 and 300 °C is assumed to possess 85 percent of its original strength. When the temperature increases to 300–600 °C, concrete begins to exhibit significant strength loss, up to 60 percent strength loss can be expected. At temperatures between 600 and 900 °C, concrete becomes weak and friable and it may be subject to 100 percent strength loss.

Steel reinforcing bars are usually assumed to have lost 30 percent of their effectiveness, mainly on account of possible loss of bond between the concrete and the steel. Bars exposed to fire because of concrete spalling are regarded as having attained a temperature of 800 °C. Steel bars may lose up to 90 percent strength after being subject to temperatures between 550 and 700 °C. It should be noted that unless heated to temperature above 600 °C, low-carbon hot-rolled steels will not suffer substantial looses in their yielding strength. However, temperatures in excess of 600 °C can cause permanent changes in the grain structure (spheroidization, grain growth), which will result in loss of strength, the finer the grain structure, the more severe the loss. The high-carbon, cold-worked steel used in pre-stressed concrete elements own their high strengths partly to their elongated grain structures. Above 450 °C, the grains tend to resume equi-axed shapes, and thereby the steel loses the excess strength imparted by cold work. Upon the temperature reaching 700–900 °C, the normal reinforcing steel will totally lose its strength.

3.8 Bridge assessment

Bridges are very important infrastructures. The structural damage and failure of a bridge can have severe consequences because bridges often provide vital links in a transportation system. The primary purpose of bridge inspection is to maintain public safety, confidence, and investment in bridges. Ensuring public safety and investment decision requires a comprehensive bridge inspection. Some of the major responsibilities of a bridge inspector are as follows (Chen and Duan 2000):

1   identifying minor problems that can be corrected before they develop into major repairs;
2   identifying bridge components that require repairs in order to avoid total replacement;
3   identifying unsafe conditions;
4   preparing accurate inspection records, documents, and recommendation of corrective actions;
5   providing bridge inspection program support.

National interest in the inspection and maintenance began to rise considerably in the 1960s. In the USA, the National Bridge Inspection Standard (NBIS) sets the national policy regarding bridge inspection procedures, inspection frequency, inspector qualifications, reporting format, and rating procedures. In addition to the establishment of the NBIS, other agencies, such as the Federal Highway Administration (FHWA) and the American Association of State Highway and Transportation Officials (AASHTO), have developed and published standards and/or manuals for bridge inspection. The frequency, scope, and depth of the inspection of bridges generally depend on several parameters such as age, traffic characteristics, state of maintenance, fatigue-prone details, weight limit posting level, and known deficiencies (Chen and Duan 2000). Today, by federal law, every highway bridge structure must be inspected periodically by a qualified engineer or bridge inspector. NBIS requires that each bridge that has been open for traffic be inspected at regular intervals of not more than 2 years.

Compared with other common construction materials, concrete is especially suitable for bridges as evidenced by its ever increasing use all around the world. It is known for its longevity, low maintenance costs and ability to be formed into different desired structural shapes. Besides, the raw materials for making concrete, such as water, fine aggregate, coarse aggregate, and cement, can be found in most areas of the world, which enables concrete to be produced world-wide for local markets and thus avoids transport costs. On the other hand, bridges can be built in many different ways. There are two major categories in classifying types of bridges.

1   In accordance with structural types:
    Beam
    Arch
    Truss
    Segmental
    Cable-stayed
    Suspension.

2   In accordance with structural materials and construction procedures:
Reinforced concrete bridges
Pre-stressed concrete bridges
Steel bridges
Steel-concrete composite bridges
Timber bridges.

To be able to effectively design either new bridges or retrofit measures for existing bridges, a clear understanding of potential problem areas is essential. There is no better way of developing this understanding than by a systematic examination and categorization of failures and damage that have occurred to bridges. It has often been said that those who ignore the lessons of history are doomed to repeat the mistakes.

3.8.1 *Two levels of degradation*

Bridges suffer environmental attack while carrying loading. Two factors in particular contribute to the deterioration of concrete bridges: severe environment and overloading. The severe environment includes bad weather, chemical attack, and physical attack. In cold climates, salt is routinely applied to bridge decks in winter to prevent ice formation and bridges will be subject to severe chloride attack. In a marine environment, bridges are subject to chloride attack from the spray of the sea water. These salts penetrate through the concrete cracks and pores to the reinforcing steel and cause corrosion due to the presence of water and oxygen. Without doubt, the chloride ion is the most destructive element for concrete bridges, especially decks. There has been a dramatic increase in this hazard regarding the use of de-icing salt (sodium or calcium chloride) in the USA, in Canada and in the UK. If de-icing salt is combined with frost–thaw loading, more severe damage can result.

De-icing salts have a number of effects that influence the frost–thaw deterioration rate: (1) salts penetrate the pores of the concrete and increase the average water content of the pores due to the hygroscopic character of the salts; (2) the salts lower the freezing point which causes scaling off of concrete surfaces that are subject to frost–thaw de-icing salt deterioration; and (3) the concrete surfaces where de-icing slats are used are generally horizontal and often wet and therefore more prone to damage (Bijen 2003). Exposure of the various parts of a bridge to corrosive action depends very much on the geometry of the whole structure and its relation to salt spray, the prevailing winds and tides, and the efficiency of drainage. The degradation of a bridge can be classified into two levels of damage.

(A) MATERIALS DETERIORATION

Damage of bridge structures or structural components due to materials deterioration is a very slow process: properties of materials decay due to interaction with the surrounding environment. Most concrete deterioration begins with the appearance of cracks. The decks of concrete bridges are more susceptible to deterioration, including surface wear, scaling, delamination, spalling, longitudinal flexure cracks, transverse flexure cracks in the negative moment regions, corrosion of the deck rebars, cracks due to reactive aggregates, and damage due to chemical contamination. The pavement of the bridge deck may also be subject to splash from vehicles, which can cause physical deterioration due to the freezing of the water in the surface layers of the concrete. Also, the corrosion, caused by chloride penetration, is accompanied by expansion of the corrosion product, which leads to high tensile stress in the surrounding concrete and hairline cracks develop. These cracks will become large as a result of continued corrosion, freeze and thaw or sometimes traffic, and eventually will lead to concrete spalling. In a marine environment, concrete structures may be subject to severe deterioration caused by salt attack. The common deterioration phenomena include section losses in concrete abutments, piers, piles, and the underside of decks.

The aim of studying the damage related to materials deterioration is to understand the deterioration mechanisms. With a clear understanding of deterioration mechanisms, proper repair methods can be developed or selected. It has been found that low-slump concrete is suitable in retarding salt contamination in bridges. In concrete for bridges, quality can be further enhanced by the use of entrained air, the use of non-porous, durable aggregates, proper vibration and compaction, and good curing techniques. Asphalt concrete (AC) overlay can provide a smooth driving and wearing surface for bridge deck. A layer of latex-modified concrete, silica fume, or polymer concrete is effective in restricting the intrusion of salt. As a general requirement, the provision of a carefully detailed membrane for the bridge deck should prevent penetration of water and chloride down into bridge deck. In order to improve corrosion resistance, reinforcing steel used for concrete bridge can be protected by galvanization or epoxy coating or cathodic protection.

(B) STRUCTURAL DAMAGE

Structural damage on a bridge is mostly caused by serious continuous damage at the material level by earthquake, wind, and scour. Structural damage caused by material deterioration often relates to concrete spalling and reduction of cross-section area of reinforcing steel due to corrosion. Earthquake-induced damage often results from design deficiencies. Earthquakes can cause damage to bridge superstructures, bearings, and substructures including columns, beams, joints, abutments, foundations. In general, the likelihood of damage to bridges, caused by earthquakes, increases if the ground motion is particularly intense, the soils are soft, or the bridge configuration is irregular. Most of the severe damage to bridges, caused by earthquakes, has taken one of the following forms: (1) unseating of superstructure at in-span hinges or simple supports attributable to inadequate seat lengths or restraint; (2) column failure attributable to inadequate ductility; or (3) damage to shear keys at abutments.

We will study the damages related to earthquakes to understand how a specific design deficiency leads to one or several forms of structural damage. With a clear understanding of the earthquake-related damage, proper retrofit procedures can be developed and applied to existing bridges. Wind can cause a different mode of vibration that may result in collapse of a bridge, especially for lightweight flexible bridges such as suspension bridges and cable-stayed bridges. Bridges spanning over rivers and streams are especially susceptible to scour of the riverbed. Scour around the bridge substructures poses potential structural stability concerns and can cause uneven settlement of bridges and then lead to collapse. The process of scour estimation is discussed in *Evaluating Scour at Bridges* (FHWA 1990). It is stated that when the scour depth is within the limits of the footing or piles, a structural stability analysis for the bridge is required; when the scour depth is below the pile tips or spread footing base, monitoring of the bridge is required.

3.8.2 Bridge rating

Before rehabilitation and renovation, a structural rating analysis may be necessary, which is related to computing the current live-load-carrying capacity of the structure or of individual members. Sometimes, a vehicle may carry much more load than the designed values and thus the bridge owner may need to determine the current live-load-carrying capacity of the bridge to determine whether to allow the vehicle to pass across the bridge or not, which is closely related to loading rating analysis of bridges. The bridge may have lost live-load capacity as a result of aging, deterioration, damage to members, or its self-weight has been significantly increased due to later added weight such as a new deck, or wearing course (Parsons Brinckerhoff 1993). It is load-carrying capacity of the superstructure that generally governs the rating of a bridge. The load-rating analysis needs to properly account for the strength of members and materials of construction in their current state. To determine the load-carrying capacity of a highway bridge, the allowable capacity of each member is first computed. Then it is reduced by the actual dead load carried by the member to produce the capacity for live load. The live-load-carrying capacities of the members are then compared to the critical live load in the member calculated by considering various truck types within the lanes on the bridge. The live-load-carrying capacity can then be determined for each member, and the whole bridge.

For different members (beams and columns), the force type used for RFs (rating factors) are different. For slabs, beams, and girders, bending moments are usually used. For columns and truss members, axial force (stress) is frequently used. In stress calculation, two methods are widely used: allowable stress and load factor (Xanthakos 1996). Concrete bridges can be rated at two levels: inventory rating and operating rating (Chen and Duan 2000). The load that can be safely carried by a bridge for an indefinite period is called the inventory rating while the rating that reflects the absolute maximum permissible load that can be safely carried by the bridge is called an operating rating. Inventory rating determines the load capacity of the bridge for normal service conditions and the fatigue strength of materials is considered. The procedures to calculate the RF are as follows:

Step 1: Calculate the member forces (or stresses) resulting from the service loads $F_{service}$ or F_{design}.

Step 2: Calculate the member forces (or stresses) resulting from dead loads, F_{dead}.

Step 3: Calculate the load-carrying capacity of the members based on their actual section properties and allowable stresses, F_{total}.

Step 4: Calculate the load-carrying capacity available to carry the live load, F_{live}.

$$F_{live} = F_{total} - F_{dead} \qquad (3.30)$$

Step 5: Calculate the RF:

$$RF = \frac{F_{live}}{F_{design}} = \frac{F_{total} - F_{dead}}{F_{design}} \qquad (3.31)$$

From the above procedures, it can be seen that individual members are rated first for their full design section. This will determine the maximum capacity of the bridge if it is rehabilitated to the original design condition. Individual members should then be rated based on their current condition to determine their present load-carrying capacity. The member with the lowest structural rating controls the eventual decision for the post load of the bridge. After rating analysis, a sign indicating the safe load limit (bridge weight limit) can be posted near the approach of the bridge, called bridge posting. This is a policy decision made by bridge owners, not the engineer. Bridge posting should not be confused with bridge evaluation and rating. It may be based on either inventory or operating ratings, but often bridges are posted for reasons other than ratings.

3.8.3 *Four levels of inspection for bridges*

Inspection is essentially a set of activities intended to determine the physical condition of bridges. Before making a detailed inspection at the site, the inspector needs to collect and review information from maintenance documents for the individual bridge and perform a case study on the condition of the bridge. These documents normally include items such as structure information, structural data and history, description on and below the structure, traffic information, load rating, condition and appraisal ratings, service history, rehabilitation and strengthening records, and former inspection findings. After that, a tentative inspection schedule should be made so that the inspector team leader can determine the equipment and personnel needs. During the condition survey and detailed inspection at bridge site, care should be taken to write clear, accurate notes from the very beginning of inspection work, as these will be the record of the current inspection. Some agencies require certain forms to be filled out during the inspection, which leads to a standardized inspection and is easier to be traced in the future.

The bridge inspection usually includes four levels: a superficial inspection, a general inspection, a principal inspection and a special inspection. The superficial inspection is conducted whenever there is a chance to pass the bridge, without a predetermined schedule as the maintaining authority is always encouraged to observe the bridge occasionally whenever passing it. A general inspection is a visual examination of the representative parts of the bridge structure and adjacent associated earthwork or waterway and conducted at two-year intervals. A principal inspection is a close visual examination of all inspectable parts and associated works. It is carried out at intervals less than six years. The special inspection is initiated by a particular event or cause of concern and would probable involve special equipment and test techniques, for example, an acoustic emission (AE) measurement for detection of an active crack.

3.9 Reinforced steel corrosion assessment

The corrosion of steel is an electrochemical process. The process can be measured and the information can be used to access the amount of corrosion that is taking place. Corrosion of embedded reinforcement steel is a major cause of deterioration of concrete structures, which can weak structure load-carrying capability due to the reduction of steel cross-section, surface staining, cracking or spalling of concrete and, in some instances, internal delaminations. Several methods can be used to assess the severity of corrosion of reinforcing steel on site.

3.9.1 Visual inspection and delamination survey

The deterioration of reinforced concrete structures is usually accompanied by rust staining, concrete cracking and spalling. These phenomena can be easily identified by visual inspection by the well-trained eye. The rust staining appears as a yellow stain on the surface of a concrete member and can be easily distinguished as a sign of corrosion. However, there are many reasons that can cause cracks to form. It is important to determine the cause of cracks first. For this reason, the width, position, and direction of a crack should be carefully examined. All these factors can help to decide the cause of a crack and what action should be taken for repair. Some typical crack types are show in Figure 3.18. The most common causes leading to concrete cracks are reinforcement corrosion, sulfate attack, frost action or alkali–aggregate reactions. As shown in Figure 3.18, reinforcement corrosion is usually indicated by splitting and spalling along the lines of bars, whereas sulfate attack may produce a random pattern accompanied by a white deposit on the surface. Alkali–aggregate reaction is sometimes (but not necessarily) characterized by a star-shaped crack pattern, while a frost attack may give patchy surface spalling and scabbing. Due to similarities in crack patterns, it will often be impossible to determine causes by visual inspection alone, but this systematic "crack mapping" is a valuable diagnostic exercise when determining the causes and progresses of deterioration. In such cases, other testing methods are required to identify the cause of deterioration, for example, cover measurements are helpful when reinforcement corrosion is involved.

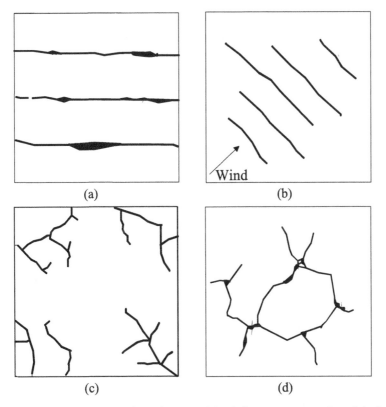

Figure 3.18 Some typical crack types: (a) reinforcement corrosion; (b) plastic shrinkage; (c) sulfate attack; (d) alkali–aggregate reaction.

Concrete delamination of a bridge deck or garage slab, which is separation of a cover layer from body concrete, usually occurs at the level of the upper surface of reinforcing steel. Concrete delamination can be determined by several methods described as follows:

1   *Chain drag method.* This is probably the simplest method. In this method, a heavy chain is dragged on the surface of concrete and the sound is generated during the process of the drag. The sounds from the solid or the hollow area are very different and can easily be distinguished by ears. Hence, a hollow, and therefore delaminated, area of concrete can be identified.

2   *Ultrasonic pulses or radar method.* In these methods, waves are generated by the equipment and sent to the concrete member to be inspected. The reflected wave from delaminations received by transmitter is different from the sound concrete in traveling time and fashion. This information can be used to identify the delaminated areas, internal defects and voids.

3   *Infrared thermography method.* This is a relatively new method used for delamination detection. In this method, an infrared camera is used to record the temperature distribution of the concrete surface while a video camera records the visual condition of the top of the deck or slab. If there is delamination below the surface of the concrete, the air or water in the void will cause a slight temperature difference from solid concrete on the surface. By identifying such temperature differences using the infrared camera, the concealed delamination under the concrete surface can be detected.

3.9.2 Half-cell measurement

Reinforcing steel corrosion is an electro-chemical process requiring the presence of moisture and oxygen and can only occur when the protective layer on the steel surface, formed under high alkaline environment surrounding the steel, has been destroyed, most commonly by carbonation or chloride diffusion (Concrete Society 1984). During the corrosion process, electrical potentials are generated, which can be detected and categorized by a half-cell measurement. The equipment and testing methods are described in ASTM C 876-91 (1999). This method has been developed and widely used with success in recent years for confirming reinforcing steel corrosion. In this method, the potential of embedded reinforcing steel relative to a reference electrode in a half-cell placed on the concrete surface is measured, as shown in Figure 3.7. As the name of the method implies, the equipment contains only one electrode, forming half of a cell or battery. Another half of the cell is provided by the reinforcing steel to be investigated as can be seen from the measurement set up shown in Figure 3.7.

In half-cell measurement, the internal connection is through the electrolyte in the half-cell and the pore solution within the concrete, while the external connection is through a high-impedance meter that is used to measure the potentials generated due to corrosion (Kay 1992). The half-cell consists of an electrode of a metal contained in an electrolyte consisting of a saturated solution of one of its own salts. Copper in copper sulfate and silver in silver chloride are commonly used in reference half-cell. The test procedure of half-cell measurement can be described as follows: (1) place a reference electrode (e.g. Cu in $CuSO_4$ solution or Ag in AgCl solution) in contact with the concrete structure; (2) set up a galvanic cell (battery) between the steel bar to be inspected for corrosion and the reference electrode; (3) measure the change in the potential generated by the copper/steel or silver/ steel cell due to the corrosion activity of reinforcement steel; (4) move the reference electrolyte (half-cell) to different points on the concrete surface, usually in a grid pattern, note the potential in the meter at each point. In this way, a potential map can be composed and locations of potentially high activity can be identified; and (5) interpret the testing results based on an appropriate standards. For accurate measurement, there should be good electrical continuity through the reinforcement over the area being surveyed. To achieve it, the concrete between surface and steel needs to be sufficiently wet to be conductive and, sometimes, the concrete structure investigated may need to be sprayed with water the night before, and tested in the early morning. It is normally required to break down the concrete cover to enable electrical contact to be made with the steel reinforcement. To increase the accuracy of half-cell measurement, the contact resistance between the concrete surface and the half-cell should be minimized as much as possible by cleaning the contact surface, and removing the surface coating.

Measured voltage in half-cell method depends on activity of the steel corrosion process. The relationship between half-cell readings and the potential for corrosion can be well interpreted following ASTM C 876-91 (1999). Normally, with a small potential (0 to −200 mV against $Cu/CuSO_4$), there is a greater than 90 percent probability that steel can be considered passive and there is no reinforcing steel corrosion; with a half-cell potential in the range between −200 mV and −350 mV, corrosion activity in the area surveyed is uncertain; with a half-cell potential more negative than −350 mV, there is a greater than 90 percent probability that corrosion considered can be regarded as significant. If an Ag/AgCl half cell is used, if the potential is greater than −100 mV, the probability of corrosion is less than 10 percent; if the potential between −100 mV and −250 mV, probability of corrosion is uncertain; if the potential is less than −250 mV, the probability of corrosion is greater than 90 percent. Sometimes, highly negative potential may be found for fully saturated concrete containing no oxygen. In this case, no corrosion can occur due to the shortage of oxygen. As stated before, half-cell potential results are often assessed by producing contour maps of the structure surface and steep gradients of potential may indicates corrosion activity occurring. It should be noted that absolute potential measurements do not make sense and can even be misleading. The half-cell method cannot indicate the actual corrosion rate or even whether corrosion has already commenced. The half-cell measurement only indicates areas requiring further investigation, and an assessment of the likelihood of corrosion may be improved by other techniques, such as resistivity measurements (Bungey and Millard 1996).

3.9.3 *Linear polarization resistance measurement*

As previously mentioned, the half-cell measurement cannot indicate the corrosion rate. In order to directly determine the rate of corrosion, a number of perturbative electrochemical techniques have been developed and the linear polarization resistance (LPR) measurement is one such technique. LPR is an electrochemical technique that allows the corrosion rate to be measured directly in real time. LPR is most effective in aqueous solutions and less accurate in concrete. The principle of LPR measurement lies in that when a steel bar is corroded, a corrosion current, I_{corr}, will be generated. The I_{corr} generated by the flow of electrons from anodic to cathodic sites could be used to estimate the corrosion rate based on the principle of Faraday's Law. However, it is very difficult to detect I_{corr} directly from a corroded steel member. In this case, LPR technique has been developed. In the LPR method, a small externally-imposed potential perturbation, ΔE, is usually applied to the steel reinforcement via an auxiliary electrode placed on the surface of concrete (Figure 3.19). The applied ΔE, usually a DC voltage charge of approximately ±10 mV to 20 mV. Under such a condition, a measurable current flow, ΔI, will be produced. The behavior of ΔI is governed by the degree of difficulty with which the anodic and cathodic corrosion process takes place and is related to corrosion current, I_{corr}, In fact, at small values of ΔE, ΔI is proportional to I_{corr}. The change in potential, ΔE, divided by the current density, ΔI, is defined as the polarization resistance, R_p (Ω-m^2).

$$R_p = \Delta E \,/\, \Delta I \qquad\qquad (3.32)$$

The real corrosion current, I_{corr}, the flow of electrons from anodic and cathodic regions in the corroded steel bar, can be derived based on Stern–Geary equation,

$$I_{corr} = B \,/\, R_p \qquad\qquad (3.33)$$

where B is a constant and normally lies between 25 mV (active) and 50 mV (passive) for steel in concrete. I_{corr} reflects the activity of steel "under" the electrode and thus it can be used to estimate the corrosion rate. The common criteria are shown in Table 3.4.

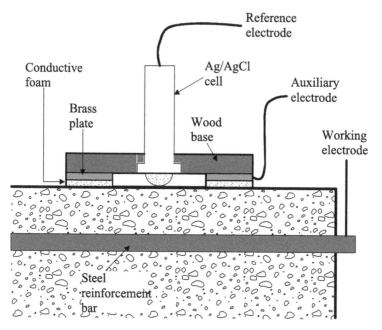

Figure 3.19 Reinforcement corrosion detection by using linear polarization resistance.

Table 3.4 Interpretation of linear polarization resistance measurement results

I_{corr} (mA/cm²)	Corrosion rate	Corrosion penetration (μm/yr)
<0.1 ~ 0.2	Passive	<1
0.2 ~ 0.5	Low to moderate	1 ~ 10
0.5 ~ 1.0	Moderate to high	10 ~ 100
>1.0	High	100 ~ 1000

There are a number of difficulties when employing this polarization resistance measurement. First, it is time-consuming. Second, this method may result in significant errors in measuring polarization resistance, R_p. Third, the choice of a correct value for the constant B requires foreknowledge of the corrosion state of the steel. Fourth, the exact area of steel measured is sometimes not clear and not easy to be accurately evaluated due to such problems as localized pitting. In order to evaluate the area of measurement accurately, some studies recommend using a large auxiliary electrode and consider that the surface area of measurement is the "shadow area" of reinforcing steel lying directly beneath the auxiliary electrode (Dawson *et al.* 1990). Broomfield *et al.* (1995) have suggested another method. By using a guard ring placed around the auxiliary electrode to confine the area of impressed current, it is accepted that the perturbation current will not try to spread laterally to steel outside the shadow area and thus the shadow area can be regarded as the area of measurement.

Both techniques show considerable promise in the field measurements, but their accuracy has yet to be independently verified (Bungey and Millard 1996). The linear polarization resistance measurement method determines the instantaneous corrosion rate which can change with temperature, relative humidity and other factors. In order to improve the interpretation accuracy, measurements at different times are required for full assessment of the corrosion process. The measurement result is only an averaged corrosion rate over the measurement area since it is tacitly assumed that corrosion occurs uniformly over the measurement area. When pitting occurs, the local rate can be much higher than the results of the linear polarization resistance measurement.

3.9.4 Acoustic emission method

The application of the AE technique to detect corrosion is relatively new (Li *et al.* 1998). The means of using the AE technique to detect corrosion is to detect the formation and propagation of microcracks at the steel–concrete interface caused by the expansion of the corroded reinforcing steel. The phenomenon of formation and propagation of the microcracks will generate a stress wave. The stress wave will spread along the medium and reach the outer surface. By placing the AE transducer on the surface, the occurrence of microcracks can be detected and used to interpret the activity of corrosion.

The mathematical model for calculating the stress caused by rebar corrosion at rebar–concrete interface can be simplified as a shrink-fit model for a smooth surface reinforcing steel case. The expansion of corrosion products of the steel is constrained by the surrounding concrete. As a result, a radial displacement will be produced in both the reinforcing steel and the surrounding concrete. The compatibility between steel and surrounding concrete displacement requires that

$$|U_r^c| + |U_r^s| = \Delta a \qquad (3.34)$$

where U_r^c is radial displacement of surrounding concrete, and U_r^s the radial displacement of steel; Δa the incremental of radius of the steel. From this condition, the stress produced at the surrounding concrete interface due to expansion of corroded steel can be derived as

$$\sigma_{\theta\theta} = \frac{4\mu_s\mu_c}{2\mu_s + (k_s - 1)\mu_c} \frac{\Delta a}{a} = C \frac{\Delta a}{a} \qquad (3.35)$$

where μ_s is the shear modulus of steel, μ_c the shear modulus of concrete, k_s the Kolosou constant of steel. It can be seen from Eq. 3.35 that $\sigma_{\theta\theta}$ is proportional to the ratio of steel radius increase due to corrosion to its original radius. For steel, the shear modulus is about 81 GPa and the Kolosou constant is around 2. For concrete, its shear modulus is about 12 GPa. Thus, the value of C is 2.23 × 10¹⁰ Pa. For $\Delta a/a$ equaling 0.0001, the stress produced is 2.23 MPa. Note that the stress is in fact the shear stress in the interface and this value is large enough to create a microcrack. And the stress wave

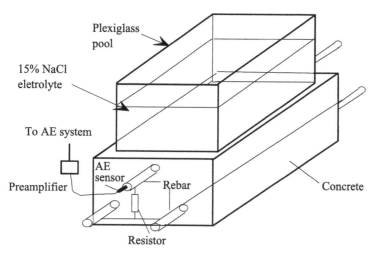

Figure 3.20 Corrosion test of reinforced concrete.

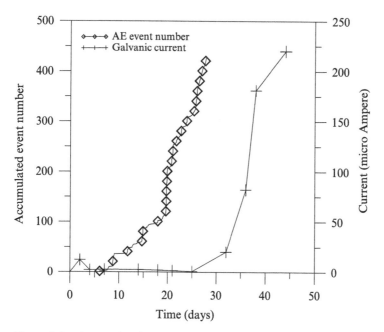

Figure 3.21 Accumulated AE events as a function of time superimposed with a measurement of galvanic current.

generated by this microcrack can be detected by an acoustic emission transducer. Thus, it is proved that the sensitivity of the detection of the AE technique to rebar corrosion is very high (0.0001 of radius of 10 mm is only 1 μm).

The sensitivity has been proved by experiment. As shown in Figure 3.20, a reinforced concrete block, 300 by 300 by 175 mm, is used to perform an accelerated corrosion test. There are three rebars in the block. The top rebar acts as corroded steel and the two bottom rebar are used to increase the cathode area. Three AE transducers are placed at the two ends of the top rebar. A shunt resistor is used to connect the top and bottom rebars. On the top surface of the concrete block, an acrylic tank is built up to hold a solution of NaCl for accelerated corrosion test.

Figure 3.21 shows the plot of accumulated AE events as function of time superimposed with a measurement of galvanic current. It can be seen that there is a sharp increase in AE events at around 20 days that is most likely due to microcracking caused by the expansion of corrosion product. On the other hand, the measurement of galvanic current shows an obvious increase at around 32 days. It proves that AE technique can detect corrosion earlier than galvanic current measurement.

4 Conventional repair and strengthening techniques

During their lifetime, civil structures are expected to provide safety against failure as well as appropriate serviceability under expected workloads and conditions of use. As the load-carrying characteristics and serviceability of the structure may diminish with time due to ageing and to environmental influences, the structures may lose their designed functions partially or wholly. Appropriate actions should then be implemented to improve the performance of structures and restore the desired function of structures. Toward this end, repair and strengthening are the two major approaches. There are several definitions of repair and strengthening of structures given by different professional organizations.

According to ACI 546R-04, to repair is to replace or correct deteriorated, damaged, or faulty materials, components, or elements of a structural system. Strengthening is the process of restoring the capacity of damaged components of structural concrete to its original design capacity, or increasing the strength of structural concrete. ACI 364.1R-07 defines strengthening as the process of increasing the load-resistance capacity of a structure or portion of the structure, which is different from ACI 546R-04. It should be mentioned that strengthening of a structure may need to be made when the load applied on the structure is to be increased even if the deterioration of the structure is not severe.

According to FIP 1991, repair refers to modification of a structure that is damaged in its appearance or serviceability; repair processes restore, partly or wholly, the pre-existing characteristics of serviceability, load-bearing capacity and, if necessary, improve its durability. From this point of view, repair may be divided into structural repair and serviceability repair. The former refers to the restoration of lost sectional or monolithic properties to damaged members, while the latter refers to the restoration of structural surfaces to a satisfactory operational standard. In comparison, strengthening is the modification of a structure – not necessarily a damaged one – with the aim of increasing its load-carrying capacity or stability with respect to its previous condition. It also refers to the case of structures which have to be adapted to a higher load category than that originally designed.

Obviously, poor design, poor construction, poor maintenance, incorrect usage, new environmental influences or an intended increase of the loading or extension of the structure's lifespan can make repair and/or strengthening necessary. In practice, inspection, maintenance, repair and/or strengthening are related processes and must be conceived of as a whole system. Frequent inspection of the structure and careful evaluation of the need for repairs and/ or strengthening can lead to fiscal savings. The extent of the repair and/or strengthening must be evaluated in accordance with the actual state of the structure, the expected environmental conditions, loading conditions and the quality of maintenance of the structure during its remaining life expectancy.

Because of the large number of existing concrete structures and increasing cost of repairing the existing structures, more attention has been paid by the professional community to the development of related industrial standard to ensure high quality repair work. In 2006, the ACI Technical Activities Committee (TAC) approved the formation of ACI Committee 562, Evaluation, Repair, and Rehabilitation of Concrete Buildings. The committee's major task is to develop a code and commentary for evaluation, repair, and rehabilitation of existing concrete buildings (Kahn 2008). In fact, this has not solely been an ACI initiative. With leadership from the ACI Strategic Development Council and strong input from the International Concrete Repair Institute (ICRI), a technical plan entitled *Vision 2020: A Vision for the Concrete Repair, Protection, and Strengthening Industry* was published in June 2006 (ACI 2006). In the 2003 International Building Code (IBC) (ICC 2002), Chapter 34 is devoted to design and detailing of repair work of existing structures. In 2006, the International Code Council (ICC) published the *International Existing Building Code* (ICC 2006).

4.1 Principal considerations of repair and strengthening techniques

Quality and durability of a structural repair and/or strengthening depend on pre-repair considerations, which are as important as the repairs and/or strengthening. First, before any repair and/or strengthening can be applied, the cause of or the mechanism of damage must be identified as clearly as possible through field inspection, investigation and evaluation. Normally there is more than one mechanism of damage involved, and the probable sequence of events resulting in the damage must be established during inspection and evaluation. In order to identify the damage mechanism, a preliminary inspection and investigation should be done prior to the detailed investigation, which should be done before the start of any plan for the repair and/or strengthening. The principal purpose of the preliminary investigation is to record the nature and extent of observed problems, identify the affected regions in the system, and pinpoint the deterioration mechanisms of the structures. These include but are not limited to the following:

1 *Review plans, specifications, and construction records.* These documents may be found in the owner's files, the city archives, files of the original designers or original contractors, testing agencies, building management firms and large subcontracting companies. For historic buildings, design documents may also be found in university and local libraries, historic societies, state preservation offices, the Historic American Building Survey (HABS), National Park Service, and the US Department of the Interior in Washington, DC. ACI 364.1R-07 and ACI 437R-03 provide details on the review processes.

2 *Check which code requirements were applicable at the time of design.* This is very important because some of the code requirements used for the design may have changed and thus be different from the present code requirements. The differences in the design codes should be identified and compared. Alteration plans and change orders should also be examined.

3 *Site observations of structural conditions.* This may include measurement of geometry, recording of crack maps and other damage of the structure or structural members. Any variation of the actual structure from the original design should be recorded, because the alteration may lead to possible problems. Field observation record sheets should be developed prior to the actual inspection. Photographic records and/or videotapes are valuable aids for the structural evaluation. ACI 201.1R-92(97) provides detailed photos and descriptions of various surface damages, such as diagonal cracks, pattern cracking, etc.

4 *Non-destructive testing.* This may include the use of non-destructive testing equipment, such as a rebound hammer, ultrasonic devices, and magnetic detection instruments. Sometimes, use of a hammer knocking on a concrete wall or a chain dragged along a concrete slab can unveil important material and structure defects. ACI 224.1R-07 provides some details on non-destructive evaluation of existing structures. In order to obtain the current stiffness of a structure, loading testing on the structure may be necessary. In this case, deflections and displacements at various locations of the structure need to be measured with a given amount of load applied on the structure.

5 *Exploratory removal, sampling, testing, and analysis.* Sampling of concrete from an existing structure has been described in detail in previous chapters. One of the basic requirements for sampling is that the samples taken from the structure should represent the current status of the structure. Since the damaged concrete areas are often localized in a certain area of the structure, concrete samples must be taken from the area and from outside of the area in order to compare the samples and to obtain a complete observation of the structural condition. Usually it is common to defer removals until the detailed investigation. Only in some special cases (e.g. no design documents are available), will sampling, testing, and analysis be performed.

There are several possible outcomes of a preliminary investigation:

- The structural system is adequate for its intended use.
- The structural system is adequate for its existing use, but may not be adequate for intended future use.
- The preliminary investigation is inconclusive, and further investigation is needed.
- It is not desirable to proceed with a further detailed investigation. This may be due to excessive damage where the structural integrity cannot be economically restored.
- The owner's objectives cannot satisfactorily be met.

For each of the possible outcomes listed above, there are usually several possible action plans from the owner:

1   Accept the structure as-is.
2   Repair and/or strengthening of the structure to correct deficiencies identified.
3   Change the use of the structure.
4   Phase the structure out of service.

The owner of a structure to be repaired should realize that repair work for a structure could be very costly. Depending on the structural type and the severity of the damage in concrete, the repair cost ranges widely. For example, repair cost of parking garages with concrete spalling damage due to reinforcement corrosion easily exceeded $10/ft^2 ($111/m^2) in 2008; and the repair cost for post-tensioned concrete slab could be $30 to $40/ft^2 ($333 to $444/m^2) (Shiu and Stanish 2008). In addition to the repair cost, other costs need to be considered, including the loss of income due to the disruption of operation of the structure and rental cost for a temporary replacement structure.

Once the owner decides on repair and/or strengthening of the structure or structural members, the next step is to prepare project plans and specifications (Newman 2001). The success of a repair and/or strengthening project will depend upon the accuracy of the project plan and specifications to address the causes of the existing damage and upon the degree to which the repair work or strengthening is executed in compliance with plans and specifications.

Existing guide specifications should be used whenever possible. These codes include the Uniform Building Code (UBC); the BOCA (Building Officials & Code Administrators International, Inc.) National Code; the Standard Building Code; the International Building Code; various state building codes covering renovations. The American Concrete Institute (ACI) and the International Concrete Repair Institute (ICRI) have published the *Concrete Repair Manual*, which provides the most comprehensive collection of concrete repair methods and technologies. It contains seven areas: General, Condition Evaluation, Concrete Restoration, Contractual, Strengthening, Protection, and Special Cases. Some of local agencies have their own specifications, such as the Rhode Island State Building Code (1997). For repair work of historic structures, the Uniform Code for Building Conservation and other design documents dealing with renovations are also useful codes and specifications, such as the General Services Administration (GSA) or the REMR Notebook of the US Army Corps of Engineers. As mentioned earlier, previous versions of codes and specifications are often useful when evaluating the design of existing structures, such as "Minimum Design Loads for Buildings and Other Structures" by ASCE 7 (1995). There is an annual conference on Structural Faults and Repair, which has been held in Europe 12 times up to 2008 (www.structuralfaultsandrepair.com). The conference provides an excellent opportunity for the participants to exchange ideas and learn about new case studies in addition to new repair material and technologies.

If the repair materials and/or repair methods selected for a project are not covered by existing guide specifications, a detailed specification based upon experience gained from similar projects and guidance should be prepared. In fact, almost every repair job has unique conditions and special requirements. The choices of methods of repair and/or strengthening of a structure often depend not just on one factor, but on a combination of several factors; some of the factors are technical, some economic, and others are project-specific practical factors. Because of the large number of influential factors involved in the decision-making process, there is usually more than one repair and/or strengthening approach that is technically acceptable. Excluding technical considerations, the ultimate choice of method of repair and/or strengthening of a concrete structure may also be influenced by the following factors (Kay 1992, p. 145):

1   cause of damage and the results of the assessments described in Chapter 2;
2   future life requirement of the structure;
3   the overall quality of repairs and the size of individual repairs;
4   access for repair;
5   requirement for continued use of the structure during repair and the time available for repair;
6   relative costs;
7   client requirements including future maintenance and economic considerations;
8   the ease of application;
9   available labor skills and equipment.

After preparing project plans and specifications, repair materials and repair methods must be selected. Proper materials selection and surface preparation are essential to high quality, durable, and functional repair and/ or strengthening. The success of repair and/or strengthening relies not only on the superior properties of the advanced repair and/or strengthening materials, but also on the influence and interaction of rehabilitation materials with neighbouring members. It should be emphasized that evaluation of repair and/or strengthening should focus on the comprehensive behavior of the entire structure after repair and/or strengthening as well as on the structural portion being repaired and/or strengthened. Very often, the evaluation is made solely on the repaired portion of the structure, and the impact of the repaired portion on the entire structure is neglected. By taking the whole structure into account as a system, the repaired and/or strengthened portion or member is only a sub-system. Three properties need to be considered, including durability, of a repaired/or strengthened structural system: the properties of repair and/or strengthening materials, the properties of existing materials, and the interaction between the repair and/or strengthening materials and existing materials.

4.2 Repair materials

4.2.1 Compatibility requirements of new and old materials

One of the important considerations in the selection of repair materials is the compatibility between the new and old (existing) materials. The compatibility of a repair system is defined as the balance between repair materials and the existing concrete in terms of physical, chemical, and electrical-chemical properties and the dimensional stability of the materials involved in the structural system. The balance ensures that the structural system can carry the load and resist attack by aggressive environment after repair. Such compatibility should be considered in the following four areas: compatibility in dimensional stability, compatibility in chemical stability, compatibility in transport properties, and compatibility in electrochemical properties. Taking dimensional stability as an example, the successful application of a repair concrete material depends largely on overcoming the tendency of new/old concrete to shrink after placement as well as on securing a good bond between the new and old concrete. To this end, a number of methods have been used to overcome the mismatch of the shrinkages between the new and old concrete and to promote a better bond. One of the methods is to use expansive additives in the repair mix, which will be explained in more detail in later sections.

There are project-specific requirements for selecting a good repair material. The repair material to be selected must meet both the project objectives and the owner's requirements, address the causes of the damage, and must be suitable for the application conditions (ICRI 1997). Taking the application conditions as an example, surface orientation of the area to be repaired is an important factor, which could be horizontal, vertical, or overhead. Thickness of repair and spacing of reinforcing bars are other application conditions needing to be considered.

In addition to the project-specific requirements, some general requirements for a good repair material are (Mailvaganam 1992):

- wide range of tolerance to accommodate dimensional changes due to temperature and moisture variations;
- appropriate chemical compatibility to avoid detrimental reactions between the new and old materials;
- good bond between the new and old materials in terms of both strength and durability;
- minimum site preparation requirements.

4.2.1.1 Compatibility in dimensional stability

Dimensional stability between new and old concrete requires that the two materials respond in similar ways to ambient temperature and moisture variations. Many repaired sections, such as shallow patches of exterior wall and concrete slabs, do not carry large structural loads; as a result, environmental loadings (e.g. temperature and moisture) become very important and often are control factors in the selection of repair materials. A compatible repair system must have monolithic action between the repaired section and the original structure as well as a sound bond between the repair material and the existing concrete. The dimensional changes of the repair material and old concrete may be considerably different due to the differences in their physical, mechanical and chemical properties as well as in their ages. The main dimensional stability problems caused by the application of new repair material are discussed below.

(A) SHRINKAGE OF CONCRETE

In practice, cracks that occur soon after the concrete has been placed are often described in general terms as "shrinkage cracks". Since concrete shrinks, the appearance of cracks is often considered to be inevitable. This conclusion is not true in many cases. First, there are several types of shrinkage during the hardening process of cement paste and during the life span of concrete:

- *Plastic shrinkage* is the early stage shrinkage of concrete before it hardens, and it is due to the loss of surface water on a fresh concrete mixture. Plastic shrinkage occurs from the time the concrete is placed to the time the concrete has hardened. Usually, it will take a few days for concrete to harden depending on the curing condition. In principle, there should be no moisture exchange between the concrete and the environment during a proper curing process; therefore there should be no plastic shrinkage. However, moisture loss during the curing process for repaired concrete occurs very often due to the use of improver curing methods, as a result, plastic shrinkage may occur, leading to shrinkage cracking.

- *Autogenous shrinkage* is the shrinkage of fresh concrete during the curing process, and it is due to self-desiccation in concretes with low water-cementitious ratios (below 0.42 or 0.3). The self-desiccation results from hydration reactions between the cement and water. Autogenous shrinkage occurs mostly during the curing process. The major difference between autogenous shrinkage and plastic shrinkage (as well as drying shrinkage) is that autogenous shrinkage may occur without any loss of moisture. The value of autogenous shrinkage varies in a large range from as low as 40 to as high as 2000 microstrains, depending mainly on cement content and cement fineness (Davis 1940; Ai 2000).

- *Drying shrinkage* is defined as the shrinkage of hardened concrete as a result of the loss of internal moisture from the pores of concrete after the curing process. Since the plastic and autogenous shrinkages are dimensional changes associated with early age concrete, drying shrinkage is the most important one for existing concrete structures. The following discussion will focus on drying shrinkage of concrete.

One can see that proper curing methods play a very important role in preventing shrinkage cracking. Proper curing methods minimize plastic shrinkage and thus the amount of shrinkage cracking at an early age. On the other hand, improper curing may cause excessive plastic shrinkage, which leads to severe surface cracking of concrete at early stages.

Assuming proper curing methods are implemented, the compatibility of dimensional stability means that the repair materials and the existing concrete will undergo a compatible drying shrinkage under the influence of temperature, moisture, loading, and other environment changes. If the compatibility requirement is met, the appearance of cracks is not inevitable.

In this sense, dimensional compatibility is the single most important factor in a successful repair. Dimensional incompatibility will cause severe problems for repair patches. For example, if a mortar is used as the repair material, the shrinkage of the fresh mortar will create stress concentration at the point of contact with the existing concrete. The stress concentration could be normal stress or shear stress depending on the orientation of the repaired area. The normal stress concentration at the interface may cause separation, and the shear stress concentration may cause debonding between the two materials. Hence, the shrinkage of repair mortar must be limited. Some commonly used shrinkage limits are shown in Table 4.1.

ICRI (1997) suggests several classifications on drying shrinkage of repair materials: very low for drying shrinkage <250 $\mu\varepsilon$; low for $250–500$ $\mu\varepsilon$; moderate for $500–1000$ $\mu\varepsilon$; and high for >1000 $\mu\varepsilon$.

In the best case scenario, the drying shrinkage of repair material matches

Table 4.1 The maximum allowable shrinkage values for repair mortar

Agency	Testing condition	Sample size	Maximum allowable shrinkage
HKHA	27 °C, 55% RH	25 × 25 × 285 mm	7 day $\varepsilon_{sh} < 300\ \mu\varepsilon$
ASTM	23 °C, 50% RH	25 × 25 × 285 mm	28 day $\varepsilon_{sh} < 500\ \mu\varepsilon$
C157–1989 AS1012	23 °C, 50% RH	75 × 75 × 285 mm	28 day $\varepsilon_{sh} < 450\ \mu\varepsilon$

the shrinkage of existing concrete. Therefore, it is advantageous to know the shrinkage value of existing concrete. However, drying shrinkage of concrete depends on many factors, such as mix design parameters (cement content and water to cement ratio), cement types, and aggregate sizes; and thus, it is difficult to estimate the shrinkage of existing concrete. One method is to take concrete samples from the existing concrete structure and carry out a drying shrinkage test using ASTM C 157 (Length Change of Hardened Hydraulic Cement Mortar and Concrete) to measure drying shrinkage of the existing concrete. If it is not feasible to conduct the shrinkage test, a rough estimation of the range of drying shrinkage of normal weight concrete will be helpful. In general, drying shrinkage of normal weight concrete ranges from 400 to 800 microstrains. Factors that affect the drying shrinkage are listed in Table 4.2.

The factors listed in Table 4.2 can be used to increase or decrease the range of the drying shrinkage for normal weight concretes. If the parameters of the concrete under consideration are not included in the table, the following general guidelines can be used:

- A high water to cement ratio leads to high shrinkage. More water in the original concrete mixture may result in high permeability of the concrete and thus loss of more water in existing concrete.
- A longer curing period leads to lower shrinkage. Longer curing of concrete results in lower permeability and thus less moisture loss, and less shrinkage.

Table 4.2 Factors affecting drying shrinkage

Factor	Reduced shrinkage	Increased shrinkage
Cement type	Type I, II	Type III
Aggregate size	1½″ (38 mm)	¾″ (19 mm)
Aggregate type	Quartz	Sandstone
Cement content	550 lb/cy (325 kg/m³)	700 lb (415 kg/m³)
Slump	3″ (76 mm)	6″ (152 mm)
Curing	7 days	3 days
Placement temperature	60 °F (16 °C)	85 °F (29 °C)
Aggregate state	Washed	Dirty

- A large proportion of aggregate in the overall volume of concrete leads to low shrinkage. In the components of concrete as a composite material, only cement paste shrinks upon the loss of pore water, while aggregate retains its volume. Therefore, more aggregate in the concrete results in less shrinkage. However, some aggregates are "shrinkable" especially some of the synthetic aggregates. In this case, more aggregates actually leads to higher shrinkage (Allen *et al.* 1993).
- A high cement content leads to high shrinkage. Again, this is because cement paste is the component in concrete that results in drying shrinkage, and thus a large fraction of cement paste in the overall volume of the concrete results in high shrinkage.
- Low environmental relative humidity results in high shrinkage. It is important to consider this environmental factor for any repair job in a dry environment.
- Older concrete shrinks less than younger concrete. This is because most of hydration reactions are completed in older concrete, which results in lower permeability of the concrete. This factor is important if the existing concrete is relatively young, say, only a few years old.

Boundary conditions of the structure member are also important, such as restrained vs. unrestrained structure members. For an unrestrained concrete member, the shrinkage results in a shortening of the member, which needs to be considered in a different way from a restrained member. Some members are restrained from moving in one direction (for example, from the bottom of a concrete slab for pavement). Drying shrinkage causes tensile stress on the top part of the slab. When the tensile stress exceeds the tensile strength of the concrete, crack will develop on the top surface (Emmons 1994).

When a shrinkage-induced crack occurs along the interface between new and old concrete, the failure of the repair work is called bond failure. There are some special methods to correct bond failure due to shrinkage (Warner 1984). One of them is to use repair materials that can expand when they are mixed and placed, such as shrinkage-compensating cement (ASTM Type K cement). This type of cement expands upon hardening, and the amount of expansion can be adjusted by adding various quantities of additives in the cement.

(B) THERMAL EXPANSION OF CONCRETE

Temperature variation changes the volume of concrete. Coefficients of thermal expansion and contraction depend on the constituents used in concrete structures. The coefficient of thermal expansion/contraction is the proportionality parameter between volume change of concrete and temperature variation. An average value for the coefficient of thermal expansion of concrete is about 10 millionths per degree Celsius (10×10^{-6}/°C or 6×10^{-6}/°F), although the value ranges from 6×10^{-6}/°C to 13×10^{-6}/°C. This amounts to a length change of 5 mm for a 10 m long concrete beam subjected to a rise or fall of 50 °C. Thermal expansion and contraction of concrete vary with many factors such as aggregate type, amount of aggregate, cement content, water to cement ratio, temperature range, concrete age, and relative humidity. Among these influential parameters, the type and amount of aggregate are most important ones.

Table 4.3 shows some experimental results of the thermal coefficient on the expansion of concretes made with aggregates of various types. These data were obtained from tests on small concrete specimens in which all factors were kept the same except aggregate type. In each case, the fine aggregate was of the same material as the coarse aggregate.

The coefficient of thermal expansion of concrete can be determined by several methods. ASTM C531 (Linear Shrinkage and Coefficient of Thermal Expansion of Chemical Resistant Mortars, Grout, and Monolithic Surfacings) is for polymer-based mortars and grouts. ASTM D696 (Coefficient of Linear Thermal Expansion of Plastics) is for polymers without any filler. For concrete, CRD-C39 can be used, which is the standard means of measuring the coefficient of linear thermal expansion of concrete.

Notably, unrestrained uniform thermal expansion does not cause problems to repaired concrete. If, however, the thermal expansion is not uniform, cracking problems similar to the shrinkage-induced damage may occur in repaired concrete. For example, if the thermal expansion of existing concrete differs significantly from the thermal expansion of the new concrete, cracks may occur due to a change in environmental temperature in the surface of repaired concrete or along the interface of the old and new concrete.

As a result, it is necessary to choose a repair material with the same coefficient of thermal expansion as the old concrete, or as close to the coefficient of the old concrete as possible. Taking polymer mortar repaired concrete infrastructures as an example, although the initial bond between polymer concrete and existing normal concrete is excellent, the cyclic expansion and contraction due to temperature changes can eventually lead to the separation of the two materials because of the large difference in their coefficients of thermal expansion. The polymers used in repair materials have a thermal expansion coefficient about ten times more than that of normal concrete (the range is about 4 to 18 times according to ACI 503.5R). When

Table 4.3 Effects of aggregate type on thermal expansion coefficient

Aggregate type	Coefficient of expansion, millionths per °C
Quartz	11.9
Sandstone	11.7
Gravel	10.8
Granite	9.5
Basalt	8.6
Limestone	6.8

fillers of natural sand and gravel are mixed with the polymers, the coefficients of thermal expansion of polymer mortars and polymer concretes are reduced significantly. But, they are still much higher than that of concrete. Therefore, when using polymer mortar or polymer concrete to repair normal concrete, special attention should be paid to their coefficients of thermal expansion.

Negative temperature changes result in shortening of concrete structural member and it can crack concrete members that are highly restrained by another part of the structure or by ground friction. Its effect is similar to the shrinkage of concrete. Calculations show that a large enough temperature drop will crack restraint concrete regardless of its age or strength, provided the concrete is fully restrained (CAC 2006).

For high temperature applications of repair materials, one should keep in mind that the coefficient of thermal expansion of concrete is no longer a constant. Under ambient temperatures, the coefficients are constants; while under high temperatures, they become functions of the temperature, which is very important for fire resistance of concrete and the effect will be discussed in more detail in Section 4.3.5 for fire damage of concrete. As shown in Figure 4.1, the coefficient of thermal expansion of concrete increases with temperature all the way up to about 600 °C. This curve was obtained from the thermal strain of Nielsen *et al.* (2004) for concrete made of quartzite aggregate.

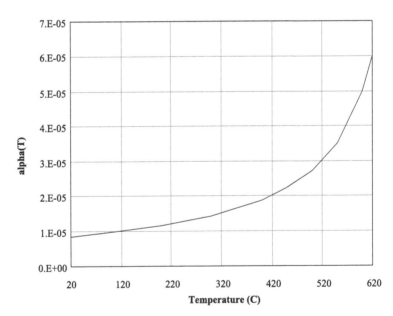

Figure 4.1 The coefficient of thermal expansion of concrete vs. temperature (Nielsen *et al.* 2004).

(C) UNEVEN DISTRIBUTION OF STRESS

Uneven distribution of stress is often caused by differential stiffness in repair systems. Stiffness of a material is measured by its modulus of elasticity. A material with low modulus of elasticity generally deforms more under the same load than a material with a higher modulus of elasticity. The volumetric compatibility is reflected in uneven distribution of stress due to the different Young's moduli of new and old concretes. When materials with widely different moduli are in contact with each other and under loading, the large difference in the stiffness will result in uneven distribution of the stress and thus lead to failure of either the material with low modulus of elasticity or the material with high modulus of elasticity, depending on the strengths of the materials.

The failure mechanism in this case is illustrated in Figure 4.2. In this figure, a new concrete is used to repair an old concrete. The old concrete could be a concrete wall, a column, or a slab, and the new concrete could be a thick surface layer coated on the outer surface of the old concrete. E_{new} and E_{old} denote the modulus of elasticity of the new material and old material, respectively. Under the loading condition shown in Figure 4.2 (in plane loading parallel to the surface of the old concrete), the deformation (or the strain) in the two materials will be the same. If $E_{new} > E_{old}$, the stress in the new concrete, σ_{new}, will be higher than the stress in the old concrete, σ_{old}. The stresses in the new and old concrete can be calculated as

$$\sigma_{new} = E_{new}\varepsilon \quad \text{and} \quad \sigma_{old} = E_{old}\varepsilon \tag{4.1}$$

Since

$$E_{new} > E_{old}, \sigma_{new} > \sigma_{old} \tag{4.2}$$

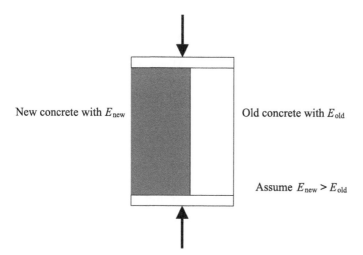

New concrete with E_{new} Old concrete with E_{old}

Assume $E_{new} > E_{old}$

Figure 4.2 Stress distribution in two materials with different moduli of elasticity.

This means that the stress distribution will not be uniform in the repaired section. Before the repair, the stress in the column was uniformly distributed. After the repair, the load can be considered as being transferred from the old concrete to the new concrete. If the strength of the new material with higher modulus is not sufficient to bear the load, the new materials may fail earlier than the old concrete. This is one of the important mechanisms that are responsible for pre-matured failures of repair work in the new material.

On the other hand, if a repair material (e.g. a polymer mortar) with low modulus of elasticity but high compressive strength is used to repair an old concrete, it could lead to the failure of the old concrete due to the stress transfer to the old concrete, assuming that the old concrete has higher modulus of elasticity than the new concrete. The failure mechanism can be explained by the same model shown in Figure 4.2. In this case, the calculation of the stresses remains the same as Eq. (4.1), and since $E_{new} < E_{old}$, we have $\sigma_{new} < \sigma_{old}$, which means more stress would be transferred to the old concrete. So, the load carried by the old concrete would be even higher than before.

Another possibility of repair failure due to the non-uniform stress distribution is the bond failure between the new and old material. It should be noted that, except for external loads, shrinkage or thermal expansion and contraction can also cause loss of bond between the repair material and old concrete when the moduli of elasticity of the two materials are significantly different.

Table 4.4 shows moduli of elasticity of various repair materials (Emmons 1994). It will be wise to select a repair material with the same or similar modulus of elasticity compared with the old concrete.

Table 4.4 Moduli of elasticity of repair materials (Emmons 1994)

Materials	psi ($\times 10^6$) Mpa ($\times 10^4$)	Materials	psi ($\times 10^6$) Mpa ($\times 10^4$)
Portland Cement Mortar	3.4 2.3	Magnesium Phosphate Cement Concrete	3.2 2.2
Preplaced-Aggregate Concrete	3.8 2.6	Microsilica Modified Portland Cement Concrete	4.0 2.8
Portland Cement Concrete	3.8 2.6	Polymer Modified Portland Cement Mortar with Non-sag Filler	2.5 1.7
Methyl-methacrylate Concrete	3.0 2.0	Latex Modified Portland Cement Concrete	2.5 1.7
Epoxy Mortar	2.2 1.5	Shotcrete	3.8 2.6

4.2.1.2 Compatibility in chemical stability

If the repair material and the existing material have different chemical properties, the repaired concrete may not be durable. For instance, if the existing concrete is made of low alkaline cement and reactive aggregate, and the repair material is made of high alkaline cement, the so-called alkali–silica reaction (ASR) between the new cement and old aggregate may take place. ASR will be discussed in detail in a later section. Another example is that shrinkage-compensating mortar or concrete is used as repair material while existing concrete is normal concrete. In this case, special care should be taken to ensure the amount of expansion of the repair material is properly selected and controlled. Otherwise, the expansion may cause cracking in the old concrete. This is especially important for underwater repair work. If the expansion of the shrinkage-compensating cement is not well controlled, cracks may appear in the repair system, resulting in leakage of the concrete structures.

Careful consideration on the chemical contents of the repair material must be made when the repair job deals with reinforcing steel or other embedded metals. Usually the chloride content in the old concrete is relatively high for the concrete exposed to de-icers and the chloride content in the new concrete is very low. A large chloride concentration gradient exists between the new and old concrete, which may form a "concentration cell" for the onset of steel corrosion in concrete. This "concentration cell" will accelerate the rate of corrosion and cause premature failure of the patch or adjoining concrete. The chloride concentration gradient is a driving force in the accelerated diffusion of the chloride in the old concrete into the new concrete, which causes the early onset of steel corrosion in the new concrete. For these cases, there are two types of methods that can be used to prevent the premature failure of the repaired system. One is to lower the chloride in the old concrete by using available chloride removal techniques (which are usually expensive). The other is to install a cathodic protection system in the new concrete. These methods will be discussed in detail in Section 4.3.6.

4.2.1.3 Compatibility in transport properties

Transport properties of the new and old materials must also be compatible. There are usually two transport properties of concrete that are worthy of consideration: permeability and diffusivity. Permeability refers to the ease that fluids, both liquids and gases, can enter into and move around in the concrete. According to Darcy's law, the flux of a fluid is proportional to the pressure gradient in the fluid, and the proportionality constant is the permeability. Diffusivity refers to the ease that species (such as chloride ions) penetrate into concrete. According to Fick's law, the flux of a diffusing species is proportional to the concentration gradient of the species, and the proportionality constant is the diffusivity.

The permeability and diffusivity of a repair material should be as close to that of exiting concrete as possible. If there is a large difference in the transport properties, some long-term durability problems may occur. For instance, when the repaired structure undergoes wet-dry cycling (e.g. at the tidal zone), the differences in the permeability and diffusivity of the new and old concrete result in concentration gradients of the oxygen and chloride in the repaired concrete, which may result in the corrosion of reinforcing steel in the concretes.

The transport properties of concrete depend on its internal micro-structure, which in turn depends on the concrete mix design. Usually the repair material has lower permeability and diffusivity than that of the existing concrete.

4.2.1.4 Other considerations regarding compatibility

In addition to the compatibility considerations discussed above, there is another compatibility related to the electrical impedance difference between the repair material (such as macromolecule compound mortar) and the host material (such as normal concrete). In the research community, there is still some disagreement over whether a large difference in impedance may reduce or accelerate the steel corrosion in concrete. In other words, it is not clear if a large impedance difference or a small difference is better for enhancing the corrosion resistance of the repaired structure. In general, there are similar electrical impedances between the repair material such as the Portland cement-based materials, and the host material such as normal concrete.

It is also worthwhile pointing out that the anticipated service conditions of the repaired system and the weather conditions when the repair material is applied play an important role in the final choice of the right repair material. The following factors relating to application and service conditions should be considered in choosing the repair material (Mailvaganam 1992):

1   The moisture content in the substrate. This will be used to determine the mix design of the repair material.
2   The temperature at the time of application of the repair; for instance, the hydration and curing rate of cementitious materials may be influenced by the environment temperature. For the same repair work, winter construction and summer construction make a significant difference in terms of selection of materials.
3   The maximum and minimum service temperature of the structure, the range of service temperature will induce possible thermal movements and build up stresses.
4   The location of the repair section should be considered; for instance, vertical or horizontal surface for repair should be considered for the mix design of the repair material.

5   Turn-around time; for instance, if the repair is subjected to early loading, the material with high early strength may be used (e.g. type I vs. type III cement).

6   Chemical exposure; for instance, if there are acids, alkalis, sulfates and so on in the environment, special cement, polymers, or additives should be considered.

7   Exposure to traffic, a repair material with good abrasion resistance may be needed for overlay exposed to heavy traffic.

8   Importance of appearance; for instance, for repairing a stamped concrete slab, the color matching between the new material and old concrete is important.

9   Life of repair, for a temporary repair work, some of the compatibility requirements may be relaxed.

4.2.2 Standard testing methods and requirements for repair materials

Depending on the type of repair material selected for a repair project, several specific testing methods can be used to evaluate the properties of the material. For cementitious repair materials, ASTM C928 (Standard Specification for Packaged, Dry, Rapid-hardening Cementitious Materials for Concrete Repairs) is the current industrial standard in the US for the evaluation of the performance of cementitious repair materials. The required tests in ASTM C928 include:

* compressive strength: ASTM C 109/C39
* slump of hydraulic cement concrete: ASTM C 143
* length change of hardened hydraulic-cement mortar and concrete: ASTM C 157
* scaling resistance of concrete: ASTM C 672
* bond strength: ASTM C 882.

The performance requirements of repair materials in ASTM C928 are listed in Table 4.5. One can see that the requirements include most of the compatibility conditions discussed earlier in the chapter. Some specific limits for the material properties are also given in Table 4.5. In addition to the listed required tests, there are three additional properties of repair materials that can be tested. But no specific limits are given for the three optional properties in ASTM C928:

* Flexural strength: ASTM C 78.
* Time of setting: ASTM C 403.
* Resistance of concrete to rapid freezing and thawing: ASTM C 666.

One can see that the above required tests are not very difficult to perform.

Table 4.5 Performance requirements of repair materials in ASTM C928

Materials' performances	R1	R2	R3
Compressive strength	3 h: 500 (psi) 1 d: 2000 (psi) 7 d: 4000 (psi) 28 d: ≥7 d	3 h: 1000 (psi) 1 d: 3000 (psi) 7 d: 4000 (psi) 28 d: ≥7 d	3 h: 3000 (psi) 1 d: 5000 (psi) 7 d: 5000 (psi) 28 d: ≥7 d
Slant shear bond strength	1 d: 1000 (psi) 7 d: 1500 (psi)		
Length change	From 3 h to 28 d in water: ≤+0.15% From 3 h to 28 d in air: ≤−0.15%		
Scaling resistance of 25 cycles	Max visual rating: 2.5/Max scaled material: 5 kg/m^2		
Consistency of concrete or mortar	Concrete slump: 3 (in). Mortar flow: 100% after 15 min	Concrete slump: 3 (in). Mortar flow: 100% after 15 min after 5 min after addition of mixing liquid	

Therefore, manufacturers of repair materials are currently using ASTM C928 as acceptance criteria of their products. If all the requirements of ASTM C928 are met, the repair material should meet the desired service life. However, premature damage of many repaired concrete structures has frequently been reported. This simply means that either the testing standard was not carried out accurately, or the required tests in ASTM C928 are not sufficient to characterize the durability properties of the repair materials. In the first case, better quality control of the testing process should be implemented, and in the second case, more advanced or more reliable testing methods should be developed in future to meet the need.

There are several different bond test methods, in addition to ASTM C882, that may be used for various repair materials other than cementitious repair materials (ICRI 1997). These bond tests include ACI 503 R (Use of Epoxy Compounds with Concrete), the Michigan DOT Shear Bond Test, ASTM C 1042 (Bond Strength of Latex Systems Used with Concrete), and AASHTO T 237 (Testing Epoxy Resin Adhesive). Xi and Li (2004) made an extensive review of testing methods for concrete repair materials.

British and European Standards for the protection and repair of reinforced concrete structures are under development (http://www.concreterepair.org.uk/cra/StdsBSRPC.pdf). A comprehensive package of standards will be developed, including test methods for specific properties, specifications for important repair materials, coatings, mortars, bonding agents, and injection materials. The package of standards is drafted by Technical Committee 104 of the European Standards body, CEN. TC 104 is the European Standards committee responsible for specifications of concrete. Sub-committee 8 of TC 104 has drafted a series of standards for repair and protection of concrete structures, "The New Approach to Concrete

Table 4.6 European Standard EN1504: the new approach to concrete protection and repair

EN 1504 Part 1	Definitions
EN 1504 Part 2–7	Products
	2 Surface protection for concrete
	3 Repair mortars, structural and non structural
	4 Structural bonding materials
	5 Concrete injection materials
	6 Anchoring of reinforcing steel bar
	7 Reinforcement corrosion protection
EN 1504 Part 8	Quality control and evaluation of conformity
EN 1504 Part 9	General principles for the use of products and systems
EN 1504 Part 10	Site application of products and systems and quality control of the works

Protection and Repair". Table 4.6 lists all the parts of this standard, EN1504.

4.2.3 Laboratory testing methods for bond strength

Of the five material properties required to be tested by ASTM C928 for cementitious repair materials, the bond strength is the one specifically important to repair materials, and it is also the most difficult one to measure accurately, thus we will discuss in detail the test methods for bond strength.

There are many testing methods available for testing the bond strength of cementitious materials (Xi and Li 2004). The testing methods can be divided into five groups according to the type of stress carried by the interface (Halicka and Krol 1999). As shown in Figure 4.3, these five groups are tension, shear, shear and compression, torsion, and others (Saccani and Magnaghi 1999; Xiong *et al.* 2002; Soares and Tang 1998; Austin *et al.* 1999; Wall and Shrive 1998; Knab and Spring 1989; Yang and Zhu *et al.* 2000; Yang *et al.* 2000). In addition to the five groups, the ultrasonic test method (Liu *et al.* 1998) is also used to evaluate bond strength.

Each testing method is influenced by different combinations of factors and on its own cannot give a comprehensive description of the required material properties. On the other hand, it is also important to know how to apply and interpret the results obtained from the bond tests to practical situations, where the interface may be in a different stress state from the tested stress state. Among the available test methods, shear, tension, and shear and compression are the frequently used methods for testing the performance of an interface between a repair material and a substrate. The other two types of testing methods are occasionally used in research laboratories or in the practice. The following is a brief discussion of the advantages and disadvantages of the three types of frequently used testing methods.

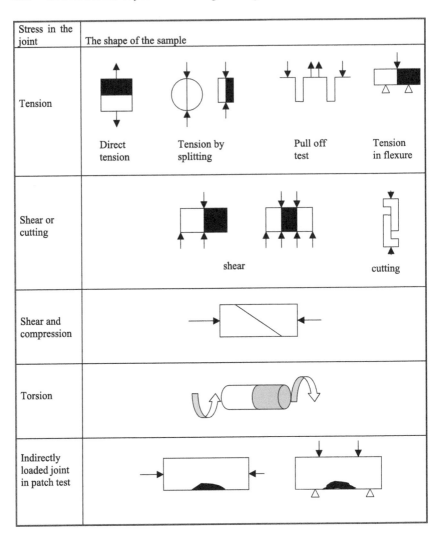

Stress in the joint	The shape of the sample
Tension	Direct tension Tension by splitting Pull off test Tension in flexure
Shear or cutting	shear cutting
Shear and compression	
Torsion	
Indirectly loaded joint in patch test	

Figure 4.3 Bond strength test methods.

(A) SHEAR TESTS

Of all the available testing methods for the bond strength, the shear test is the most frequently used method for testing the performance of an interface between a repair material and a substrate under different stress states. If the bond surface is straight and smooth, such as a surface from a saw cut, the shear failure line can pass along the bond interface and cause a pure shear failure. If the bond surface is not straight and smooth, such as the bottom interface between a patch and substrate, there will be a strong mechanical interlock due to the surface texture, resulting in an incorrect bonding capacity for the repair material. The insensitivity of the shear test to surface roughness can be overcome by testing in tension and torsion, to generate a combination of tensile and shear stresses.

(B) COMPRESSION AND SHEAR TESTS

This test method was adopted by ASTM C882 (called the slant-shear test), in which the bond interface is under a combined stress state of compression and shear. The compressive slant-shear loading configuration is shown in Figure 4.4. This test is used by repair material manufacturers to evaluate the bond strength of repair materials. This test, however, has some shortcomings:

• Failure is crucially dependent on the angle of the plane.
• It is insensitive to surface roughness and condition, only producing bond failures with smooth surfaces (Robins and Austin 1995).
• The test is sensitive to the difference in elastic moduli of the repair and substrate materials.

Considering these shortcomings, some researchers have proved that the effect of surface roughness depends upon the inclination of the bond plane in the compression shear test, and there is a critical angle associated with a particular surface roughness (Austin *et al.* 1999). A critical bond angle can be defined as the inclination at which the load required to produce a bond failure is a minimum. Fortunately, the critical bond angle corresponding to smooth surface is about 30°, which exactly meets the specification of ASTM C882. In real repair projects, the concrete to be repaired is often sawed by a diamond saw and the interface of new and old concrete is quite smooth.

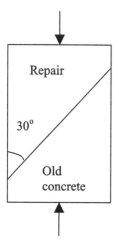

Figure 4.4 Compressive-shear loading configuration in ASTM C882.

Therefore, ASTM C882 is suitable for determining the bond strength. Moreover, the testing method is comparatively simple and easy to perform in practice.

(C) TENSION TESTS

Tension tests are more difficult to perform than the compressive shear test and the shear test. Although the pull-off test (see Figure 4.3, the first row) is a more direct simulation of an actual repair situation than the slant shear test, significant variation in pull-off measurements within a specimen was observed. The main cause of the variability was a delamination at the edges between the patch and substrate due to the shrinkage in the repair mortar (Marosszeky *et al.* 1991).

Flexural tests can also be categorized as tension tests. There are several different flexural testing methods. The one shown in the last of the first row of Figure 4.3 is the three-point bend test. Flexural tests can be considered an indirect testing method for measuring the interfacial bond strength of repair materials with old concrete (Yang *et al.* 2000), in which the bond property is evaluated by using a special beam made of new and old concretes. In the flexural test, the size effect on flexural bond strength is important and should be considered. Some researchers prefer to use flexural tests instead of the direct tensile test to evaluate the bond capacity, mainly because the direct tensile test is very difficult to perform. As illustrated in Figure 4.5, a four-point bending test was carried out on a beam with a prefabricated crack (a notch) at the bottom of the beam. The applied load and the crack mouth opening displacement (CMOD) were recorded, and then by analysing the load–CMOD curve, one can obtain the tension softening diagram (see Figure 4.5), which is the relationship between crack opening w and the stress σ in the interface of the new-old concrete (Kunieda *et al.* 2000). The σ–w relationship can be considered a measure of the bond strength, which is not a constant. In this sense, the bond strength measured by using other testing methods is actually an averaged value for the interface bond strength. It should be realized that the flexural testing methods are not standard test

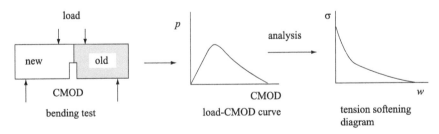

Figure 4.5 Determination of tension softening diagram from the bending test.

methods for evaluating bond strength but sometimes used in research laboratories.

4.2.4 Field testing methods for bond strength

Bond strength of repair materials has been measured both in the laboratory and in the field. Rizzo and Sobelman (1989) evaluated the selection criteria for repair materials. A number of adhesion/bond test methods were discussed, including direct tension, pull-off, direct shear, flexure, and slant shear. A compilation of studies of each method was summarized by Knab (1988). Several field test methods have been proposed to evaluate bond properties and the performance of repair materials in general. Tensile bond tests are gaining in popularity for field testing because of their relative simplicity and the ability to meet the requirements imposed on in-situ bond strength test. Tensile test methods can be divided into indirect and direct techniques.

The pull-off test method is one of the tensile test methods. Unlike the other bond test methods that are used for laboratory testing, the pull-off test can be used in the field to evaluate the bond strength between repair material and parent concrete in a structure (Figure 4.6). The first modern development of the pull-off concept for strength testing of in-situ concrete was undertaken independently in the United Kingdom at Queens University, Belfast (Long and Murray 1984), and in Austria, where it was called tear-off test (Stehno and Mall 1977). This led to "Limpet" test equipment being

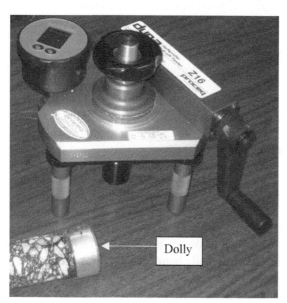

Figure 4.6 Pull-out tester made by Proceq. The dolly provides the connection between the loading device and the concrete surface.

commercially available in the United Kingdom. Further test equipment has since been developed in several countries, leading to a wide range of test configurations and procedures now being available (McLeish 1993).

A number of different pull-off tests have been reviewed by CIRIA (McLeish 1993), the majority involving cutting of the repair material interface before applying a tensile load. Mathey and Knab (1991) studied the bond strength of concrete overlays by using in-situ uniaxial tensile tests (pull-off tests with partial coring). Two types of equipment were used in the tests: a hydraulic, uniaxial tensile test apparatus, which was a modification of the ACI 503R field-test apparatus, and a pneumatic apparatus developed at the National Institute of Standards and Technology (NIST). In the pull-off tests, cores were drilled through the overlay concrete to about 25 mm past the interface. A steel disk was then glued on the top surface of the core with a high-strength quick-setting epoxy. Then, a load can be applied to test for the bond strength. The important issue associated with pull-off tests is the depth of the core drilling into the existing concrete. It is suggested that the influence of the steel dolly and reaction tie on test results depend on the depth of coring into the substrate concrete. Ignoring the effect of drilling depth may be one of the main causes of difficulties in reproducing and comparing test results. It is very important to have a standard for the core depth beyond the repair–substrate interface for the pull-off bond test. The depth of core drilling below the interface should be a minimum of 25 mm (1 in.) or one-half of the core diameters, whichever is larger.

Although there are different testing methods and equipment for carrying out the pull-off tests, the general procedures can be described as follows:

1   Marking and preparing the test area.
2   Partial coring into the existing substrate perpendicular to the repair surface. In some cases, partial coring is done around the attached loading disc.
3   Attaching the disc to the core, using an epoxy resin.
4   Attaching a loading frame to the disc. A frame around the disc provides the reaction force to the load.
5   Pulling the disc until the specimens fails.

At the interface between the old and new concrete, the stress distribution is not uniformly tensile, because of the complicated failure surface, which is not perfectly perpendicular to the loading direction. As a result, the stress state is more complex than uniaxial tension, and there are stress concentrations in the interface. Therefore, most of the pull-off test results do not provide an accurate value for the tensile bond strength; they instead provide relative comparisons among the test data obtained from different types of interfaces. In many practical applications, the pull-off tests are conducted to determine if the bond strength between repair and concrete substrate meets the specified failure criteria.

Bond failure is usually classified by failure patterns as well as specified by critical bond strength values. There are several bond failure patterns. One is called adhesive failure, which is the failure in the repair–substrate interface. Another failure pattern is called cohesive failure, which includes the fracture failure within the repair material or within the substrate concrete. The other failure pattern is called adhesive/cohesive failure. It is a mixture of the first two types of failure patterns, partial adhesive and partial cohesive. In general, it is not desirable to have adhesive failure. In another words, it is not desirable for the failure to occur at the repair–substrate interface. Cohesive failure and adhesive/cohesive failure are acceptable, providing that the bond stress is equal to or greater than the specified bond strength. If failure occurs at the steel disc–repair interface, then the pull-off result represents minimum bond strength, and the test should be repeated if the strength is not acceptable.

4.2.5 Repair materials

From the material component point of view, the repair materials used for concrete structures can be divided into two types: cementitious repair materials and polymer modified repair materials. In addition, some other materials, such as coatings, are also used for repair work. There are several commercial companies supplying general concrete materials, such as W.R. Grace and Master Builder in the US who provide various types of products for concrete repair work. There are also some companies specializing in materials for fast concrete repair, such as Conproco (http://conproco.com/applicationCharts.htm) and Sika (www.sikausa.com). The websites and brochures developed by the material suppliers are good references for understanding basic properties of the repair materials.

4.2.5.1 Cementitious repair materials

There are several types of cementitious materials that are often used for repair concrete structures, including Portland cement-based and gypsum-based concrete, magnesium phosphate concrete, and high alumina concretes.

(A) PORTLAND CEMENT-BASED CONCRETE

Regular Portland cement concrete (PCC) is the most commonly used repair materials for spall repair. Usually, PCC used for patching should meet the following requirements:

- PCC (ASTM C 150), Type I and II, 362 kg/m^3 (or higher);
- Maximum w/c ratio: 0.42;
- Total air content: 6 percent to 8 percent;

- Minimum compressive strength: 3500 psi (24 MPa) at 7 days.

Several different mix designs can be used for repair concrete. Perkins (1986) has suggested mix proportions of concrete and mortar for surface patching repair as follows:

Concrete: 1 part ordinary or rapid hardening Portland cement;
2.5 parts concreting sand;
2.5 parts 10 mm coarse aggregate;
The amount of mixing water should be such as to give a slump of 50 mm ± 25 mm which should be adequate for the compaction of small areas of concrete.

Mortar: 1 part ordinary or rapid hardening Portland cement;
3 parts coarse concreting sand.

To prevent excessive shrinkage, the total water content should be kept to a minimum. If an increased slump is desired to improve workability, a high-range water reducer should be added. High-range water reducers should not be used without first testing trial batches. If the repaired structure must be opened to use relatively quickly, rapid-setting or high early-strength materials, such as set-accelerator, may be used. Rapid set hydraulic cement/mortar should meet the following requirement:

- Minimum of 30 minutes time to initial set as tested by ASTM C403, Standard Test Method for Time of Setting of Concrete Mixtures by Penetration Resistance. The mixture should have sufficient workability to allow placement and consolidation before initial set.
- Field-cured cylinders should have a minimum compressive strength of 1,200 psi (8.25 MPa) at 2 hours.
- Expansion/shrinkage should be tested in accordance with ASTM C157, Standard Test Method for Length Change of Hardened Hydraulic Cement Mortar and Concrete. The acceptance values for shrinkage are listed in Table 4.1, and the thermal expansion of concrete is discussed in section 4.2.1.1.
- Freeze–thaw durability should be tested in accordance with ASTM C666, Standard Test Method for Resistance of Concrete to Rapid Freezing and Thawing. The specimens should maintain at least 80 percent of their relative dynamic modulus of elasticity after 300 cycles of testing.
- The material shall not contain any other form of chloride.
- The product shall be used before its recommended shelf life expires.

Most structure owners want to reduce the period of repair. Therefore repair materials that can gain early strength have become very popular for many applications. Superplasticizer may be needed as an additive when the time available for the repair is limited. For surfaces exposed to freeze–thaw/salt cycles, an appropriate air void content in the repair mortar or concrete must be achieved. It was shown by many researchers that very early strength (VES) concretes can be achieved by using conventional materials, normal mix designs, regular placing and curing practices. Two kinds of VES concrete (A and B) were developed by Virginia DOT. For VES A concrete, a minimum compressive strength of 2000 psi is reached 6 hours after water is added to the concrete mixture using Portland cement with a maximum water–cement ratio of 0.40. For VES B concrete, a minimum compressive strength of 2500 psi is reached 4 hours after water is added to the concrete mixture by using Pyrament PBC-XT cement, with a maximum water–cement ratio of 0.29. Both types of concrete can achieve a minimum durability factor of 80 percent at 300 cycles of freezing and thawing.

Depending on the thickness of repair patches, mortar or concrete can be used for various repair work. The maximum aggregate size should not exceed $1/3$ of the required thickness. For heavy traffic areas, a cement mortar or concrete with good bond should be used with a minimum thickness of 35 mm. For normal traffic area, this requirement reduces to a minimum 20 mm in thickness.

Concrete mortar and concrete are applied by casting, trowelling and spraying. Casting is used for large repair jobs with horizontal and vertical surfaces. A possible problem with casting application is the adhesion to the old concrete. Spraying application of concrete is similar to shotcrete and will be discussed later.

(B) GYPSUM-BASED CONCRETE

Gypsum-based (calcium sulfate) patching materials gain strength rapidly and can be used in situations where the ambient temperature is near freezing point (FHWA 1999). However, gypsum concrete does not appear to perform well when exposed to moisture or moist weather. In addition, the presence of free sulfates in the typical gypsum mixture may promote steel corrosion in reinforced concrete structures.

(C) MAGNESIUM PHOSPHATE CONCRETE

Magnesium phosphate concrete sets very quickly, and therefore has high early-strength. It can be used to make impermeable patches that bond to clean and dry surfaces. Many research studies have been carried out on phosphate cement-based materials for rapid repair of concrete. Phosphate cement-based material usually is prepared by mixing the MgO (M) and $NH_4H_2PO_4$ (P) powder with borax (B). Some researchers (Yang and Zhu *et al.* 2000; Abdelrazig *et al.* 1988; Seehra *et al.* 1993) studied the chemical compositions and mechanical properties of the materials, and their study shows that MPB has good bond and compatibility with old concrete as well as low shrinkage. Yang (2002) studied the de-icer scaling resistance of MPB and found that MPB mortars have quite high de-icer frost resistance. However, these materials are extremely sensitive to water on the surface, and even a very small amount of extra water in the mix severely decreases its strength. They also cannot be used with limestone aggregates. MPB is a neutral or weak acidic material that may not protect the steel reinforcement from corrosion. In addition, this type of material contains a high level of Na$^+$ ions, which may cause alkali–silica reaction in the concrete.

(D) HIGH ALUMINA CONCRETE

Calcium aluminate concretes gain strength fast, bond well and shrink very little during curing (FHWA 1999). They are often used as refractory concrete. However, they may lose strength over time because of the chemical conversions that take place.

4.2.5.2 *Polymer modified repair materials*

Polymer modified repair materials are a combination of polymer resin, aggregate, and a set initiator. Polymer modified repair materials are very useful not only for repairing concrete buildings but also for overlay applications on reinforcement concrete slabs and bridge decks mainly because they set very fast and have low shrinkage. Fast setting allows the repaired section of the structure to be used earlier, and low shrinkage ensures the durability of the repaired section. However, some of the polymer modified repair materials are sensitive to moisture. Magnat and Limbachiya (1997) studied three typical repair materials: high strength non-shrinkable concrete, a mineral-based cementitious material with no additives or coarse aggregate, and a cementitious mortar containing styrene acrylic polymer with fiber additives. The research results show that the shrinkage of polymer concrete was very sensitive to the relative humidity of the environment as compared to the other two concretes. The total shrinkage of the polymer mortar is the highest among the three repair materials despite the presence of some fiber additives. Furthermore, the compressive creep strain of polymer concrete is also the greatest as compared to the other materials.

Moreover, there is significant discrepancy between the coefficient of the thermal expansion and the elasticity moduli of some polymer modified concretes and conventional Portland cement concrete. The difference in the coefficient of thermal expansion will generate a significant interface shear stress between polymer modified concrete and existing (regular) concrete upon temperature variations. The difference between the elasticity moduli will also generate interface shear stress between the new and old concrete.

There are several types of polymer modified concretes commonly used for repair applications, including epoxy, methyl methacrylate, and polyurethane modified concretes.

(A) EPOXY CONCRETE

Epoxy usually has short curing time, high early strength, and high resistance to chemical attacks. Therefore, epoxy modified concretes or epoxy concretes are often used for repair work when a short curing time, a high early strength and a high resistance to chemical attack are required. Epoxy concretes are especially suitable for repair of surface damage and edge damage at joints. But epoxy resin systems have some disadvantages. As discussed earlier, they have larger coefficients of thermal expansion than that of concrete. Besides, they are relatively expensive. Resin systems usually have mechanical properties that differ from those of the concrete substrate, which result in the build-up of stresses at the interfaces as the resin cures. So a clean and sound substrate surface is very important for a successful surface repair (Saccani and Magnaghi 1999).

Epoxy resins usually consist of two components, one is the active ingredient and the other the catalyst or hardener. It is important that the component materials of resin systems be mixed in the proportion suggested by the manufacturer. Epoxy resin repair materials should conform to ASTM C881, Standard Specification for Epoxy-resin-based Bonding Systems for Concrete. The commonly used mix design for epoxy mortar/concrete is four to seven parts dry silaceous aggregate to one part resin by weight. The maximum aggregate size should be less than 0.50 in. Aggregate passing the #100 sieve is generally excluded. The aggregate should preferably be under oven-dried condition. If not, its moisture content should not be greater than 0.5 percent. More information on the use of epoxy mortar/concrete can be obtained from ACI 546.1R, Guide for Repair of Concrete Bridge Superstructures, and ACI 503.4, Standard Specification for Repairing Concrete with Epoxy Mortars.

Epoxy concretes are impermeable and excellent adhesives. They have a wide range of setting times, application temperature, strength, and bonding conditions. Epoxy gives off a large amount of heat during its hardening process, and thus, for deep epoxy repair work, the repair material should be placed in several lifts to control heat development in the concrete structure.

(B) METHYL METHACRYLATE CONCRETE

Methyl methacrylate concrete has high compressive strengths and good adhesion (FHWA 1999). The binder consists of a high-molecular-weight methyl methacrylate (HMMA) monomer that is polymerized in place. Initiators and promoters are added to start polymerization and to accelerate polymerization, respectively. HMMA is preferred over other methyl methacrylate monomers as it is less flammable. Other advantages of HMMA are low viscosity, high bond strength to concrete, and relatively low cost.

The quantity of HMMA monomer to be mixed should be divided into two equal portions. The entire required quantity of initiator should be thoroughly mixed with half the monomer, and the entire required quantity of promoter mixed with the other half of the monomer. Then the two portions of monomer should be blended. The monomer should be mixed in clean, electrically grounded containers. Mixing should be conducted in a well-ventilated area out of direct sunlight. A class B fire extinguisher should be available during mixing and placing. The aggregate should conform to ASTM C33. The maximum aggregate size should be less than 0.50 in. The aggregate should preferably be oven dry, but in no case should the moisture content exceed 0.50 percent. More information on the use of HMMA mortar/concrete for repair of bridge decks can be obtained from ACI 546.1R, Guide for Repair of Concrete Bridge Superstructures, Chapter 6(19). In addition, many methyl methacrylates are volatile and may pose a health hazard from prolonged exposure to the fumes.

(C) POLYURETHANE CONCRETE

Polyurethane concretes generally consist of a two-part polyurethane resin mixed with aggregate. Polyurethanes generally set very quickly (FHWA 1999). Some manufacturers claim their materials are moisture-tolerant; that is they can be placed on a wet surface with no adverse effects.

(D) LATEX MODIFIED CEMENT SYSTEMS

Latex is a dispersion of very small particles of an organic polymer in water. Latex modified cement systems should be proportioned in accordance with the specified mix design. The amount of water used in the mixture should be kept as low as possible to reduce shrinkage. Some research studies showed that the total water/cement ratio may fall within the range of 0.25–0.35 by weight (Meinheit and Monon 1984). To prevent latex losing its stability, air-entraining agent should be avoided in the mixture to prevent the formation of small air bubbles in the mix. Latex modified concrete is often used for repairing bridge decks and parking garage deck overlays. The main drawback of latex modified cement system is that it is sensitive to high temperature (i.e. 29.4 °C or above) due to the film-forming characteristics of the latex. Besides, latex modified cement system is not suitable for underwater applications since it needs to dry in the air.

The hardening mechanisms of the latex modified cement system are similar to regular concrete. During the curing process, the Portland cement hydrates as usual. In addition, a network of polymer film is formed within the mixture of aggregate and hydrates. This system provides high early strength, better flexural strength, better bond, better freeze–thaw durability, lower permeability and better durability than those of regular concrete. For instance, the bond strength of latex modified cement system may exceed that of the unmodified cement system by factors of 2–3. The latex itself tends to seal the micropores in the cement system, resulting in a major reduction of permeability of cement system. Also, due to the dispersing effect of components in the latex combined with water; latex modified cement system has better workability than normal cement system. Immediately prior to applying latex modified cement systems for surface repair, the substrate concrete must be pre-wetted to a saturated surface dry (SSD) condition in order to delay film formation of the latex. As one can see from the above descriptions, latex actually has several different functions when added in the concrete mixture: (1) increasing tensile and flexural strengths of the concrete; (2) as a water-reducing agent (plasticizer) for concretes with lower water/cement ratios, improving workability of fresh concrete and decreasing shrinkage of hardened concrete; (3) improving the bond between the repair concrete and the existing concrete; and (4) reducing the permeability of the concrete by increasing its resistance to aggressive chemicals. Therefore, latex modified cement systems are usually used for repair jobs which require special properties that regular concrete does not have: high adhesion, low shrinkage, high durability, and low permeability.

There are many different latex products available for repair applications. They have different requirements for curing procedure. In general, moist curing is required for latex modified cement systems in surface repair for 24 hours to let the cement hydrate properly, then the moist curing is usually followed by 72 hour of dry curing. Because of its excellent adhesion characteristics, latex modified mortar has no minimum thickness requirement in general concrete repair, provided that the size of sand particles in the mortar is small enough.

4.3 Repair techniques

The word "concrete" has become a symbol of strength and durability in most people's minds, but in reality, concrete structures come under attack from both natural and manmade forces. The rate of degradation of concrete depends on a number of factors; only some of them are controllable. Fundamental understanding of these factors is essential to determine how to repair the distressed structure. Some controllable factors that affect the rate of degradation of concrete are quality of concrete placement, mix design of concrete, drainage condition of the structure, application of de-icing chemicals, etc. Non-controllable factors include the environmental conditions, such as moisture fluctuations, freeze–thaw cycles, and chemicals in surrounding water sources. Man-made forces affecting degradation of concrete generally fall into two categories: (1) design deficiencies or neglect; and (2) overuse and overload of the structures.

There are several types of concrete degradation that can be found in many concrete structures, which will be discussed in detail in later sections. One of the common degradations of concrete structures is the damage caused by corrosion of the embedded reinforcing steel. The corrosion of embedded steel reduces the cross-section area of the steel and thus reduces the load-carrying capacity of the structure. At the same time, corrosion results in various types of damage in the concrete surrounding the reinforcing steel. The damages include cracking, delamination, and spalling. Because reinforcing steel is virtually found in all reinforced concrete structures, corrosion of reinforcing steel is a very critical and widespread problem. The corrosion of steel and the deterioration of concrete are two individual durability problems for reinforced concrete structures and they tend to occur together. The accumulation of rust formed in the corrosion process of a reinforcing bar results in cracks in the surrounding concrete cover. The first crack may be very small but it invites easier intrusion of moisture and aggressive chemicals such as chloride ions from the de-icing salts, in turn, the combination of the moisture and the chemicals accelerates the corrosion process. Eventually and inevitably, the outward symptoms of scaling, cracking and spalling gradually begin to appear in concrete structures. Figure 4.7 shows a bridge pier experiencing severe corrosion damage.

Figure 4.7 Example of steel corrosion.

The most important task in planning a successful repair of concrete structures is to find the cause (damage mechanism) of the deterioration of the structure and apply effective methods to stop the damage mechanism. This, apparently, is a very difficult task even for experienced professionals. In the following sections of this chapter, we will first introduce general methods that have been used to repair concrete structures such as how to repair concrete surface damage and how to repair concrete cracking, without going into details on the cause of the damage in the concrete. Then, we will further discuss in greater depth several commonly seen deteriorations in concrete and the specific repair methods for the damage, including fire damage, corrosion damage and the damage related to alkali–silica reaction in concrete. Finally, we will discuss some specific repair methods for special structures such as bridges and pavements, which are different from buildings.

Concrete repair provides extended service life to the structures by integrating new materials with existing materials to form a composite structure that can withstand environmental and mechanical loadings. There are many advanced solutions, far beyond the simple concrete patch, that can be utilized to implement an effective concrete repair. Basic understanding of concrete repair options, such as surface repair, protection, stabilization, strengthening and waterproofing, will successfully provide extended service life of the structure (Emmons 2006).

It is important to mention that, in addition to the commonly used repair methods, some of the repair methods described here may not be commonly used methods and they are still in the experimental or developmental stage. For these repair methods, the readers should obtain more detailed information or consult experienced professionals before actual implementation.

4.3.1 General repair strategies

In repairing deteriorated concrete structures, the primary object is to restore the concrete structural member or the whole structure to its original shape and condition, to ensure structural integrity, durability, and composite behavior and, sometimes, to match the existing concrete in color and appearance. Repair is not only to recover the original condition of an infrastructure, but also to extend its service life to satisfy the design requirement. A successful repair is not only involved in the portion of the structure to be repaired, but also the reinforced members connected to the repaired portion. Basically, the repair procedure for reinforced concrete structures consists of some or all of the following four steps:

1   Removal of damaged concrete to expose all damaged portions of the structure. The damaged concrete portion may contain cracks, delamination, and etc.
2   If there is reinforcing bars in the damaged portion of the structure, the surface of the steel should be cleaned to remove any rusting layers, and

additional steel should be incorporated if necessary. A protective coating may also be applied to the existing and new steel surfaces.

3   Application of repair mortar or concrete to replace removed concrete, which serves two purposes: (a) the repair material prevents the ingress of corrosive agents (physical protection); and (b) to re-passivate the steel (chemical protection). Primer may need to be applied first to improve bond between existing concrete and repair material.
4   To enhance protection for steel, external membranes may be applied over the repaired section, or the whole concrete surface.

Like the selection of a proper repair material, the selection of a repair method is often a difficult process which involves consideration of a large number of factors, some of which are technical, some economic, and others may be practical. A successful and durable concrete repair can result only from correct choices for repair method and for repair materials. In choosing an appropriate repair material for concrete structures, there are a number of considerations such as those discussed in earlier sections: strength, durability, compatibility, ease and safety of application, and last but not the least, cost. Cost is a very important issue. In a limiting case, the cost of repair should not be more than the cost of replacement. But in some cases, even if the cost of repair is more than the cost of new construction, people preferred to choose repair work because of other considerations such as environmental preservation and historic restoration for some landmark buildings.

Repair work not requiring strengthening of the structure may be classified in different ways: either by geometric feature of the area to be repaired such as vertical and horizontal surfaces and cracks or by the failure mechanisms of the structure or structural member such as fire damage and corrosion damage. For the first type of classification: the repair methods for concrete surfaces and concrete cracks will be described. For the second type of classification, the repair methods for fire damage, corrosion damage, and ASR damage will be discussed. The following sections will be presented in the remaining part of this chapter:

1   Repair of the concrete surface
2   Repair of concrete cracks
3   Repair of fire damaged concrete
4   Repair of corrosion damage
5   Repair of ASR-induced damage.

4.3.2 *Preparation methods*

Depending on the deterioration mechanisms and damage conditions of concrete surfaces, different methods may be available for repair. For shallow and minor damage in the concrete surface, surface coating, surface impregnation and surface sealers may be applicable. For deep and serious damage, the damaged and deteriorated concrete should be removed and replaced by new repair material. For all cases, repair work begins with a proper surface preparation of the damaged or deteriorated concrete surface to ensure good bonding development between the new material and the old material. Many surface cleaning and preparation procedures are available and the choice will depend on the condition of the substrate and the degree and nature of contamination. The ability of repair materials to wet out the concrete surface varies, and consequently, adhesion to concrete relies greatly on the mechanical bond.

4.3.2.1 Preparation of the surface

In all cases where concrete surfaces are repaired, the condition of the existing concrete in the exposed damaged area is of primary importance for the durability of the repair. The repair work can seriously be compromised if the adhesion between the new repair material and the existing sound concrete surface is not strong. Surface preparation includes all the steps taken after removal of large amounts of deteriorated concrete. It also includes the steps taken to prepare surfaces on which the new repair material is to be placed, even if no concrete is removed. In order to fully understand the extent of damage to the structure, all damaged concrete should be removed. Concrete removal techniques should be selected on the basis of safety, economy and their effect on the remaining sound concrete. They also depend on the situation, especially on the extent and thickness of the layer which has to be removed, as well as on the type, location and position of the damage in the structure (FIP 1991).

There are many industry guides and standards related to surface preparation for repair. The guides and standards that should be referenced, include but not limited to the following:

ASTM D 4258	Surface Cleaning Concrete for Coating
ASTM D 4259	Abrading Concrete
ASTM D 4260	Acid Etching Concrete
ASTM D 4261	Surface Cleaning Concrete Unit Masonry for Coating
ASTM D 5295	Guide for Preparation of Concrete Surfaces for Adhered (Bonded) Membrane Waterproofing System
ACI 546R-04	Concrete Repair Guide
ACI 224.1R-07	Causes, Evaluation, and Repair of Cracks in Concrete Structures
NACE RP0591-91	Coatings for Concrete Surfaces in Non-Immersion and Atmospheric Service
NACE TCR6G166	Surface Preparation of Concrete Coatings

For different types of damage, there are different technical guidelines for surface preparation. For example, ICRI guideline 03730 (ICRI 2002) covers surface preparation for repair work of corrosion damage.

Surface preparation should introduce as little damage as possible to the concrete remaining in place. On the other hand, it is difficult to determine whether all the damaged material has been removed, because the zones of damaged or deteriorated concrete are usually difficult to define. Surface imperfections should be well treated during the surface preparation process. For instance, sharp edges should be ground smooth and blowholes should be filled with a suitable filler.

Proper surface preparation provides a dry, even and level surface free of dirt, dust, oil and grease. Removal of surface contaminants allows primers and repair materials to have direct contact with the substrate, increasing the surface area of high quality existing concrete, increasing roughness of the surface, and providing increased anchorage of the applied new repair material. Table 4.7 lists some of commonly used surface preparation methods.

In order to choose the best concrete removal method or combination of methods, the following safety, environmental and job-related information should be considered in addition to the industry guides and standards:

1   The cleaning and surface preparation instructions supplied by the material manufactures must be carefully studied. Some specific surface preparation methods are required for special repair materials.
2   The location of repair work, e.g. inside a building or outside. The

Table 4.7 Standards of surface preparation

Surface conditions	Methods	Expected tensile adhesion
Dry and free from grease, free from dust, flakes, salts and laitance, with a sharp even finish	Application of emulsified degreaser and washing clean with water, grit blasting, high-pressure water jetting and acid etching	2.5–3.0 N mm^2
Dry and free from grease, free from dust, flakes, salts and superficial laitance	Application of emulsified degreaser and washing clean with water scraping off any loose matter, power wire brushing and vacuum cleaning to remove dust	1.5–2.0 N mm^2
Dry and free from grease, free from dust flakes and salts	Application of emulsified degreaser, washing clean with water, scraping and hand wire brushing to remove loose matter and brushing clean	0.5–1.0 N mm^2

restrictions with respect to noise, dust, vibration, exhaust fumes and disposal of wastewater should be considered. This is important especially when the repair work is inside a building.

3   The additional load due to construction equipment and construction materials. This is important when a large amount of repair materials and equipment must be piled up temporarily during the repair process. It is crucial to make sure there is no overload applied to the structure.

4   The thickness of the damaged concrete to be removed. This is important when choosing the concrete removal equipment.

(A) MECHANICAL METHODS

A wide variety of mechanical equipment is available for the surface preparation of concrete, such as pneumatic drill, electrical saw, electrical hammer, compressed-air hammer, mill machine, sand blasting, and so on. In general, these devices are of two types: rotary and impact. Rotary equipment includes rotary discs and grinders which are usually used on the substrate containing concrete of relatively low compressive strength that does not have a steel trowelled finish. They are not effective on hard, dense concrete. Abrasive blasting is a fast and effective method of surface preparation using rotary equipment. A widely used abrasive tool is copper slag grit. The grit is propelled against the concrete surface using large volumes of compressed air. But the aggressive abrasion may cause numerous blowholes which have to be filled before repair work. Impact tools such as electric hammers, scabblers, and needle guns will effectively remove several millimeters of concrete. Scabblers use compressed air to hammer piston-mounted bits into the concrete surface. They tend to roughen the concrete surface more than either abrasive blasting or shotblasting. Scabbling operations are dusty, noisy, and produce some vibration, so vacuuming or spray-pressured water is always required to improve the working condition. Impact tools pulverize the concrete and can cause fracture of the concrete below the surface. Consequently, it may be necessary to use another means (such as water jetting or wet sandblasting) for the final cleaning of the surface. Heavy hand labor is needed when using mechanical methods, such as hammer and hand tools to prepare the surface for repair. If a concrete layer with a small thickness is to be removed, chipping-off is recommended. A milling device is helpful in removal of concrete to a flat surface. Several milling passes may be required for a thick layer of concrete. Sand-blasting is suitable for roughening of concrete surface, but it is not suitable for removal of a thick concrete layer.

For cleaning concrete surfaces, high pressure water jetting are often used, but in case of heavy pollution or ingrained dirt, it may be necessary to use wet sand blasting and manual scrubbing with detergents to clean the concrete surface. A power wire bush is quite often used to clean the concrete surface. Concrete treated by this technique exhibits a regular and relatively blowhole-free surface.

A hydraulic water jet with 10–40 MPa pressure is able to remove loose concrete particles. A great advantage of this technique is that there is no dust produced. A hydraulic water jet with a higher pressure up to 40–120 MPa is an efficient way to remove areas of soft concrete surface. With a higher pressure to 140–240 MPa, the hydraulic water jet may be used to cut concrete.

Whenever deteriorated concrete is removed using impact tools, the surface of the remaining concrete may be damaged. If this damaged layer is not removed, the repair material may debond from the substrate. Therefore, the remaining concrete should be further prepared using wet sandblasting or high pressure water jetting to remove this damaged surface material. Usually the removal of limited areas of concrete to permit a repair requires the saw-cutting of the perimeter of the areas to minimize further damage to the area to be repaired.

(B) CHEMICAL METHODS

Many deterioration processes result from penetration of the concrete by gases or salts in solution. In such cases, surface treatments are designed to reduce the passage of potentially deleterious substances and slow down the rates of deterioration. Concrete that is contaminated with oil, grease, or dirt requires chemical cleaning piror to the application of repair materials. Acid and alkalines are suitable for this type of work, for instance, acid etching is a particularly efficient method of surface preparation. Acid etching is especially suitable for areas demanding clean conditions since no dust is produced, but its most effective use is restrticted to floors due to practical difficulties with vertical or inclined surfaces (Allen *et al.* 1993). A 10 percent solution of muriatic (hydrochloric) acid is normally used for concrete surface cleaning in acid etching. The concrete should be pre-wet, and all oil, grease, paint, sealers, gum, tar and other foreign materials must be removed before etching.

Chemical etching should be followed by vigorous scrubbing and thorough rinsing with water to remove all residues. If necessary, the surface should be water-jetted to ensure that all water-soluble salts have been removed. If thorough cleaning cannot be done after chemical etching, a neutralizing agent can be used. In this case, the neutralizing agent should be dispersed in water and then the solution can be used on the concrete surface. Diluted ammonia (4 percent solution) in water mixed at a rate of 0.9l concentrate to 19l water flooded over the surface of the concrete will be sufficient (Holl and O'Connor 1997).

It is normally good practice to check the pH level of the prepared concrete surface. If the pH level is below 10, it is recommended to carefully wash the concrete surface with a weak solution of sodium hydroxide to raise the pH level above 11. Conversely, if concrete containing reactive aggregates has been attacked by alkalis, the surface can be washed with water to lower the pH level below 12 (Newman 2001).

It should be noted that muriatic acid, commonly used to etch concrete surface is relatively ineffective in removing grease or oil (CSA 1989). Besides, due to the potential risk of corrosion of the rebars in concrete, acids should not be used for reinforced and prestressed concrete (FIP 1991).

(C) FLAME CLEANING

Flame cleaning is generally used to clean concrete surfaces that are to receive coatings or resinous overlays. This method is particularly useful for oil-stained floors because it permits the application of coatings to the concrete immediately below. A special multi-flame oxy-acetylene blowpipe is passed over the concrete surface at uniform speed. The thickness of the concrete layer removed depends on the speed at which the blowpipe is moved and the properties of the concrete. The most suitable blowpipe speed lies between 0.02 m/s (0.066 ft/s) and 0.03 m/s (0.099 ft/s). Concrete and coating removal involves both the spalling and melting off of the surface. The laitance layer is usually removed to a depth of 1 or 2 mm (0.04″ or 0.08″) and in a few instances up to 4 mm (0.16″). The moisture content of the concrete has the greatest effect on the concrete removal – completely dry slabs do not produce much spalling, while slabs soaked in water prior to flame cleaning produce uniform concrete removal.

European experience indicates that flame cleaning does not promote the migration of deep-seated oil to the surface, does not remove the alkalinity of the matrix – the surface gradually attains alkalinity similar to that of new concrete – and does not promote the development of any visible cracks in the surface. The method has proved useful for such applications as the recoating of concrete floors or the removal of defective elastomeric waterproofing membranes from parking decks (Mailvaganam *et al.* 1998).

4.3.2.2 Concrete surface requirement

The desired condition of the concrete surface depends on the type of repair and the condition of the substrate. After the surface preparation, the strength of adhesion and moisture content of the remained concrete have to be evaluated to ensure a good surface repair quality. The strength evaluation can be done by pull-out test. In this method, a 50 mm diameter steel plate at least 10 mm thick is glued to the concrete surface after a corresponding circular-core drill cut is made in the concrete surface. After the glue hardens, the steel plate is pulled off the concrete surface and the tensile stress achieved in the pull-off test is referred to as the strength of adhesion. It is required that the measured strength of adhesion should be greater than 1.5 MPa. If the adhesion strength is less than 1.5 MPa, the concrete is required to be removed to a further depth.

The moisture content can be evaluated by mass measurement method by using a moisture meter, or using the method specified in ASTM D 4263, Indicating Moisture in Concrete by the Plastic Sheet Method. In this method, a concrete specimen is sealed by a plastic sheet about 24 in^2 (15,500 mm^2). If the concrete appears dark, damp, or wet, the presence of excessive moisture in the concrete is indicated. The concrete should be allowed more time to dry until the excessive moisture is removed (Holl and O'Connor 1997). When electrode-type moisture meters are utilized, care must be exercised, if readings are taken on surfaces recently wetted, since the concrete interior could be dry. It is wise to chip off some of the surface and repeat the test on the underlying surface. Normally, readings of less than 5 percent moisture content by mass are required before coating, and specific requirement by local building codes should be referred to regarding the residual moisture content before coating.

The required moisture condition can be achieved by water spray or air dry out. The permissible moisture content depends on the selected repair method and the applied materials. A cement bond system requires wetting of the concrete surface, so that the existing concrete will not absorb water from a repair coating. However, an excessive water content may also be detrimental to adhesion. Therefore, the concrete surface should be moist but not saturated. For a plastic bond system, a dry surface is required, which means the moisture content is less than 6 percent by weight to a depth of approximately 20 mm. Good adhesion cannot be achieved if the water in old concrete prevents the penetration of liquid repair materials. Surface preparation also needs the removal of reaction products such as laitance that cover the surface by various methods described above. The removal of the surface contaminates allows primers, and the repair materials themselves, to have a strong bond to the substrate. For most inorganic materials, the concrete should be saturated and surface dry to prevent rapid loss of water from the repair material to the substrate so that shrinkage and cracking in the repair material are prevented. Maximum adhesion for organic (resin-based) materials is achieved by ensuring that the concrete surface is dry.

For reinforced concrete, surface repairs may include proper preparation of the surface of reinforcing steel to ensure good bond development with the replacement concrete so that the desired behavior in the structure is obtained. The deteriorated concrete surrounding the reinforcing steel should be removed so that the original reinforcing steel is exposed and then cleaned. A pachometer may be needed to determine the depth of the steel in the concrete. A pachometer is an electromagnetic device used for determining the location and cover of reinforcement bars in concrete.

4.3.2.3 Bonding agents

There are mainly two kinds of bond between the old concrete and new concrete: (1) physical bond through adhesion and cohesion; and (2) chemical bond through reactions among the materials in contact in the interface. In most cases, both types of bonding exist at the same time. Although adequate concrete bond strength can be achieved when patching, mortar or repair concrete is placed against a prepared surface that is water-saturated but surface dry, a bonding agent is recommended to use in repair work to improve the bond between the old and new concrete. It is essential to obtain the best possible bond at the interface with the help of bonding agents. Bonding agents may increase the cost of repair. The commonly used bonding agents for surface repair of concrete include:

1   Cement paste with relative low water–cement ratio. The cement paste can be brushed onto the surface to be repaired. Cement-based bonding agents usually require some initial protection (i.e. proper curing), because a rapid drying of the materials could lead to shrinkage cracking and delamination. The bonding agents should be applied soon after surface preparation is completed.

2   Cement mortars with a cement and sand ratio of 1.0. Proper curing is also needed for this type of bonding agents.

3   Polymer modified cements, such as latex modified mortars or pastes. In this type of bonding agent, the polymers are mixed with the cement paste or cement mortar via the mixing water. Additives, such as dispersion agents and emulsions, may be added to the mixture to improve the bond strength, workability and water retention capacity. Latex modified mortars have increased workability characteristics and a tendency to entrain air in the mix. Therefore, they should be mixed by a slow speed mixer or by hand to reduce the amount of air entrainment. With latex materials, it is advisable to wet the concrete surface beforehand, and allow the surface to become dry prior to installation.

4   Resin systems, such as epoxy resin and hardener with or without fillers. Resin-based bonding agents do not generally need any protection during their curing period. In the case of epoxies, however, the concrete surface should be dry to allow the epoxy to penetrate into the surface pores. Damp concrete surfaces inhibit the proper adhesion of epoxy material and may resulting in debonding (Mailvaganam 1992). In many cases, filling materials are used with resins mainly because they are less expensive than epoxy resin, they can reduce shrinkage as the resin cures, and they can reduce temperature as the exothermic reaction takes place when the active ingredients of resins and the hardener are mixed together.

4.3.3 Repair of concrete surface

4.3.3.1 Shotcrete

Repair of concrete surface normally uses shotcrete or gunite, and, sometimes, this technique is also referred to as sprayed concrete. Sprayed concrete is a mixture of cement, aggregate and water projected at high velocity from a nozzle by pneumatic pressure into a location in an existing structure or formwork where it is compacted by its own velocity to produce a dense homogeneous mass (Allen *et al.* 1993). The Concrete Society of the UK (1979, 1980 and 1981) has defined gunite as a sprayed concrete with a maximum aggregate size of less than 10 mm; whereas shotcrete is defined as a sprayed concrete where the maximum aggregate size is 10 mm or greater.

This technique was first developed as long ago as the 1930s and is also suitable for concrete replacement and for strengthening structural elements. Within the category of shotcrete, a distinction can be made between a dry mix process and a wet mix process. In the dry mix process, the solid constituents, cement and aggregate, are mixed without the addition of water. This dry mixture is then delivered along the hose to the discharge nozzle where water is added in sufficient quantity to hydrate the mixture and to provide the right consistency so that the produced mixture can be projected at high velocity onto the structure to be repaired. The impact due to the high velocity compacts the mixture to form the required quality of the concrete. The dry mix process can produce sprayed concrete with very low water–cement ratios and almost no slump that enable it to be placed to a limited thickness on vertical and overhead surfaces. In the wet mix process, all the ingredients including the water, are batched and mixed and then pumped along flexible hoses into a discharge nozzle. At the nozzle, compressed air is introduced and projects the mixture into position at high velocity as shown in Figure 4.8. Nowadays, the wet mix process has become more and more popular because the working conditions are better than that for dry mix process and there is better quality control.

When sprayed concrete for surface repair is used, in either a dry or wet mix process, pre-treatment of the old concrete surface is important. Sandblasting is efficient for surface treatment and the existing concrete surface should also be pre-moistened. If reinforcement steels are exposed after removing the deteriorated concrete, the rust on the steel must be removed and the steel must be cleaned down to the bare metal. Bonding agents, such as epoxy bonding agents, latex-modified cement slurries, or pure cement slurries can be used as a base coating to improve the bond between the sprayed concrete and the existing concrete substrate. But there are some arguments that there is no need for these bonding agents since they may be removed from the concrete surface during the initial pass of spraying concrete. Besides, during the initial application of spraying concrete, rebound is

Figure 4.8 Application of shotcrete for repairing a concrete arch.

high, resulting in a cement-rich layer left, which may fulfill the same function as the bonding agents (Crom 1986).

One important note in the application of shotcrete is that the thickness of the concrete layer should be considered. If the thickness is large, multi-layer sprayed concrete is needed, which requires the preceding layer to achieve a sufficient hardness. If the thickness is more than 50 mm, minimum reinforcement is required. Evaporation protection is helpful in curing sprayed concrete after application to prevent a rapid drying out.

The purpose of using sprayed concrete in surface repairs is to provide a relatively thin, high quality, protective layer on the area to be repaired. If prepared and applied properly, sprayed concrete can provide excellent bond, high strength, and a protective cover of low permeability to the existing steel reinforcement and concrete. Each process, both dry mix and wet mix, has its own advantages and disadvantages, and the choice of process depends on the cost and availability of equipment and crews, operational features and the particular circumstances of the application. The major advantages of both processes are that no formwork is needed for the placement of the concrete and that a variety of shapes of surface can be formed. The repair of concrete by shotcrete has very wide applications for thin walls, certain roofs, slabs, reinforced concrete tanks, tunnels, canals, coatings on brick, steel and masonry, encasement of steel for fireproof and repairs to earthquake damages. The disadvantages for sprayed concrete include: the quality of sprayed concrete is largely dependent on the skill of the operator; the uniformity of the sprayed concrete is not good; the placing process of sprayed concrete is relatively dusty; it is generally more suitable for thin-layer repair, not good for thick section; and the control and inspection of the material and repair work by sprayed concrete are difficult. The dry mix process produces more dust and rebound than the wet mix process. Therefore, the wet mix process is favored for work in a confined environment such as tunnels and in environmentally sensitive areas. Since no water is added to the mixture until just before placing, the dry mix process is appropriate where sprayed concrete is to be used intermittently and spraying can be stopped and started at will. On the other hand, the wet mix process is more suitable for continuous spraying and it is often used for large-scale repair project, typically in new construction work.

Most commonly used mixes for the dry mix process have an aggregate–cement ratio in the range of $3.5:1$ to $4:1$ by weight and the sprayed concrete in place shows good density, high strength (typically 40–50 N/mm^2) and very good bond to a suitable substrate. In the wet mix process, cement contents are usually in the range of 350–450 Kg/m^3 with a water–cement ratio down to 0.4. The mix, including cement contents, aggregate–cement ratio, maximum aggregate size and grading, has to be designed to be pumpable. Normally higher water–cement ratios are found in the wet mix process than those in dry mix process in order to achieve pumpability. Compared with the sprayed concrete made by the dry mix process, the in-place strength of the wet mix process sprayed concrete is lower, typically 20–40 N/mm^2.

In some cases, no-fines concretes needs to be used. No-fines concrete is a concrete that only contains normal Portland cement, water and coarse aggregate. The use of non-sand concrete has expanded into many areas such as: drainage pipes, wells, small retaining walls, pavements (for local roads and parking lots), and repair work. One of the disadvantages of such concrete is the difficulty in mixing and placing (pumping) due to the stiff consistency of the mixture caused by the low water–cement ratio required to prevent segregation.

It was shown that non-sand concrete can be made pumpable by adding several admixtures (Amparano and Xi 1998). Pumping characteristics are based on the mixture's ability to provide good slump and excellent cohesiveness. The purpose of the research was to obtain a pumpable non-sand concrete mixture, within the specified design parameters, which achieved two goals, that is, exhibits excellent pumping characteristics and yields equal or higher strength values than that of mixtures made without admixtures. Their results indicated very good performance of mixtures containing silica fume concurrently with the hydroxyehtyl cellulose admixture for non-sand concrete mixtures with 12.7 mm and 6.35 mm maximum size aggregates.

4.3.3.2 Replacement

When the damaged concrete is in a large volume, especially when the thickness of the damaged concrete is large, the damaged concrete may be replaced by new cast of repair concrete. Strictly speaking, the shotcrete described in the previous section is also a means of replacing damaged concrete. It is a special repair method that can be used for large areas of damaged concrete with a small thickness, and it is basically used to repair surface damage on a concrete structure.

When damaged old concrete is intended to be replaced, the compatibility between old and new concrete should be taken into account. The surface of the old concrete requires adequate preparation, careful cleaning and pre-moistening. Placing the new concrete should avoid the entrapment of air. Therefore, the formwork must be sufficiently rigid and tightly fitted to the existing concrete in a manner to minimize leakage of cement paste. The new concrete should be re-compacted to improve contact to the old concrete before initial setting. For large-volume replacement of old concrete, sometimes minimizing the temperature difference between old and new concrete may require special procedures (cooling of new concrete and/or heating of old concrete). In general, the replacement of concrete should have final properties, such as strength and modulus of elasticity, that match the existing concrete as closely as possible. Thus, the mix design of the repair concrete, such as type of cement, cement content and the water–cement ratio should be carefully chosen.

4.3.3.3 Grinding

Surface grinding can be used to repair some bulges, offsets, and other irregularities that exceed the desired surface tolerances. This method is often used to repair concrete driveways, sidewalks, and roadways when a concrete slab is not levelled very well with the surrounding slabs. Excessive surface grinding, however, may result in weakening of the concrete surface and exposure of easily removed aggregate particles. If the concrete slab or the concrete member is to carry a heavy load, then the load-carrying capacity of the member should be examined with regard to the grinding process to make sure that the load-carrying capacity of the reduced cross-section is sufficient for the load. For these reasons, surface grinding should be performed subject to the following limitations:

- Grinding of surfaces subject to cavitation erosion (hydraulic surfaces subject to flow velocities exceeding 40 feet per second) should be limited in depth so that no aggregate particles more than $1/16$ inch in cross-section are exposed at the finished surface.
- Grinding of surfaces exposed to public view should be limited in depth so that no aggregate particles more than ¼ inch in cross-section are exposed at the finished surface.

- In no event should surface grinding result in exposure of aggregate of more than one-half the diameter of the maximum size aggregate.

4.3.3.4 *Surface patching*

Patching refers to the restoration of relatively small areas of damaged concrete to the profile of the surrounding concrete. This technique has been used for many years and is typically used to restore concrete sections damaged by delaminating and spalls. It is also an appropriate repair method when the concentration of aggressive chemicals is low to moderate and when the areas of visible deterioration are relatively few (Newman 2001). The disadvantage of surface patching is that the appearance of the repaired surface often becomes a problem if too many small patches are installed, as shown in Figure 4.9.

Similar to other repair techniques, the surface preparation of the substrate is critical to a successful patching. The boundaries of the areas requiring patching should be clearly defined with a saw cut to a depth of about 6 mm and contaminated and unsound concrete within this area should then be removed. For easy and proper patching, the patch geometry should be simple, with minimal edge length and perimeter. Disc cutting at right angles to the surface (to a depth of 5–25 mm) is usually employed to achieve the required geometry. Feather edges should be avoided when cementitious

Figure 4.9 Many small patches applied on a bridge deck.

repair are used, but resin-based materials, such as epoxies which have considerably better bonding characteristics, can be feather-edged.

The repair systems normally include bonding agents. The appropriate bonding agents include sand/cement, cement/latex slurries and epoxies. These bonding agents help in reducing loss of moisture from the repair and also improve the bond between the new and old concrete. Certainly, the use of bonding agents increases the cost of repair, with cement slurry being the cheapest and epoxy the most expensive. The bonding agent should be applied soon after surface preparation is completed. The repair system may also include reinforcement primer, polymer modified mortar, pore filler levelling mortar and protective coatings. When choosing repair materials, normally the mechanical and transport properties of the repair material should resemble as closely as possible those of the surrounding concrete. Portland cement mortars and grouts are the most frequently used materials for surface patching because of their good compatibility with the existing concrete.

Latex modified mortar and epoxy mortar are often substituted for cementitious patching materials when a fast cure time, higher bond strengths, and feather edging are required. The addition of polymers to repair mortars improves their properties in different stages. For the mix design of the repair concrete, when polymers are used, the water dosage needed can be reduced, and thus also does the long-term shrinkage of the repair mortar. For the interface bond, the polymers help to improve the bond between the repair mortar and the substrate, and they can increase flexural and tensile strengths of the modified mortars and reduce permeability to the diffusion of moisture and carbon dioxide (Kay 1992).

Rapid setting materials based on magnesium phosphate and high alumina cement are also used to patch damaged flat structures, especially when rapid temporary repairs are necessary. The current process to prepare repair materials is that all of the repair materials are provided in a packaged form as a system from one manufacturer. As far as mortars are concerned, preblended dry sand and cement may be supplied in one package, and the latex and the appropriate quantity of water in another. The ingredients are simply mixed together at the construction site to produce the repair mortar.

There are several steps in the patching repair process:

1   *Cut out the area to be repaired.* The first step of a surface patching repair is to cut a groove around the anticipated edges of the repair. The edges of deteriorated areas should be cut perpendicular to the surface to a depth of at least 10 mm (Mailvaganam 1992). The shape of the repair should be delineated by straight lines and abrupt changes in width and depth should be avoided. The main surface-breaking work is usually undertaken by pneumatic or electric tools. Cutting by high pressure water jet is an alternative for surface breaking which causes less damage to the substrate and less noise. When heavy tools are used for surface preparation, the damage in the form of loosened aggregates should be

removed with lighter tools. In the case of reinforced concrete with corroded reinforcement steel, concrete is usually cut out beyond the reinforcement steel to give a clear gap of approximately 20 mm below the steel bar, so that the backs of the reinforcement bars can be properly cleaned during the preparation of surface patching repair. These exposed reinforcement bars are then cleaned by grit blasting to satisfy the specified qualities of surface preparation. Sometimes, grit blasting followed by washing with fresh water and further blasting is needed in order to get a clean surface for patching. In severe cases it may be necessary to cut out the corroded reinforcement steel and install a new bar as a replacement. The new steel bar should be grit blasted to the same high standards as the existing reinforcement within the repair. Once the reinforcing steel has been prepared by grit blasting, the first coat of reinforcement primer should be applied within 3 hours and it should totally encapsulate the exposed reinforcement including locations where bars cross (Kay 1992). A good surface patching is shown in Figure 4.10(a), and a good quality of bonding between patched and old concrete is shown in Figure 4.10(b) by a core taken out of the patch.

2   *Surface preparation.* The next step is to prime the surface of the concrete in the repair area. If cementitious repair mortars are used, the substrate should be thoroughly soaked with clean water before applying a bonding aid. When the bonding aid is applied, the surface should be damp but free of standing water. When polymer mortars are used as repair material, the concrete surface must be soaked with water. After that the first layer of repair mortar may be applied to the concrete surface. For cementitious mortars, the mortars must be applied when the bonding aid is wet. For polymer mortars, normally the bonding aid should reach a tacky condition before the application of repair mortars. The repair mortars have to be painted by hand, layer by layer, so that the repair mortars are in good compaction and uniform density. The thickness of individual repair mortar layer depends on the repair material as well as the orientation of the surface being repaired, typically 25–50 mm for vertical repairs and 20–30 mm for overhead repairs with normal weight mortars (Kay 1992). In order to obtain a good bond to the next layer, the surface of intermediate layers is usually left in a rough condition. As a general rule, the following layer is painted when the previous layer has stiffened sufficiently to carry the applied weight but before final setting. In cases when the installation of the following layer is delayed, the surface of the previous layer should be scratched rough and damped with water and a bonding aid should be applied before the following layer is painted. It is recommended that the final layer should be at least 10 mm thick and finishing work is required to match the surrounding concrete. The finishing on flat areas can be achieved by a wood or steel trowel to give the required texture. After completion of trowelling, the repaired

Figure 4.10 (a) A good surface patching on a concrete slab; (b) a good quality of bonding between patched and old concrete.

areas must be either sprayed with a resin-based curing membrane, or covered with polyethylene sheets held down around the edges and kept in position for 4 days (Perkins 1986).

3 *Curing for the repaired concrete.* The last step is curing the newly cast repair concrete. The curing time and condition depend on the properties of the repair materials. Cement-based repairs usually do require some

initial protection during curing. When ordinary Portland cement is used, all traffic should be kept off the repaired areas for at least 7 days; this can be reduced to 2 days if rapid hardening Portland cement is used. In cold weather, these periods normally should be extended by about 50 percent (Perkins 1986). If emergency repairs have to be carried out, high alumina cement can be used (Perkins 1986). Sprayed-on curing membranes cannot be used between layers of a multilayer repair because they prevent bonding between layers. Since the curing of resin-based repair materials involves heat evolution, i.e. the faster the curing, the greater the rate of heat evolution. The build-up of high thermal stresses is possible for large-scale repair work, and in this case, the thermal stresses will be released when the repair material cools down. As stated before, spray-on curing membranes or sheets of polyethylene tapes are helpful in protecting resin-based repair materials from cold, heat and rain. For the similar reason, a large volume of resin-based materials should not be mixed and applied in one batch.

4.3.4 Repair of cracks in concrete

There are usually some cracks on the surface of a concrete structural member, such as a concrete wall or a reinforced concrete beam. If the crack width is very small, the crack may not have to be repaired. However, when the crack width becomes large, it is very important to evaluate the effect of the crack on structural performance and durability of the structure, and repair the crack if necessary. Depending on the location and orientation of a crack in a structural member, a large crack can significantly reduce the load-carrying capacity of the structure. For example, if a crack significantly reduces the strength of anchor bolts embedded in concrete, it should also be repaired. In general, a professional structural engineer should be consulted for the structural evaluation. In addition to the assurance for load-carrying capacity, crack repair is also necessary for long-term durability concerns of the structure, such as corrosion protection and bond strength protection. Therefore, crack width is an important factor in making a decision as to whether the crack should be repaired and how to repair.

As a general rule, cracks in reinforced concrete wider than those allowed by building codes or specifications should be repaired or sealed. A number of codes of practice specify maximum permissible crack widths for various service conditions. Tolerable crack widths according to varying exposure conditions specified by ACI 224R-90 are shown in Table 4.8 (Emmons 1994).

Under loading, the crack width in concrete members depends on loading conditions. In the latest version of the ACI code (ACI 224.1R-07), the following formula was given for calculating the maximum crack width under bending:

Table 4.8 Exposure conditions and tolerable crack width, ACI 224R-90

Exposure condition	Tolerable crack width	
	(in)	*(mm)*
Dry air, protective membrane	0.016	0.41
Humidity, moist air, soil	0.012	0.30
De-icing chemicals	0.007	0.18
Sea water and sea water spray; wetting and drying	0.006	0.15
Water-retaining structures	0.004	0.10

$$w = 2 \frac{f_s}{E_s} \beta \sqrt{d_c^2 + \left(\frac{s}{2}\right)^2} \qquad (4.3)$$

where

w = maximum crack width, in. (mm);

f_s = reinforcing steel stress, ksi (MPa);

E_s = reinforcing steel modulus of elasticity, ksi (MPa);

β = ratio of distance between neutral and tension face to distance between neutral axis and centroid of reinforcing steel (taken as approximately $1.0 + 0.08d_c$);

d_c = thickness of cover from tension face to center of closest bar, in. (mm); and

s = bar spacing, in. (mm).

For structural members under direct tension, the following formula has been suggested by ACI 224.1R-07 to estimate the maximum crack width:

$$w = 0.1 f_s^3 \sqrt{d_c A} \times 10^{-3} \qquad \text{(in.-bl)} \qquad (4.4)$$

$$w = 0.0605\, f_s^3 \sqrt{d_c A} \times 10^{-3} \qquad \text{(SI)} \qquad (4.5)$$

According to an NCHRP Study (NCHRP Report 380) and various FHWA publications, the acceptable crack width from a corrosion and durability standpoint is between 0.004 in and 0.008 in (0.1 and 0.2 mm) (Xi *et al.* 2003). British Standard 8110 suggests that the maximum crack width of 0.3mm may be acceptable or aesthetically acceptable. ACI 318 (1995), *Building Code Requirements for Structural Concrete*, has limited the acceptable concrete crack width to 0.013 in (0.33 mm) for an exterior crack and 0.016 in (0.41 mm) for an interior crack.

A fundamental weakness of this approach lies in the fact that the crack width at the surface of the concrete will always be greater than the interior cracks, or greater than the width at the reinforcement, which may not always be the case. The difference will depend largely on the loading conditions and the thickness of cover.

Furthermore, cracks may be considered unacceptable in the following two situations:

1 *Aesthetically unacceptable.* In this case, since crack widths are usually very small, care must be taken as inadequate crack repair techniques often result in poor appearance upon completion (as shown in Figure 4.11).
2 *Non-watertight for the structure.* Some structures require watertightness, such as reinforced concrete water tanks. Small cracks on the surface of concrete result in accelerated water penetration into the concrete, which reduces the service life of the structure and more importantly leads to possible leakage of the structure.

There are many crack repair methods as listed in ICRI Guidelines and ACI documentations, such as epoxy injection, gravity filling, routing and sealing, near-surface reinforcing and pining, grouting, drilling and plugging as well as crack arrest. These methods may be grouped into three basic types (Newman 2001):

1 "Glue" the cracked concrete back together by epoxy injection or grouting.
2 "Stitch" the cracked concrete with dowels.

Figure 4.11 Appearance of a repaired concrete by improper crack repair techniques.

3 Enlarge the crack first and then "caulk" it with a flexible or semi-rigid sealant.

The choice of repair methods for cracks should take into account several factors, such as the cause and extent of cracking, the present condition of the crack, the location and environment of the cracks, and whether the crack is still actively moving or not. Typical examples of live cracks are those in movement joints. For example, cracks under wet-dry, industrial and marine conditions will require materials and methods quite different from those found acceptable for cosmetic repairs. Techniques that can be successfully used on horizontally cracked surfaces may not be effective on vertically cracked surfaces. For a crack induced by a one-time load application and which has stopped spreading can be repaired right away. For the cracks induced by shrinkage or settlement, the repair may be delayed so that further deformation can be minimized. In the case of live cracks, one must be aware that completely filling those cracks by injection will always lead to new cracking within the crack filler, on the interface with the cracked concrete, or within the old concrete (Bijen 2003).

Repair of dead cracks wider than about 1 mm can usually be achieved by cement grouting although some traces may be left. Finer dead cracks are to be sealed by the injection technique (Allen *et al.* 1993). Inactive cracks can be repaired by epoxy injection or by sealing with mortar. For cracks which are deep or pass right through the member, crack injection can provide a satisfactory solution.

4.3.4.1 Resin injection

Resin injection is used to repair concrete that is cracked or delaminated and to seal cracks in concrete to stop water leakage. Because of the high costs (generally about $200.00 per linear foot of injected crack), resin injection is not normally used to repair shallow, drying shrinkage, or pattern cracking. Two basic types of resin and injection techniques are used to repair cracked concrete.

(A) EPOXY RESINS

Crack repair usually adopts the injection technique, because the resin used for the repair of cracks has high mechanical strength and resistance to most chemical environments encountered by concrete. Two methods of mixing and injection of resins for crack repair are commonly used. In the first method, the components are combined in a separate mixer and then transferred to the application equipment. In the second method, the components are pumped separately in carefully controlled proportions to a mixing head where mixing takes place automatically as the two components flow together. The latter method is suitable for large volume work in the situation where continuity of the process over a long period is assured, while the former method is more suitable for a small repair job where the work is intermittent (Kay 1992).

There are wide variations in the properties of different resins and it is therefore necessary to carefully choose to match the individual job requirements in relation to such matters as temperature of application, capability of bonding to moist concrete, shrinkage, thermal and elastic properties of the hardened resin and other special needs such as fire-resistance and high temperature stability (Campbell-Allen and Roper 1991; Trout 1997). ASTM C-881 provides more details on the types of resin that can be used for repair work. Also, the correct formulation of the resin is of vital importance and the requirements for the repair resin are likely to vary from job to job. The desirable qualities for repair resin include low viscosity, ability to bond to damp concrete, suitability for injection in as wide a temperature range as possible, low shrinkage, and toughness rather than high strength (toughness is the area under a stress–strain curve of the material) (Perkins 1986).

Resin injection can be used to fill cracks and to bond together the concrete surfaces. The crack must be cleaned prior to the injection. Appropriate cleaning of the crack can be achieved by vacuum cleaning, or flushing of water or other solvents. When reinforcement rusting and concrete spalling have occurred, the best repair method may be removal of the defective concrete, cleaning of reinforcement, resin injection and then repair by mortars. The consistency of the materials used for crack repair must be adequate for the resin to penetrate into cracks and to provide durable adhesion to surfaces of cracks. Fluid resins are the most widely used repair materials for cracks. Important properties of any injection resin are its resistance to moisture penetration, to alkaline attack, a similar tensile strength to concrete and a similar modulus of elasticity with concrete. The tensile strength of the repair material should approach that of concrete as closely as possible.

Before a resin injection can begin, the crack that appears on the surface of concrete member must be sealed, so that the liquid resin will not leak and flow out of the crack. During the injection process, epoxy resin is injected into cracks and micro-fissures in concrete and into voids between concrete and reinforcing steel. The bond between the epoxy and the concrete should be stronger than the concrete itself, so the repaired concrete should be as good as new concrete. It has been found that the injection of a low-viscosity epoxy is a possible repair method for cracks between about 0.02 mm and 6 mm in width (Warner 1977). For a crack width smaller than 0.02 mm, the epoxy cannot flow into the cracks. For small cracks, compressed air and solvents can be used for the cleaning work and for removal of water in the crack.

(B) POLYURETHANE RESINS

Polyurethane resins are used to seal and eliminate or reduce water leakage from concrete cracks and joints. They can also be injected into cracks that experience some small degree of movement. Such systems, with the exception of the two-part solid polyurethanes, have relatively low strength and should not be used to structurally re-bond cracked concrete. Cracks to be injected with polyurethane resin should not be less than 0.005 inch in width.

Polyurethane resins are available with substantial variation in their physical properties. Some of the polyurethanes cure into flexible foams. Other polyurethane systems cure to semi-flexible, high density solids that can be used to re-bond concrete cracks subject to movement. Most of the foaming polyurethane resins require some form of water to initiate the curing reaction and are, thus, a natural choice for use in repairing concrete exposed to water or in wet environments.

Polyurethane resin used for crack injection should be a two-part system composed of 100 percent polyurethane resin as one part and water as the second part. With appropriate water to resin mixing ratios, the resulting cured resin foam can attain at least 20 psi tensile strength with a bond to concrete of at least 20 psi and a minimum elongation at tensile failure of 400 percent. The manufacturer's certification that the product meets these minimum requirements should be required before the injection resins are accepted for use on the job (ADWR 2006).

(C) INJECTION EQUIPMENT

Resins can be injected using several types of equipment. Small repair jobs employing epoxy resin can use any system that will successfully deposit the epoxy in the required zones. Such systems could use a pre-batch arrangement in which the two components of the epoxy are batched together prior to initiating the injection phase with equipment such as small paint pressure pots.

Large epoxy injection jobs and polyurethane resins injection jobs generally require an injection pump in which the two epoxy components are pumped independently of one another from the reservoir into the mixing nozzle. At the mixing nozzle, located adjacent to the crack being repaired, the two epoxy components are brought together for mixing and injecting. The epoxy used in this injection technique must have a low initial viscosity and a closely controlled set time. Several companies have proprietary epoxy injection systems, which allow them to make satisfactory repairs under the most adverse conditions. Contact information of these companies can be obtained from the Materials Engineering and Research Laboratory, Code D-8180, Denver, Colorado.

(D) APPLICATION

Sealing is commonly achieved by providing a surface dam which completely bridges the crack. Resin injection requires a high degree of skill for satisfactory execution and is therefore normally carried out by specialist contractors. As a general principle in sealing cracks by injection, it is suggested to start at one end and work progressively along the crack. A series of injection points are formed at intervals along the length of the crack and grout is injected into each point in turn until it starts to flow out of the next one. The point in use is then sealed off and the injection equipment can be moved to the next point, and so on until the full length of the cracks has been treated. Complete filling is assured when resin appears at all the injection points and vents. Sometimes, re-injection is necessary, because the injected resin may flow from the main crack into fine capillaries. This is especially true for high-pressure injection. Re-injection has to be accomplished prior to hardening of the previously injected resin. For large structural elements or structural elements with deep cracks through the structural thickness, both sides of the elements should be injected by resins in order to successfully repair the cracks.

(E) QUALITY CONTROL AND EVALUATION

As a quality control procedure, concrete samples can be cored to assess the effectiveness of the repair in achieving desired objectives (see Figure 4.12). Coring concrete samples should be done through the crack plane. The concrete samples are indicators for evaluating the depth and completeness of resin penetration into the crack. The injection quality can also be checked by

Figure 4.12 A concrete core is taken from a bridge deck.

ultrasonic method. To conduct an ultrasonic test, one transducer can be placed at one side of a crack, and the other transducer on the other side, as shown in Figure 4.13. To evaluate the repair effectiveness, ultrasonic signal travel time between the locations of the two transducers can be measured prior to the repair, which provide baseline data and can be compared to those obtained after the injection. The baseline data of the signal travel time obtained prior to the repair should be larger than those obtain after the repair, because the repair work is supposed to heal the crack and the concrete between the two locations should become one solid piece after the repair (the ultrasonic signal travels faster in a solid without any cracks and voids). More than one location along the crack length should be tested for the signal travel time. The test data will reflect the average quality of the repair work.

4.3.4.2 Stitching

Instead of gluing a crack together, repair can be done by stitching a crack together as shown in Figure 4.14, in which the crack is stitched by iron or steel dogs. Stitching can be considered as one of the crack arrest techniques. Other crack arrest techniques are described elsewhere, such as ACI 224.1R-07. When stitching a crack, a series of stitches are sufficient to make the total tensile strength of the repaired concrete equal to or greater than the

Figure 4.13 Ultrasonic testing of concrete cracking before repair.

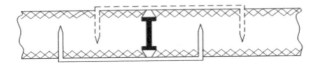

Figure 4.14 Repair of a crack by stitching of concrete.

tensile strength of the original concrete. The best method of stitching is to bend bars into the shape of a broad flat-bottomed letter U between a foot and three feet long and with ends about 6 inches long or less if the thickness of the concrete is less. Before stitching, clean and seal the crack, and use a flexible seal for active cracks. Drill holes on both sides of the crack. The holes should not be in a single plane and the spacing should be reduced near the ends of the crack (because of the stress concentration at the crack tip). Clean the holes and anchor the legs of the dogs in the holes with a non-shrink grout or an epoxy. The stitching dogs should vary in length and orientation to prevent transmitting the tensile forces to a single plane.

The following considerations should be made when using stitching: Stitch both sides of the concrete section where possible to prevent bending of the stitching dogs. Bending members may only require stitching on the tension side of the member. Members in axial tension must have the stitching placed symmetrically. Stitching does not close the crack, but can prevent its propagation. Stitching that may be placed in compression must be stiffened and/or strengthened to carry the compressive force, such as encasement of the stitching dogs in a concrete overlay. As an alternative to stitching at right angles sometimes it is necessary to use diagonal stitching since stitching at right angles does not resist shear along the crack. One set of dogs can be placed on each side of the concrete if necessary (Champion 1961).

4.3.5 Repair of concrete damaged by fire

4.3.5.1 Fire damage in reinforced concrete structures

Generally, concrete is recognized as one of the most excellent thermal-resistant building materials. Compared with other construction materials, concrete suffers less from a fire disaster, therefore concrete has been used in buildings and industrial facilities, nuclear reactor containment structures, and nuclear waste containers as a fire-resistant material. To determine the required concrete thickness for fire resistance and to determine if cracking or spalling is likely to occur in the situation of fire exposure, an understanding of the high-temperature behavior of concrete is necessary for designers.

Fire usually causes damage to a large area of concrete on beams, columns and soffits of slabs. In the case of floor slabs and beams, it is the soffit which is normally the most affected. For reinforced concrete structures, the reinforcement may also be affected. For instance, it is often found that, after fire, some of the reinforcements in the slabs buckle due to the relative smaller depth of the cover. This reinforcement must be replaced.

The effect of fire on concrete depends on the temperature reached, the exposure period in fire and the characteristics of concrete itself in terms of cement type, water–cement ratio, cement content, aggregate type, and the thickness of concrete cover to the steel reinforcement. The structural damage of concrete structure caused by fire may be (1) spalling; (2) strength reduction in concrete and steel; (3) loss of anchorage of reinforcement; (4) excessive deflection of beams and slabs; and (5) distortion of the whole structural framing. The latter three types of damage may be so severe that demolition and replacement are sometimes the only possible solution (Campbell-Allen and Roper 1991).

Generally, the strength of concrete is significantly affected by extreme temperatures occurred during a fire. Figure 4.15 shows the temperature-strength curves of concrete. In Figure 4.15, all test data were residual strengths of concrete measured after the exposure to different high temperatures (Lee *et al.* 2008). The test data were then divided by the strength of concrete under room temperature, so they are relative residual strengths. The three curves represent the residual strength of concrete cooled by different methods after the high temperature exposures. One can see that water cooling resulting in the highest reduction in the relative residual strength in all temperature ranges except 800 °C, which implies that if a fire in a structure was extinguished by using large amount of water, the damage in the concrete is higher than that in a structure under fire and then cooled down naturally. The same phenomena can be observed in Figure 4.16 for the reduction of stiffness of concrete exposed to different high temperatures (Lee *et al.* 2008). It is important to emphasize that using water to extinguish fire is a commonly used method, and we do not mean this method should

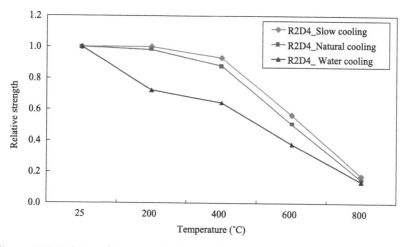

Figure 4.15 Relative ultimate residual strength vs. maximum temperature.

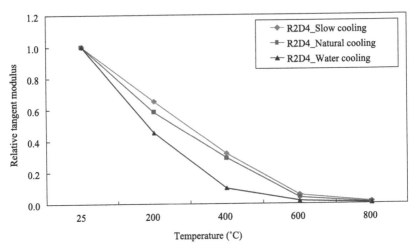

Figure 4.16 Relative initial tangent modulus vs. maximum temperature.

not be used. One should realize the consequence of water cooling, which leads to high thermal gradient in concrete members and thus high thermal stress.

Figure 4.17 shows the stress–strain curves of concrete exposed to different high temperatures, which indicates the deterioration of concrete in the full range of strains, including ascending and descending portion of the curve. The degradation of concrete under fire depends heavily on the type of aggregate used in the concrete.

Heating rate is another important factor to consider when evaluating fire

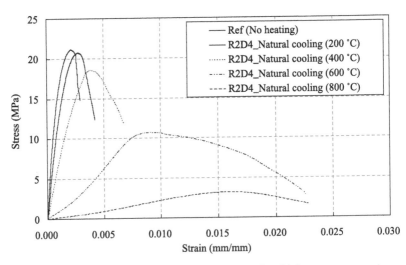

Figure 4.17 Stress–strain curves of concrete exposed to high temperatures (natural cooling).

(a) Heating rate: 2 °C/min. (b) Heating rate: 15 °C/min.

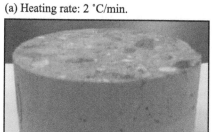

Figure 4.18 Concrete specimens exposed to maximum temperature 800 °C and cooled down by natural cooling.

damage. Figure 4.18 shows two damaged concrete specimens by fire with the same maximum temperature 800 °C and the same natural cooling method. One can see clearly that the one with slow heating rate of 2 °C/min. exhibits less surface cracks and smaller crack width than the other one with the heating rate of 15 °C/min. Under high heating rate, spalling damage of concrete occurs in the temperature range below 350–450 °C. Some severe fire disasters showed that progressive spalling of concrete may occur during a fire, when the concrete spalling takes place in the surface layer first and progresses into deeper part of concrete. As a result, the load carrying capacity of the structure can be severely reduced.

High temperature results in significant damage to concrete as well as to steel. Therefore, the repair for fire damage of concrete structures may necessitate the addition of new reinforcing steel and link the new steel to the existing steel.

It is important that a realistic assessment of the damage is made as soon as possible after the fire, since an immediate inspection will make it easier to evaluate the maximum temperatures reached in different parts of a concrete structure during a fire and the temperature gradient through the structural members. Maximum temperature reached in the structure can be estimated reasonably well by a careful examination of the debris after the fire. This can be compared with the temperature distribution and strength loss determined from test evidence. The detailed quantitative investigation using various tests is to determine the temperatures to which the member has been subjected and the loss of strength of the concrete as well as the reinforcement.

There are three main tests which could be carried out in order to determine strength loss in concrete: (a) color determination; (b) Schmidt hammer; and (c) core testing. Change in the color of concrete resulting from heating may be used to indicate the maximum temperature attained and the equivalent fire duration. In many cases at above 300 °C a pink discoloration may be readily observed. Pink or red spots are often observed on concrete exposed to a maximum temperature of 400 °C and above. Depending on the cooling method, the color of concrete subjected to slow and natural cooling is generally pink or red, while the color of concrete subjected to water cooling was dark pink or dark red. The onset of noticeable pink discoloration is important since it coincides approximately with the onset of significant loss of strength due to heating. Therefore, any pink discolored concrete should be regarded as being suspect.

There are various books and industrial codes for performance and requirements of reinforced concrete structures under high temperatures (Bazant and Kaplan 1996; Phan 2004; Felicetti *et al.* 2004), such as ACI 216.1-07/TMS-0216-07, *Code Requirements for Determining Fire Resistance of Concrete and Masonry Construction Assemblies*; and ACI 349.1R-07, *Reinforced Concrete Design for Thermal Effects on Nuclear Power Plant Structures*. Figures 4.19, 4.20, and 4.21 depict the strengths of various types of concretes and steels under high temperatures specified by ACI 216.1-07. Table 4.9 shows the temperature dependence of hot-rolled reinforcing steel specified in Eurocode3 (ENV 1992-1-2, 1995a; ENV 1993-1-2, 1995b). The stiffness and strength values have been normalized by the respective values at reference temperature (i.e. room temperature).

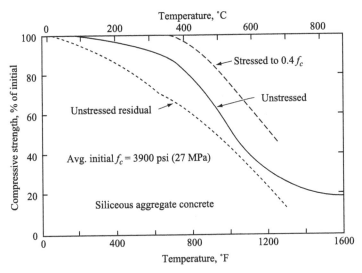

Figure 4.19 Compressive strength of siliceous aggregate concrete at high temperature and after cooling (ACI 216.1-07).

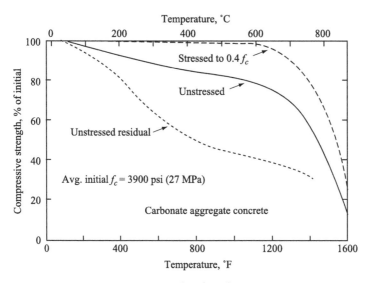

Figure 4.20 Compressive strength of carbonate aggregate concrete at high temperature and after cooling (ACI 216.1-07).

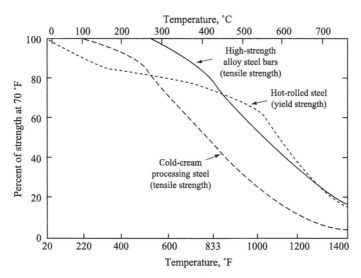

Figure 4.21 Strength of flexural reinforcement steel bar and strand at high temperatures (ACI 216.1-07).

4.3.5.2 *Repair methods of fire-damaged concrete*

Repair of fire-damaged reinforced concrete structures requires restoration of any loss in strength, durability, and fire resistance of concrete and steel. In situations where there is still sufficient strength and durability left for the

Table 4.9 Hot rolled reinforcing steel (0.2%), ENV 1992-1-2 (1995a)

Temperature (°C)	$E(T)/E(20\,°C)$	$\sigma_{pr}(T)/\sigma_{0,2}(20\,°C)$	$f_y(T)/\sigma_{0,2}(20\,°C)$
20	1.00	1.00	1.00
100	1.00	1.00	1.00
200	0.90	0.81	1.00
300	0.80	0.61	1.00
400	0.70	0.42	1.00
500	0.60	0.36	0.78
600	0.31	0.18	0.47
700	0.13	0.07	0.23
800	0.09	0.05	0.11
900	0.07	0.04	0.06
1000	0.04	0.02	0.04
1100	0.02	0.01	0.02
1200	0.00	0.00	0.00

materials, a thin hand- or spray-applied repair material could be used to restore the structure. The commonly used repair materials for a fire-damaged concrete structures are cementitious mortars, epoxy resins-modified mortar or concrete. Caution should be used when epoxy-based repair materials is applied for the repair of fire-damaged concrete because the epoxy rapidly loses its strength if heated above 80–100 °C. Consequently, the most commonly used materials for the repair of fire damage are cementitious mortars and concrete. The repair of the fire-damaged concrete structure normally consists of the following stages:

1   Remove unsound concrete.
2   Clean reinforcing steel and install new steel bars if needed.
3   Clean and roughen the exposed concrete surface.
4   Replace removed concrete.

The removal of unsound concrete and the roughening of the surface are usually carried out by hand tools and light pneumatic tools, or by sand blasting. The removal of concrete, cleaning and roughening of the new surface, and the cleaning of the reinforcing bars can also be done in one step by water jetting. After removing the cover concrete, the reinforcing steel is exposed. The rust on the corroded rebar can be removed by sand blasting device, hammers, or wire brushing. It is always a good idea to encapsulate the cleaned reinforcing steel in an alkaline coating prior to restoration of the concrete cover. If there is a significant loss of steel section, new steel bars should be used to replace the corroded bars. If the reinforcement steel has buckled after fire damage, it must be replaced.

Three methods may be used for replacement of removed concrete: (a) recasting in formwork; (b) spraying or shotcrete; and (c) hand-applied mortars. The choice of the repair method is usually determined by practical and cost considerations. Spraying and recasting are more suited for large volumes and large area applications, and when a speedy structural repair is required. However, if a high standard of surface finish is required, recasting is more preferred. The formwork must be well sealed against the existing structure, rigidly and firmly fixed. A high workability concrete should be used in order to have good placing and compacting of concrete. To replace concrete surfaces damaged by fire, shotcreting or guniting is the most frequently used technique. This method produces a dense concrete and no additional compaction is required after repair. Provided that the repairs are properly carried out, the shotcreting or guniting provides a high quality repair for fire-damaged concrete. The thickness of the sprayed concrete is largely dependent on the thickness of the original concrete cover needed for protecting reinforcing steel.

It should be noted that the sprayed concrete may be applied over areas of undamaged concrete, and the surface of this must also be prepared by the methods aforementioned to be sound, rough and homogeneous. All traces of coatings must be removed since they would prevent a strong bond between the sprayed concrete and the existing concrete. It is always necessary that the sprayed concrete should incorporate a welded steel fabric, or steel or other fibres, to increase the bond between the existing concrete and the new concrete and to minimize the risk of cracks which allow the penetration of water into the concrete.

Hand-applied mortars are more suitable for patch repairs and repair jobs of less volumes. Resin-based repair materials are normally applied by hand for small area repair. Before applying these mortars, the damaged concrete surface should be well treated by the methods described above. A thin layer of slurry grout coat of cement/latex can be applied first by brush to the prepared concrete surface. The repair mortars, normal cement-based or resin-based, are then placed on the coating layer by brush while the coating layer is still tacky. The repair should be built up in layers if the depth of the damaged concrete is large. The surface of the previous layer should be furrowed before the subsequent layer is applied. The final surface should be finished to match the surrounding surfaces.

4.3.6 *Repair of corrosion-damaged reinforced concrete structures*

In industrial countries, the total cost of repairing metallic corrosion related deterioration is believed to be in the region of 4 percent of GNP. Therefore, how to repair corrosion-damaged concrete structures has become an important topic.

For the onset of steel corrosion, there are four necessary conditions: enough moisture, air, low pH value, and enough ions (usually chloride ions from deicing salt) to form electrolyte. As one can see, moisture and air cannot be blocked from entering concrete. So, effort has been focused on how to maintain high pH values and how to slow down the penetration of chloride ions into the concrete. High pH values in concrete mean high alkalinity in the concrete pore solution. Steel is passive under high alkalinity environment, and therefore corrosion will not occur. Then, there are two ways to trigger the onset of corrosion. One is by the penetration of carbon dioxide from the environment into concrete and the other is the penetration of water containing dissolved salts through the concrete cover or through a concrete crack. In the first case, the alkalinity of the concrete surrounding the steel could be reduced by the atmospheric carbon dioxide which reacts with calcium hydroxide in cement paste to form calcium carbonate (called carbonation of concrete). The reduction of calcium hydroxide leads to a low pH value. This creates an environment for the corrosion of steel to take place. In the second case, the alkalinity of concrete is not reduced, but when the chloride ion concentration is high enough, reaching a certain ratio with the hydroxyl ions (Cl/OH), the corrosion of steel may start.

The required treatments for restoring the protective environment for steel depend on the extent and cause of the corrosion damage:

1   *Carbonation-induced corrosion damage.* Under such conditions, carbonated concrete should be removed and new concrete should be installed, re-passivation is provided by the new repair mortar or concrete.
2   *Chloride-induced corrosion damage.* Under such conditions, if chloride has penetrated to the level beyond the steel reinforcements, removal of chloride around steel bars does not guarantee re-passivation as chloride ions may diffuse back from the deeper part of the concrete to the new concrete cover. This is the so-called redistribution of chloride after the repair. In this case, the repaired concrete will become cathodic and the rebar will be the anode. The corrosion will occur in the bars immediately. Other factors may influence the re-passivation of steel, for instance, coating of the steel reinforcements, and the application of membranes or sealers to limit the moisture content.

4.3.6.1 Restoration of concrete surrounding corroded bars

The restoration of the corrosion protection system of the reinforcing steel can be accomplished by concrete, cement mortar or epoxy resin mortar. If the removal of concrete is selected, care must be taken to prevent the chloride redistribution. It is better to remove more concrete than needed to ensure all the chloride-contaminated concrete and all the carbonated concrete has been removed. Figure 4.22 shows the removal of the concrete under corroded steel bars by using hand tools. After the distressed concrete is removed, the condition of the rebars must be checked. For uncoated steel bars, the rust on the surface must be cleaned, and the fresh metal surface must be exposed, as shown in Figure 4.23 (a sensor is shown in Figure 4.23

Figure 4.22 Removal of the concrete under corroded steel bars using hand tools.

Figure 4.23 Steel bars after cleaning operation.

which was used to monitor the corrosion process). For epoxy coated bars, an inspection must be made to make sure the epoxy coating is not severely damaged. If there are damages and scratches as shown in Figure 4.24, the damaged areas on the coating must be repaired first by epoxy resins. Otherwise, the corrosion of steel bars will start from the damaged areas.

The choice of the repair system depends on the thickness of concrete cover. For a repair with concrete, the use of sprayed concrete (shotcrete) is advisable. For restoration of the corrosion protection system with cement mortar, it is better to use polymer-modified cement mortars with high quality mix as the repair material, which can be placed in individual layers whose thickness does not exceed 5 mm. The water–cement ratio for the repair cement mortar is normally less than 0.4. If an adequate thickness of concrete cover cannot be achieved by placing the polymer modified cement mortar, an additional sealing of the concrete surface or a protective coating on the reinforcing steel will be necessary to prevent corrosion in the future.

To make a high quality repair for corrosion protection, the water–cement ratio should be kept low. A low water–cement ratio will make concrete of reduced diffusivity. For Portland cement concrete, depth of carbonation decreases almost linearly with the water–cement ratio. The diffusion coefficient of chloride ions decreases with decreasing water–cement ratio of concrete or cement mortar. At the same water–cement ratio, the chloride ion diffusion coefficient decreases when cement content increases.

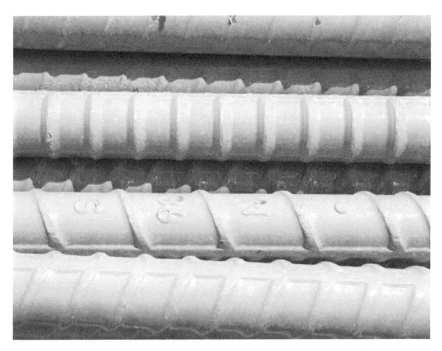

Figure 4.24 Damages and scratches on the epoxy coating of new steel bars.

The addition of pozzolanic materials such as fly ash and/or silica fume into concrete also helps improve the quality of repair concrete or cement mortar since these pozzolanic materials result in a low diffusion coefficient of the cementitious repair materials. Berke *et al.* (1988) have found that, at up to 10 percent addition by weight of cement with silica fume, the concrete exhibited significant improvements in properties directly affecting steel corrosion resistance. Compared with conventional concrete mixes, such concrete has much lower diffusion coefficient mainly due to the reduction of porosity. Sufficient curing is also of vital importance in improving quality of concrete and cement mortar, since insufficient curing will reduce durability against both carbonation and chloride ingress, and thus the ability of protection of reinforcing steel from corrosion.

Epoxy resin mortars may be useful for small repairs with thin layer thicknesses. These mortars are able to prevent carbonation or the access of corrosive agents. Prior to restoration of the corrosion protection system with epoxy resin mortar, the concrete surface should be dry and the corroded reinforcing steel should be cleaned and should receive two coats of paint consisting of epoxy resin with active corrosion-inhibiting agents.

4.3.6.2 *Cathodic protection methods*

Corrosion is an electrochemical process as discussed in Chapter 2. Anodic and cathodic areas develop on corroding materials or systems, with measurable potential differences between the anodes and the cathodes. Cathodic protection, an electrical system designed to stop corrosion by applying an electric current to the affected metal surfaces, has been extensively used to protect steel pipelines and tanks from corrosion for many years and has been applied in the protection of the reinforcing steel in concrete in recent years. The principle of the cathodic protection (CP) is to enforce the rebar to become a cathode and another metal as an anode. Since corrosion only occurs at the anode, the rebar will be protected. It should therefore be possible to reduce corrosion rates if the reinforcing steel under consideration can be shifted to a cathodic condition by some externally applied potential. This is the basic theory behind cathodic protection techniques. For practical applications, there are two basic CP systems: the surface-mounted anodes (or sacrificial anode) system and the impressed current system.

(A) THE SACRIFICIAL ANODE SYSTEM

Sacrificial anode cathodic protection is a very effective system for many reinforced concrete structures where the concrete is often in a wet condition and electricity resistivity is low. In this method, the structure or the structural member to be protected is connected to a more reactive metal as shown in Figure 4.25.

Figure 4.25 A sacrificial anode cathodic protection system, called Galvashield cathodic protection system was installed in SH 85 SB in Greeley, Colorado.

Anodes of magnesium or zinc can be used in this way to provide protection to buried reinforcing steel. In this method, no additional power supply is needed, the small potential difference between the cathode and anode is sufficient to drive the electron and ion flow. The anodes corrode and are consumed in the process of providing protection and, for this reason, they are known as sacrificial anodes.

(B) THE IMPRESSED CURRENT SYSTEM

An external direct current supply can be installed as shown in Figure 4.26(a) for a concrete slab (such as a bridge deck). This type of CP system is called the impressed current CP system. The impressed current is DC with the positive side of the output connected to a purposely made anode, and the negative to the steel being protected. The anode can be of any material that can conduct electricity, but those materials that ensure system longevity are preferred, such as special metal alloys, those materials based on graphite, high silicon-iron, titanium, tantalum, niobium or the lead/silver alloys. Anodes in this type of CP system are comparatively inert and are designed to last much longer than sacrificial anodes. When correctly selected and used in a well-designed CP system, such anodes may have a useful life of 20 years and more (Perkins 1986).

An additional power supply is necessary for this type of CP system exposed to the atmosphere where a higher potential difference is required. When this type of CP system is applied to reinforcement in structures above ground, to get effective cathodic protection, the reinforcement should be electrically continuous and the concrete between reinforcing steel and anode

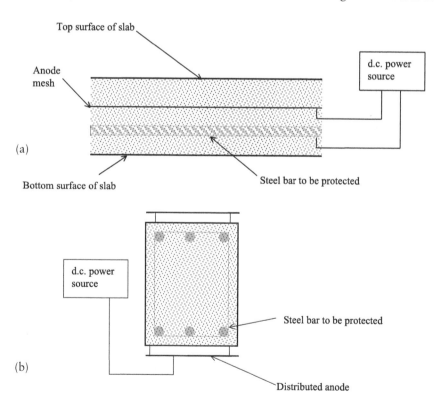

Figure 4.26 (a) Impressed current cathodic protection system installed in a concrete slab; (b) schematic for reinforced concrete cathodic protection.

should provide sufficient conductivity. The current flows through the concrete and the anode cannot be placed further away than the surface of the concrete. As a result, and due to the high electricity resistivity of concrete, it is necessary to use an anode system which is distributed across the surface if all of the reinforcement is to be protected. A schematic cathodic protection system for a reinforced concrete beam is illustrated in Figure 4.26(b). In the case of reinforced concrete superstructures, there is no appropriate electrolytic medium. The anode or anodes have to be distributed over the concrete surface if a reasonably uniform cathodic protection current is to be supplied to the reinforcement.

Also, alkali-reactive aggregate should be avoided as cathodic protection may aggravate the reaction. The second type of CP system is more widely used than the first one. The procedures to install an embedded CP system include:

1   Assessment of the location of corrosion.
2   Preparation of the concrete surface.
3   Paint the rebar.
4   Place the anode.
5   Establish an electric connection between the reinforcement and DC power source and between anode and DC source. A control box as shown in Figure 4.27 is usually installed in a nearby location.
6   Patch repair is applied.
7   The DC source is activated.
8   Verification tests are conducted to ensure proper operation of the CP system.

This type of cathodic protection system is generally used for the situation where reinforcement corrosion has happened because of the presence of high concentrations of chlorides in the concrete. If only cracked and spalled areas are repaired, a large area of reinforcement may remain in concrete with chloride concentrations sufficient to activate additional corrosion. Consequently, further cracking and spalling may occur. In such cases, the CP can only provide a limited means of prolonging the service life of the structure. If applied to the entire structure, the CP provides one of the most effective practical methods for reducing the corrosion rate to virtually zero. It is generally recognized that the CP system is a global approach to the corrosion problem and the only permanent repair of existing corroded reinforced concrete structures.

The key to a successful cathodic protection of steel in concrete is to provide a uniform current density to the reinforcement. If the current density is too low, corrosion can occur; if it is too high, deterioration of the concrete around the reinforcement could occur. In cathodic protection, normally power is supplied to the anodes in the form of a very low direct current. For

Figure 4.27 The control box for an impressed current cathodic protection system.

new reinforced structures, cathodic protection is expensive compared to the use of better concrete.

The CP method should be used with caution for pre-stressing steel because hydrogen could develop on the steel surface that is polarized as the cathode, which may cause hydrogen embrittlement of the high strength steel. Simultaneously, hydroxyl ions move away from the cathode and they may react with aggregate, leading to a reduction in concrete/steel bond. Therefore, cathodic protection is not recommended for pre-stressed concrete.

4.3.6.3 Realkalization techniques

Realkalization is a technology to provide the long-term restoration of alkalinity to carbonated but otherwise sound concrete under the passage of an electric current. All areas of concrete damaged by corrosion must first be cut out and replaced. Realkalization is achieved by applying a potential across a temporary anode, external to the concrete, and an internal cathode – the reinforcing bar. The electrolyte, which is an alkaline solution of sodium or potassium carbonate, diffuses into the concrete towards the reinforcement. The electrochemical production of hydroxyl ions caused by the electric current creates more alkalinity at the surface of the reinforcement, which repassivates the steel, and reduces the risk of further corrosion. The process normally takes between 3 and 15 days and it restores the alkalinity of concrete to a level greater than pH = 11.5, a level which supports passivity of the reinforcement. Figure 4.28 illustrates the process schematically using a shutter system to contain the electrolyte. There are several special considerations for this process, which will be described in Section 4.3.6.4 on chloride removal techniques.

4.3.6.4 Chloride removal techniques

For chloride-induced corrosion damage, the chloride contamination can be removed from concrete as a repair process. The possible methods of chloride removal are: water treatment and chloride extraction (FIP 1991).

Water treatment is based on the concept of chloride removal by water transportation. In this treatment, chloride contaminate is flushed out by a water stream as well as water dilution. However, since the permeability of concrete is very low compared to other porous media, it is difficult to remove the chloride content at greater depth.

Chloride extraction is a method of corrosion control by extracting chlorides from around the reinforcement in otherwise sound concrete under the passage of an electric current. Chloride extraction, like realkalization, is achieved by applying a potential across a temporary anode, external to the concrete, and an internal cathode – the reinforcing bar. The positive anode attracts negatively charged chloride ions and the cathode repels the same. Chloride ions will either be removed from the concrete into the electrolyte or repositioned, away from reinforcement. In addition, the electrochemical production of hydroxyl ions creates more alkalinity at the surface of all the reinforcement, which repassivates the steel, and reduces the risk of further corrosion. Process run times are of the order of 3–15 weeks and electrolytes such as water or saturated calcium hydroxide are most often used. Figure 4.29 illustrates the process schematically using a shutter system to contain

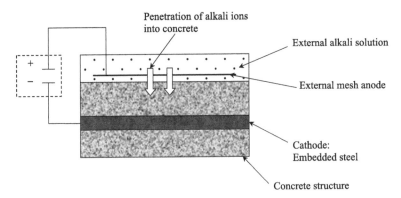

Figure 4.28 Electrochemical realkalization.

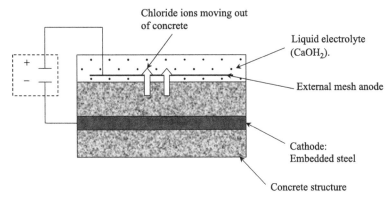

Figure 4.29 Electrochemical chloride extraction.

the electrolyte. To avoid chlorine gas production at the anode, an ion exchange resin is mixed in the electrolyte $(CaOH_2)$. In this method, the concrete may be heated to 70 °C.

Before applying realkalization or chloride extraction, any cracked, spalled or delaminated concrete needs to be cut out and replaced with material of similar electrical resistivity to the parent concrete. The exposed reinforcement should have been blast cleaned, but should not be primed.

The processes involve the application of a temporary impressed current between anodes fixed to or near the surface of the concrete and the reinforcement. The anodes are normally of mild steel or activated titanium mesh. The anode is connected to the positive terminal of a direct current electric supply with the negative terminal connected to the reinforcement which thus becomes the cathode. Electrical low resistance continuity is important for both anode and cathode systems. The voltage used is normally between 10 and 40V. The electrolyte used is a solution of sodium or potassium carbonate for realkalization and water, sometimes with calcium hydroxide added, for chloride extraction. If alkali–aggregate reaction is a possibility, an alternative electrolyte such as lithium hydroxide can be used. The electrolytes are contained within an absorbent layer applied to the surface or within liquid tight shutters.

The current flow, measured in amperes per surface area of concrete, is carefully controlled as is the total charge passed, both of which are related to the results produced. After completion of either process, the concrete surface may require cleaning. It is then beneficial to apply an anti-carbonation coating in order to maintain a stable environment and prevent further ingress of carbon dioxide or to coat and seal the concrete to prevent the ingress of chlorides. There are some side effects to be considered which include the following.

Both processes increase alkalinity and may give rise to alkali–aggregate reaction if reactive aggregate is present. Where reactive aggregate is present, proper consideration should be given and testing carried out before, during and after treatment in order to ensure that realkalization or chloride extraction does not accelerate the alkali–aggregate reaction. They both produce hydrogen at the reinforcement, which may cause hydrogen embrittlement of the steel if it is sensitive to hydrogen. The high alkalinity generated adjacent to the steel may temporarily reduce the bond between reinforcement and concrete. Some researchers have concluded that realkalized concrete is stronger, denser, less water absorbent and shows a tendency to lower levels of alkali silica expansion. Other research has indicated that any short-term loss of ductility of normal reinforcing steel would not cause significant problems and there is little or no effect on bond strength for deformed reinforcement. Realkalization and chloride extraction of pre-stressed and post-tensioned concrete are not normally recommended.

4.3.7 Repair of ASR affected concrete structures

It is quite difficult to repair ASR damaged concrete structures. In general, blocking further moisture penetration into the concrete is the major mitigation method for repairing existing ASR damaged concrete. This is because research has shown that the gel formed during ASR is expansive only when there is a plenty of moisture involved in the reaction. Therefore, how to reduce the amount of water or moisture getting into concrete has become very important for ASR-affected structures. For the structures that are in contact with water or already have very high moisture content, it is extremely difficult to repair or to control the ASR-induced damage. Various sealants and waterproof membrane have been used to seal the surface of concrete. The problem with this type of method is that any small hole or damage to the membrane will defeat the purpose, because the amount of moisture needed for the formation of the expansive ASR gel is actually quite small.

Depending on the extent of damage induced by ASR on a structure or a structural member, different methods can be used to repair the ASR-induced damage. For moderately damaged concrete structures, electrochemical methods can be used to mitigate ASR-induced damage and prolong the service life of the structure. For severely damaged concrete structures, the affected concrete must be replaced by new concrete with non-reactive aggregate and low alkali cement. More importantly, some additives should be used in the new concrete to reduce the risk of ASR damage. Otherwise, the new concrete will be affected very soon, for its working environment is the same as the old concrete. It is important to emphasize that if a concrete surface is affected by ASR, the interior concrete will eventually be affected by ASR, and it is just a matter of time before the entire structure is affected by ASR. Therefore the replacement of part of the existing concrete (the severely damaged part) by new concrete will not solve the problem, but only temporarily alleviate the ASR-induced damage.

4.3.7.1 Replacement of ASR-affected concrete

Replacement techniques of ASR-affected concrete are similar to the techniques introduced earlier for repairing concrete surface. The difference is that special attention must be paid to the mix design of the new concrete to be used for the repair. Usually, there are two special requirements that should be satisfied. The first is that the cement to be used for the repair job should be low alkali cement, that is, the equivalent alkali content should be below 0.6 percent (ASTM C150). Low alkali cement is a necessary condition to reduce the risk of ASR in the new concrete. The second requirement is that the aggregates (sand and gravel) to be used in the new concrete should not be reactive. The reactivity of aggregates can be determined based on several ASTM standard methods, such as ASTM C1260, 1293, and 1567. In Canada, there are two standard methods for testing reactivity of aggregate, CSA A23.2-25A which is similar to ASTM C1260, and CSA A23.2-14A which is similar to ASTM C1293. In Japan, JASS 5N T-603 is a commonly used method for detecting reactive aggregates.

In addition to low alkali cement and non-reactive aggregates, it is necessary to add some admixtures to reduce ASR potential. Pozzolanic admixtures, such as fly ash, ground granulated blast-furnace slag (GGBFS), and silica fume are often used. The function of added pozzolanic admixtures is to reduce the amount of calcium hydroxide in hardened concrete, and thus reduce the amount of reactants in ASR. Fiber reinforcement such as polypropylene fibers and steel fibers have been used to reduce the ASR-induced damage in concrete. The function of the fibers is to increase crack resistance of the concrete and thus the cracking time and crack width will be reduced in the ASR-affected concrete. Since the fibers do not alter the ASR damage mechanism, the expansion due to ASR will not be changed.

4.3.7.2 Lithium technologies

Since the early 1950s, lithium compounds have been shown to be effective in mitigating ASR in concrete. Lithium compounds can be added to fresh concrete mixtures or hardened concrete, and the lithium ions do not stop ASR, but interfere with the expansive mechanism of the alkali–silica gel by changing the reaction product. Some of early experimental results (Durand and Gravel 1995) show that the use of lithium hydroxide (LiOH) affects the expansion due to ASR. Lithium combines with reactive silica to form a lithium-silica gel that does not absorb water and therefore does not expand. Hence, lithium compounds can be used for both the construction of new concrete structures and the mitigation of existing concrete structures affected by ASR (Folliard *et al.* 2003). Since 1995, there have been some projects in the US using lithium technologies to mitigate the ASR problem in pavement, bridge decks, and dams (Stokes 2002).

The lithium ion, Li^+, in several forms is capable of suppressing deleterious

ASR. Lithium hydroxide (LiOH), lithium fluoride (LiF), lithium carbonate (LiCO$_3$), and lithium nitrite (LiNO$_3$) have been used or studied for the mitigation of ASR expansion. LiNO$_3$ is safe to handle, and has the least effect on concrete properties, while it has the greatest effect on concrete to prevent deleterious ASR. LiNO$_3$ also does not generate hydroxide ions in concrete pore solutions, and therefore does not increase pH value of the pore solution. LiOH, on the other hand, is a hazardous material and difficult to handle.

For an existing structure affected by ASR, a lithium compound can be sprayed on the surface of concrete. Lithium ions will penetrate into concrete and get involved in ASR. This method is effective only for very thin concrete members, because the penetration process of lithium ions is very slow and thus they can only reach a very shallow depth from the concrete surface. In order to accelerate the penetration process, an electrochemical treatment method similar to realkalization and chloride extraction may be used for mitigation of concrete slabs such as pavement and bridge decks (see the sketch in Figure 4.30). The technique involves the application of a high DC current density for a short period of time, typically a few days to weeks, between steel reinforcement acting as a cathode and an extended anode placed in an external electrolyte in contact with the surface of the concrete. Using this system, lithium ions can penetrate into concrete at a much faster rate than the spray application and become involved in the alkali–silica reaction.

It should be pointed out that there is only a small window during the deterioration process of concrete due to ASR when the lithium technologies can be applied effectively. The window corresponds to an early stage of ASR damage in concrete. If severe ASR has already occurred in a structure, manifested by crack mapping on the surface of concrete, the concrete should be replaced instead of repaired by using lithium technologies. If the concrete structure is still in very good shape without any cracks, the penetration of lithium ions will be very slow even under the electrochemical system, and

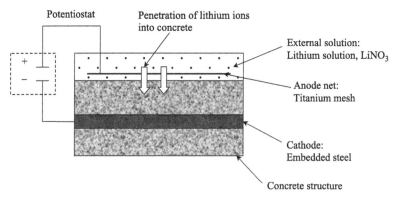

Figure 4.30 The electrochemical repair system used in the highway industry.

thus the repair will not be effective. The best time (the window) for the application of lithium technologies is when the structure is diagnosed as at an early stage of ASR and there are only minor ASR-induced cracking. The cracking does not impose severe long-term damage in the concrete but allow accelerated penetration of lithium ions.

4.3.8 Repair of reinforced concrete bridge decks

Bridge decks are usually made of reinforced concrete, while the girders and beams of a bridge may be made of steel, reinforced concrete, or pre-stressed concrete. In addition to environmental loadings such as freeze–thaw and wetting–drying cycles, bridge decks are exposed directly to traffic loading and attacks of de-icing salts in cold regions. Therefore, bridge decks are the structural members that need the most attention in terms of maintenance and repair in a bridge structure. Depending on the type of damage and the location of the damage, different repair methods can be used in practice. One important measure that should be taken to prolong service life of bridge decks is preventive maintenance or small-scale repair that can help significantly to improve the serviceability of bridges.

4.3.8.1 Preventive maintenance

Preventing concrete deterioration is much easier and more economical than repairing deteriorated concrete. Preventing concrete deterioration begins in the design of the structure with the selection of the proper materials, mixture proportions, concrete placement, and curing procedures (Air Force 1994). Even a well-designed concrete structure will generally require follow-up maintenance actions. The primary types of maintenance for concrete are surface protection, joint restoration and cathodic protection of the reinforcing bars. Surface maintenance involves the application of sealers and coatings for protective purposes. It should be mentioned that coatings are usually different from sealers in the materials used, not in their thickness. Joint problems are usually treated with one of a variety of types of joint sealers, and cathodic protection was discussed in an earlier section, which involves the use of sacrificial anodes connected to the reinforcing bars or installation of an impressed current system.

(A) SURFACE SEALERS

Surface sealers are applied to concrete for protection against chemical attacks by alkalines, salt solutions, or other chemicals. The actual need for a sealer must first be established, and then the cause and extent of any deterioration, rate of attack, and environmental factors must be considered when selecting the right sealer for the job. A variety of sealers are available for waterproofing and protecting concrete surfaces. Some of these products have been successful in protecting new concrete from contamination by de-icing salts and other harmful environmental agents. They have generally been unsuccessful at stopping the progression of already contaminated concrete (*Utah DOT Research News* 1998).

(B) TWO TYPES OF SEALER AND THEIR APPLICATIONS

Linseed oil. A mixture of 50 percent linseed oil and 50 percent mineral spirits is normally used. The mixture is applied in two applications on a dry, clean concrete surface. The surface sealers should be less than 5 mils, and a test strip should be used to help determine the rate of application. The normal rate of application is about 40 square yards per gallon for the first application and 65 square yards per gallon for the second application. This treatment could last for 1 to 3 years under normal traffic condition.

Silicone. Silicone has been used on concrete to minimize water penetration. Care must be taken where moisture has access to the back of the member and carries dissolved salts to the front face where it is trapped by the silicone. Silicone oxidizes rapidly and is somewhat water soluble. Treatments are required every 1 to 5 years.

(C) APPLICATION PROCEDURES FOR SURFACE SEALERS

Specific surface preparation instructions provided by the manufacture of the selected sealer should be followed. The description in Section 4.3.2.1 should be studied. The report by NRC (National Research Council 1993, SHRP-S-360) has a detailed description of the construction procedures related to the application of surface sealers. In general, a sealer should be applied to a clean and dry surface. The surface should be dry for maximum penetration of the sealers into concrete. Epoxy injection of visible cracks should be considered before the application of the sealer. An alternative is to first seal the cracks and then reseal the entire surface, including the cracks. To provide for proper penetration, the subsurface pore must be dry to the desired depth of penetration before sealing. This will help the penetration of sealers into a greater depth of concrete. The drying requires a sufficient period of dry, warm weather before application. Since the surface of concrete should be dry, water blasting is not recommended, and acid etching is discouraged.

Newly patched and overlaid areas should be allowed to cure a minimum of 28 days after placing, or longer if recommended by the manufacturer. In any case, penetrating sealer should not be applied if the moisture content of the concrete is greater than 2.5 percent when tested in accordance with AASHTO T239 standard test procedure.

Sealer may be applied by low-pressure pump and by flood and brush techniques. To ensure proper coverage, the area that needs to be sealed should be delineated on the deck surface. This area can be calculated from the capacity of the sprayer or container used for flooding the sealer. In addition, the sealer should contain a fugitive dye to enable the solution to be visible on the treated surface for at least 4 hours after application. The fugitive dye shall not be conspicuous more than 7 days after application when exposed to direct sunlight. Sealer should not be applied when the temperature of the concrete surface is below 4 °C. Before sealing, exposed joint sealers and painted steel joints adjacent to areas to be sealed should be masked off. The sealed areas should be protected from rain and traffic spray for six hours after application.

(D) SURFACE COATINGS

Plastic and elastomeric coatings form a strong, continuous film over the concrete surface. To be effective in protecting concrete, the coating must satisfy some basic requirements: the adhesive bond strength of the coating to the concrete must be at least equal to the tensile strength of the concrete; the abrasion resistance must prevent the coating from being removed; chemical reactions must not cause swelling, dissolving, cracking or embrittlement of the material; the coating should prevent the penetration of chemicals that will destroy the adhesion between the coating and concrete; for proper adhesion, the concrete must be free of loose dirt particles, oil, chemicals that prevent adhesion, surface water, and water vapor diffusing out of the concrete.

1   *Epoxies.* Epoxies are often used as coatings. As with most thin coatings and sealers, a protective overlay or cover is required if they are exposed to traffic wear or abrasive forces.
2   *Asphalt.* Asphalt is used as a protective overlay for bridge decks. It provides water protection and a protective wearing surface.

(E) APPLICATION PROCEDURES FOR COATINGS

The surface preparation methods are similar to those suggested for surface sealers. For epoxy coating applications, the coating should be mixed to produce a uniform and homogeneous mix. For spray application, a low-pressure spray gun should be used. The coating should be applied when the temperature is between 50 and 90 °F. Two applications of the coating should be applied to ensure even coverage and minimize the likelihood of pinholes. Each application should produce a dry film thickness of 2–3 mil. The second coat is normally applied 24 hours after the application of the first coat, but this can vary with environmental conditions and material type. The coating should be applied at the rate of coverage approved during the material acceptance tests. For asphalt coating, the thickness of the coating is usually in the range of 2–4 inches. The application is similar to asphalt pavement overlay, which will not be described in detail here.

4.3.8.2 Repair methods for crack reinforcement

Crack reinforcement is primarily used to repair cracks in the load-bearing portion of the structure for active and dormant cracks. This repair bonds the cracked surfaces together into one monolithic form. Repair procedure is as follows:

1 Clean and seal the existing crack with an elastic sealer applied to a thickness of $1/16$ to $1/32$ inch and extending at least ¾ inch on either side of the crack.
2 Drill ¾-inch holes at 90 degrees to the crack plane, fill the hole and crack plane with epoxy pumped under low pressure (50 to 80 psi).
3 Place a reinforcing bar (No. 4 or No. 5) into the drilled hole with at least an 18-inch development length on each side of the crack.

There are other repair methods for cracking in bridge decks, which are similar to those described in Section 4.3.4. Details can be found in ACI 224.1R-07.

4.3.8.3 Spall repair methods

Spalling of concrete is a localized damage in concrete structures. Spalling of concrete cover outside of reinforcement bars is usually caused by corrosion of the steel bar. Another form of concrete spalling is due to high heating rate when the concrete is under fire. Patching methods are often used to repair the spalling damages. For bridge decks, the deteriorated areas may include the top layer of steel bars or both the top and bottom layers of steel bars. If only the top steel bars are corroding, a partial-depth repair may be used. For partial-depth deck repairs, the deteriorated concrete should be removed to the depth required to provide a minimum of 1.9 cm clearance below the top layer of steel bars. Maximum depth of removal for a partial-depth repair should not exceed half the deck thickness (National Research Council 1993, SHRP-H-356). Corrosion of both the top and bottom layers of steel bars requires full-depth repairs. In this case, the concrete within the delineated area for the entire deck thickness, normally 8 in. should be removed. It should be pointed out that patch repair of bridge decks has a relatively short service life because they do not address the cause of the problem, i.e. corrosion of the reinforcing steel, but address only the symptoms: spalling and delaminations.

It is also important to note that when concrete contaminated with chloride beyond the threshold level is left in place in the area surrounding the patches, patches often accelerate the rate of deterioration of the surrounding concrete. The patched concrete area acts as a large non-corroding site adjacent to corroding sites and increases the rate of corrosion. Therefore, it is better to install a cathodic protection system as shown in Figure 4.25 for new patches. The following are the construction procedures for partial-depth repair (FHWA 1999).

Areas of unsound concrete should be located using drag chains and hammer. The unsound concrete is evidenced by a hollow sound. The area to be removed should be delineated by the site engineer as the unsound or delaminated area plus a periphery of 3–6 in. The delineated area should be outlined with a saw cut 0.75 in. deep. Care should be taken to avoid cutting existing reinforcing steel. In no case are feathered edges acceptable.

(B) REMOVAL OF UNSOUND CONCRETE

Many kinds of tools can be used to remove unsound concrete, which usually include: pneumatic breakers, milling machines, and hydrodemolition (National Research Council 1992, SHRP-S-336). These removal methods are for large scale bridge deck demolition, and they are different from those described in Section 4.3.2.1 for small-scale repair preparation of structural members in buildings. Each method has very specific strengths and weakness with regard to work characteristics, production, economics and quality. The following is a brief summary:

- *Pneumatic breakers.* They are the most expensive method but also the most flexible. They can be used for all sizes and shapes of area, to all depths and all bridge structural elements.
- *Milling machines.* These provide the least expensive method of concrete removal but they are also the most inflexible. They can only be used to remove large areas of surface and/or cover concrete on decks.
- *Hydrodemolition.* This technology lies between pneumatic breakers and milling machines in terms of cost and flexibility. Surface, cover, matrix and core concrete can be removed, but economies are only realized if work is done on large horizontal areas such as decks.

Quality, availability, flexibility, total coat and contractual risk can easily override the technical aspects. These vary with time and location and thus it is extremely difficult to make any firm rules as to which method must be used under general circumstances.

(C) PREPARATION OF REPAIR CAVITY

Once all the unsound concrete has been removed, the cavity should be blasted clean to remove all loose material and provide a dust-free surface. All exposed reinforcing steel should be blasted to near white metal. With widely used epoxy coated rebars, one precaution should apply, that is, the rebar after sand blast should be re-coated with epoxy, otherwise, the exposed steel due to the blast will be the new site for corrosion. Reinforcing bars with greater than 20 to 25 percent sectional loss as determined by the engineer should be lapped with reinforcing bar of equal diameter for 30 bar diameters on either side of the deteriorated area. Following the completion of blasting, the cavity should be air blasted to remove all dust and debris. The compressed air source used for blasting should have all oilers removed and oil traps installed. Full depth patches will require formwork constructed to prevent leakage of mortar.

Patches should be reinforced with wire mesh attached either to reinforcing bars or dowels to secure the patch to the old concrete. Loose reinforcing bars should be tied at each intersection point to prevent relative movement of the bars and repaired concrete due to the action of traffic in adjacent lanes during the curing period. If new reinforcement is required, an adequate length to attain a lap splice (30 times the bar diameter) must extend from the existing section. If a proper splice is not possible, holes must be drilled into the existing concrete and dowels or anchors installed.

(D) APPLICATION OF BONDING AGENTS

A good interface must be established between the existing and new concrete. For instance, the following options can be considered:

1   *Epoxy bonding.* Ensure the surface is clean, dry, and free of oil. Apply the epoxy agent to the prepared surface.
2   *Grout or slurry.* Clean the prepared surface and saturate with water. Remove all free-standing water with a blast of compressed air, and apply a thin coat of grout.

(E) PLACEMENT AND CONSOLIDATION

The patch material may be batched and mixed at the site or supplied ready-mixed. The patch material should be placed in a manner that will prevent segregation, and consolidated with an internal vibrator. The surface of the patch should be floated and textured to match the surrounding concrete surface.

(F) CURING

Portland cement concrete and hydraulic mortar/concrete must be moist cured for a minimum of 72 hours. Moist curing should be provided by clean, wet burlap covered with plastic sheeting anchored around the patch perimeter with sand. Shorter moist curing times may be approved by the site engineer in rapid patching conditions. For rapid repair materials, the patches should be cured until the structure is to be opened to traffic or for the minimum time recommended by the manufacturer.

4.3.9 Repair of reinforced concrete pavements

There are many types of damage in concrete pavement (NCPP 2004). Figures 4.31(a) and (b) show the so-called "D" cracking and map cracking in concrete pavements. To extend the service life of concrete pavement, preventive maintenance is very important. The preventive maintenance methods used for concrete bridge decks such as sealers and coatings can also be used for concrete pavement. However, the cost required for large-scale application of the preventive maintenance methods will be quite high. Fortunately, the serviceability of concrete pavement can usually be kept very well for a long period of time without requiring any major repair.

Pavement rehabilitation encompasses major and minor repair activities. The major repair activities are different from periodic maintenance activities. Major repairs will be viewed as any work that is undertaken to

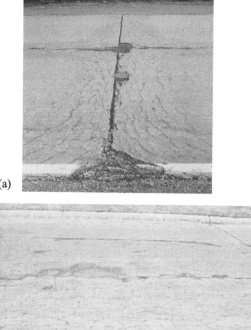

(a)

(b)

Figure 4.31 (a) Severe "D" cracking in a concrete pavement; (b) map cracking in a concrete pavement.

Source: (NCPP 2004).

significantly extend the service life of an existing pavement through the so-called 3R program: Restoration, Resurfacing, and Reconstruction. Restoration is used at an early stage when a pavement has little deterioration, and repair work is limited to isolated areas of distress. Resurfacing is used when the pavement has medium to high levels of distress and when restoration is no longer effective. Reconstruction is used when the pavement has high levels of distress and when resurfacing is no longer effective. Minor pavement repair activities include: thin asphaltic overlays, routine sealing of cracks and joints, slab sealing, pavement patching and pothole repairs.

There are no definite equations, guides, or step-by-step procedures that one can use as a recipe for proper rehabilitation of concrete pavement. There are no "right" and "wrong" solutions to pavement rehabilitation problems, but rather "better" or "optimum" solutions. Optimum solutions maximize benefits while minimizing costs, which is often not attainable due to the constraints imposed. A realistic procedure is as follows:

1   Determine cause of the distresses or pavement problems.
2   Develop a candidate list of solutions that will properly address the problem.
3   Select the preferred rehabilitation method given economic and other project constraints.

Reconstruction involves concrete pavement design (rigid pavement design), and resurfacing involves thickness design for overlay, which is based on concrete pavement design. Both of them are beyond the scope of this book. We will only briefly introduce resurfacing methods and then discuss in detail a restoration method in this section.

4.3.9.1 Concrete overlay

Based on traffic loading, the thickness of concrete overlay can be calculated. Then the construction method of a concrete overlay is similar to the construction method for a concrete pavement, in which the old concrete pavement is treated as a sub-base. Depending on the material used for the overlay, it is called white topping if the overlay is made of Portland cement concrete and black topping if it is made of asphalt concrete.

Many attempts have been made to reduce the permeability of concrete overlay on bridge decks and concrete pavement so that the penetration of chloride ions into the concrete can be slowed down. Low-slump dense concrete overlay has been used by many Department of Transportations (DOTs), in Iowa, Kansas, Minnesota, and New York. High performance concrete overlays with silica fume, microsilica, fly ash, slag, latex have been used by many DOTs. Where a bare concrete deck is desired, Region 6 of Colorado DOT has been topping the deck with 2 inches of silica fume concrete. Recently, very early strength (VES) latex-modified concrete overlay has been used by Virginia DOT.

Various fiber reinforced concretes have been used by Oregon and South Dakota DOTs for concrete overlays. In Europe, ultra-high performance fiber reinforced concrete (UHPFRC) has been used for in the overlay of concrete decks, which has very low permeability and very high tensile strength. High tensile strength helps to reduce cracking tendency of concrete. However, the initial cost of UHPFRC is quite high because of the use of large amount of fibers and fiber dispersion additives in UHPFRC.

Polymer concrete has also been used for overlays. They are mixtures of epoxies with aggregates such as polyesters with aggregates, methacrylates with aggregates, and polyurethanes with aggregates. Similar to UHPFRC, the initial cost of polymer concretes are relatively high.

4.3.9.2 Full-depth repair – a restoration method

Restoration methods are also referred to as non-overlay pavement rehabilitation methods, which include the following:

- full-depth repair;
- partial depth patching;
- joint crack sealing;
- subsealing – undersealing;
- grinding and milling;
- sub-drainage repair;
- pressure relief joints;
- load transfer restoration;
- surface treatments.

Some of above listed methods are similar to those described in the last section for bridge deck repairs, or similar to Section 4.3.3 and Section 4.3.4. They will not be repeated here. We will only introduce in detail the full-depth repair.

FULL-DEPTH REPAIR

The full-depth repair consists of five basic steps:

1 *Locating and isolating the patch area.* Good judgement is essential in defining the limits for full-depth repairs. Each repair should be large enough to replace all significant distress, resist rocking under traffic, and be easy to work in, and small enough to minimize the material cost. Typically, a patch that is of full-lane width and half-a-lane width long meets these criteria. For utility cuts, the patch length and width should be slightly larger than the planned trench length and width. This creates

a shoulder around the utility cut that keeps the excavation from undermining the existing concrete during utility repair and helps support the concrete patch after placement.

2 *Removing the damaged concrete.* There are two procedures to remove the old concrete: the lift-out method and the break-up and clean-out method. With the lift-out method, the old concrete is lifted out in large sections. With the break-up and clean-out method, the concrete is broken into small pieces and removed. The lift-out method generally is faster and causes less sub-base disturbance. The break-up and clean-out method removes severely damaged concrete that the lift-out method cannot remove. However, this method may damage the sub-base, and so may require more sub-base repair than the lift-out method.

3 *Preparing the patch area.* With the lift-out method, the base repair generally consists of shovelling up the damaged materials from the opened area. With the break-up and clean-out method, base repair includes recompacting the sub-grade and adding granular sub-base layers. If the area to be repaired includes a joint, tie bars or dowel bars should be used. Tie bars are deformed steel bars grouted into the existing concrete, while dowel bars are smooth steel bars inserted into holes drilled in the existing concrete. Both tie bars and dowel bars are installed for transferring the load between the joints. For a pavement with little truck load, the load transfer can be done by aggregate interlocking, which is the action between the roughened face of the old concrete and the face of the patch. The roughened face can be created by chips along the patch edges of the old concrete. The sub-base should be excavated out from beneath the concrete slab and replaced with new concrete.

4 *Placing and finishing the new concrete.* Generally, the patch thickness is the same as the existing slab thickness. Normal Portland cement concrete and high early strength concrete have all been used for full-depth repair. The strength of the new concrete should not be lower than the old concrete. Asphalt concrete is not a good material because it does not last as long as Portland cement concrete. For continuously reinforced concrete pavement, there is no joint along the slab. Usually the repair work should be done in the afternoon in the summer season to avoid crushing failure under large thermal expansion. Proper curing of the patch is important. It is best to begin curing operations as soon as possible after completing the finishing operation. Typically, either a clear or white-pigmented curing compound may be used. For high early strength patches, insulation mats should be placed over the repair during curing to increase the strength gain of the concrete. For utility cut, flowable fill is an ideal alternative to base repair. Flowable fill is a low-strength, self-levelling material made with cement, fly ash, and sand that flows in and around repairs and then hardens. Since it is designed not to become too hard, it is easy to remove later.

5 *Opening to traffic.* The proper time to open a patch to traffic depends on

the strength gain of the patch material. Generally, full-depth repairs can be opened when the strength of new concrete has reached at least 20 MPa (3000 psi), or the strength has reached the design value specified by the engineer. Most of concrete mixtures achieve this strength in 24–72 hours.

4.4 Strengthening techniques

Strengthening of a concrete structure may be required due to several reasons:

1 Change of usage which may cause over-stress in the structural member.
2 Serious materials and structural deteriorations which cause structural members to be no longer able to carry the imposed loads with an adequate factor of safety.
3 A combination of (1) and (2).

Strengthening of structural members can be achieved by replacing poor quality or defective material with better quality material, by attaching additional load-bearing material, such as high quality concrete, additional steel, thin steel plates, various types of fiber reinforced polymer sheets, and so on, and by the redistribution of the load such as by adding a steel supporting system. The purpose of strengthening is to increase the load-carrying capacity or stability of a structure with respect to its previous condition.

4.4.1 Design considerations for strengthening

A repair or strengthening work starts with a diagnosis, or evaluation of the structural condition, and then selection of materials and the strengthening method, preparation of the areas in the structure to be strengthened, and application of the repair or strengthening. It is necessary to conduct a careful investigation and assessment of the condition of the existing structure before any strengthening work. It is usually essential to check carefully the position, condition and amount of the reinforcement with the help of original drawings and design data. If the original documentations are not available, some non-destructive testing methods may be used, as described in previous chapters. The information obtained from the structural assessment is necessary and useful for the design of strengthening. In general, the strengthening of structures should be designed and constructed in accordance with appropriate design codes. In design of strengthening work, typical problems that should be solved are the transfer of shear forces between the old concrete and the new concrete applied for strengthening reinforcement, and the post-tensioning of the existing structure which in some respects is different from the post-tensioning of a new structure.

Interactions between new and old concrete should be always considered in a design for strengthening of concrete structures or structural members. As a general rule, the aim is to get structural parts, composed of different concretes, old and new, to act as a homogeneously cast structural component after strengthening. To achieve this, the interface between the old concrete and the new concrete must be capable of transferring shear stresses without relative movement of such a magnitude that the structural performance is significantly affected. In addition, the joint must be durable for the severe environment.

In strengthening work using a large volume of concrete, additional stress may be produced due to the heat of hydration in new concrete. Temperature differences between old and new concrete can be limited by special measures, such as by pre-heating the old structural elements and/or by cooling the fresh concrete. Differences in shrinkage and creep between old and new structural elements require careful evaluation. Suitable mortars or concrete with low shrinkage and creep properties and minimal development of the heat of hydration should be employed for strengthening. An effort should also be made to match, as closely as possible, the strength and modulus of elasticity for the two concretes. Of course, compatibilities of the old and the new concretes should also be seriously taken into consideration when strengthening a job with large volume of concrete.

Strengthening of reinforcement subjected to tensile forces can be achieved by several methods, which will be discussed in detail in the following sections:

1   Additional reinforcement placed in the old cross-section or in an additional layer.
2   Replacement of corrosion damaged rebars.
3   Addition of epoxy-bonded steel plates.

4.4.2 Addition of reinforcing steel

In the simple case, a strengthening of the concrete tension zone is possible by means of addition of rebars. Rebars should be added after unloading and after the cover has been removed. Sometimes, additional support is needed for the structures and/or structural members when a concrete structure is strengthened by adding reinforcement. After additional rebars have been placed, the concrete cover must be re-established. Effective anchoring for rebars is required.

Welding is recommended when the space between the rebars is limited. For a welding joint, attention should be paid to the heat propagation in the bar that could create serious problems in old concrete. The uneven elongation of the heated bar and concrete may lead to high local tensile stress in concrete that could result in cracking and debonding of the rebar. It is therefore suggested that the anchorage length of the bar should be increased by six bar diameters.

Strengthening of the reinforcement by incorporation of additional steel has mainly the following disadvantages: (a) it is a labor-intensive and time-consuming job; (b) it needs significant space for the retrofitting process to be performed; (c) it may affect the operation of the existing facilities; and (d) it may reduce space or headroom after retrofitting.

The following preparation work is important for the addition of reinforcing steel:

1   Concrete around corroded bar should be removed to a 50 mm distance beyond the corroded portion. The reinforcement steel bar should be exposed with at least 20 mm clearance below.
2   If concrete around non-corroded bars are damaged to an extent that will affect concrete/steel bond, the affected steel should be completely exposed.
3   Rust should be removed from corroded steel surface. Grit blasting and needle gun are effective but they will produce air and noise pollution. A high pressure water jet can be used while wire brushes are not so efficient. After rust if removed, dust on steel surface can be removed with an industrial vacuum cleaner.
4   If there is a concern about insufficient compaction of repair material, which may lead to incomplete contact of repair material with steel and ineffective re-passivation, the steel surface can be treated with alkaline slurry first. Care should be taken as slurry may thicken quickly and not bond well to the repair material. Polymer cement slurries can be used, but they may reduce alkalinity or insulate steel from the passivating repair mortar or concrete.
5   For reinforcement steel in concrete structure, when re-passivation is not possible, the electrical isolation of steel from the surrounding materials may be helpful. However, the preferable technique for treatment of reinforcement steel is the use of zinc coating to provide cathodic sacrificial protection. Epoxy coating can also be employed, but there may be the following problems: (a) the steel surface may not be completely covered in the field; (b) corrosion develops in the uncoated part which becomes an anode; and (c) penetration of water into steel/epoxy interface may lead to debonding and corrosion.

4.4.3 Replacement of the reinforcement

If reinforcement steel inside the old concrete structure has been severely damaged, or has lost a substantial proportion of its cross-sectional area through corrosion, it may need to be replaced by new rebars. Pullar-Strecker (1987) suggests that replacement of steel is necessary if it has lost more than 20 percent of its cross-section area but other researchers require replacement if more than 10 percent of the cross-section area is lost (Campbell-Allen and Roper 1991).

Replacement of the reinforcement is more complicated than the addition of new rebars discussed in the previous section, because in this operation, the existing reinforcement must first be removed. Before any existing steel is removed, it is extremely important to check that all aspects of design have been considered before this step is taken. The structure should be unloaded, and additional support provided. In replacement of steel, it is necessary to open up sufficient concrete to provide adequate lap lengths with the undamaged portion of the steel bars by lapped splices, by welding or by coupling device. The replacement steel bar should be of the same quality as the original bar. If a different steel must be used, it should certainly not be so different that corrosion cells can be set up between the two adjoining bars. For instance, stainless steel bars should not be used to replace or reinforce conventional steel. Otherwise, the conventional steel next to the new stainless steel will immediately become an anode and the corrosion will start right away from the existing steel.

If it is convenient to saw into sound concrete, new bars may be conveniently anchored in concrete placed in dovetailed slots. A longer anchorage length than that used in conventional design should be provided in all cases of replacement or addition of reinforcing material. The concrete cover should certainly be re-established after the damaged rebars have been replaced. In reinforced concrete beams or columns where the damage is confined to the steel stirrup, the new steel stirrup may be hooked round existing main corner reinforcement. In such situations, an extra thickness of concrete has to be anchored on while the structure may not allow an increase in the cross-section of the beam or column. A good solution may be to drill into the sound concrete and anchor the steel stirrup in epoxy. If this measure is adopted, the cause of reinforcement corrosion must be completely eliminated as a part of the repair procedure, since otherwise the new steel stirrup will be at risk of corrosion where it emerges from the epoxy anchorage (ibid.).

When the new rebar just replaces the old deteriorated reinforcement steel, it carries no load until the structure is reloaded and the stress is transferred to it through the surrounding concrete or other anchorage. Since it will be difficult to determine how effective the transfer is going to be, generous allowances should be made for this uncertainty when assessing the final capacity of a structure repaired by replacement of reinforcement.

4.4.4 Epoxy-bonded steel plates

The strengthening of concrete structures by means of epoxy-bonded steel plates has been used in many countries to provide additional load-carrying capacity or additional stiffness for the concrete structures, and to control flexural crack widths and deflections (McKenna and Erki 1994; Colotti and Spadea 2001). The advantages of this method can be attributed to the availability of high quality epoxy resins, and the minimum changes in geometric dimensions of structural systems. The disadvantages of the method of strengthening by epoxy-bonded steel plates are: (a) it is a labor-intensive and time-consuming work; (b) it needs significant space for the retrofitting process to be performed; (c) it may affect the operation of existing facilities; and (d) it may reduce space or headroom after retrofitting.

Also, additional measures are needed to prevent steel from corroding when applying this strengthening method. The gradual loss of strength of ambient cured resin adhesives at temperatures above 60 °C also limits the use of this strengthening method. If an epoxy-bonded steel plate is used for strengthening the soffits of beams and slabs, the steel plate could become detached from the concrete surface during fire due to the deterioration of adhesive. Under such situations, a secondary positive fixing system between the steel plate and the concrete surface should be provided. For instance, the steel plate can be bolted into concrete, providing mechanical anchors as the secondary fixing system.

When employing this strengthening method, the following parameters and main performance criteria for epoxy resin adhesives are important (Perkins 1986):

1   The concrete surface must be well prepared to receive the epoxy and the surface of the steel plate must also be well cleaned and polished to a high standard using grit blasting.
2   The epoxy must be carefully selected for both concrete and steel.
3   The entire steel plate surface in contact with the concrete must be covered with epoxy.
4   After the strengthening, if failure occurs, it will occur in the concrete because the shear strength of the cured epoxy is usually higher than that of the concrete.

After the concrete and the steel surface are cleaned, the installation of the steel plate is quite straightforward. First, the epoxy is applied within a short period of time on the surface of concrete, and the steel plate is then pressed against the epoxy-coated surface. The plate must be held in position until the epoxy has cured and gained enough strength. After the installation of the steel plate, the behavior of concrete structures strengthened by means of epoxy-bonded steel plates should be examined in two aspects: short-term behavior and long-term behavior.

(A) SHORT-TERM BEHAVIOR OF STRUCTURES STRENGTHENED BY
EPOXY-BONDED STEEL PLATES

The load-carrying capacity of the strengthened structures depends on the strength of the reinforcement, the old concrete, and the adhesive (epoxy). Yielding of the reinforcement will cause the adhesive to fail, but the utilization of high-strength reinforcement may not be helpful. As mentioned above, the shear strength of the cured epoxy is likely higher than that of the old concrete, so that the strength of the old concrete has a large influence on the efficiency of the strengthening method because the failure is usually located within the concrete. To ensure a good bond between the epoxy and the old concrete, the surface of the old concrete should be treated to make sure that it is free from any defects likely to have a detrimental effect on the transfer of stress between concrete and steel plate.

A common failure pattern of a strengthened structure is the slip between the reinforcing element and concrete. Under a one-dimensional tensile load, the slip will start at the loaded end of the reinforcing element and move, with increasing load, to the center of the element. Failure occurs suddenly, by abrupt elongation of the slip interface up to the end of the reinforcing element (peeling off of the plate). In correctly designed structures with bonded plate reinforcement, a ductile failure with yielding reinforcement can be attained.

Theoretically, higher bond stress is to be expected from an increase in the elasticity of the reinforcing element and a decrease in the elasticity of the adhesive. Geometric influences are primarily the dimensions of the steel plate. The length of steel plate has a strong influence on the bond stress intensity, which decreases with increase in the length. The bond stress will also be influenced by the thickness of the plate. Therefore, glued-on reinforcing elements behave differently from the deformed bars, which can be designed using the same permissible bond stress for all diameters. With increasing width, there is a risk of defects in the adhesive and an increase of width results in a reduction of bond strength. Therefore, the width of the steel plate is normally limited to 200 mm (FIP 1991). The thickness of the adhesive coat, within a range of 0.5–5 mm, has no significant influence on ultimate load. The concrete dimensions, according to previous tests, do not appear to have any decisive effect on the ultimate load. The surface condition of the steel is an important parameter to achieve a successful repair. Suitable conditions can be achieved by sand-blasting. As a cleaned surface corrodes rapidly, a primer coating should be applied immediately which serves as a corrosion-protection layer and as an adhesive base for epoxy resin adhesive.

(B) LONG-TERM BEHAVIOR OF STRUCTURES STRENGTHENED BY EPOXY-BONDED STEEL PLATES

Epoxy resins are a new component in the strengthened structural system compared with the original structure made of only concrete and steel. Epoxy resins are polymers whose long-term behavior such as creep and aging are considerably different from those of concrete and steel. The effects of these material properties on the long-term performance of strengthened structures must be considered.

The creep of epoxy resins is considerably higher than that of concrete. In thin adhesive layers (normally less than 3 mm), the influence of creep is restricted by the cohesion. For the sake of good long-term performance, the epoxy resin is required to be stiff enough not to creep significantly under sustained load, but flexible enough so that no high stress concentrations can arise (Mays 1985).

Aging is a change of properties resulting from long-term mechanical, physical and chemical influences, e.g. relative humidity, radiation, heat, weathering and moisture. For strengthening with steel plate, aging reduces strength so that the long-term strength is only 50 percent of the short-term strength. Epoxy resin adhesives have a certain porosity, which will allow the penetration of water and other solutions. Exposure to water over long periods of time can cause epoxy resin adhesives to lose strength. It is generally required that the epoxy resin for bonding steel-plate has long-term durability for a service life of at least 30 years at service temperatures from −20 °C to +40 °C (Mays 1985). Also, epoxy is generally much more sensitive to fire than concrete so that this strengthening technique is only suitable for concrete structures with low fire risk. In cases of high fire risk, fire protection should be applied to the epoxy as well as steel plate.

Some tests show that the fatigue strength of epoxy is about 50 percent of the short-term strength. The fatigue strength, rather than the short-term strength, should be considered when designing the epoxy-bonded steel plates strengthening method for long-term application.

4.4.5 Externally bonded fiber reinforced sheets

One of the problems with epoxy-bonded steel plates is that the steel plates are very heavy and sensitive to the moist environment (such as the corrosion damage). There are many types of fiber reinforced materials that have been developed for strengthening applications of reinforced concrete structures, such as glass fiber reinforced polymers (GFRPs) and carbon fiber reinforced polymers (CFRPs) (see ACI 440.2R-02). These fiber reinforced materials are in the form of thin sheets and can be externally bonded onto the existing concrete structure and enhance significantly the load-carrying capacity of the structures. In addition to very high tensile strength of the fibers, GFRPs and CFRPs are much lighter than steel and have very high corrosion resistance. The repair and strengthening applications of GFRPs and CFRPs will be discussed in greater detail in Chapter 5.

Similar to CFRPs and GFRPs, steel cord reinforced polymers (SCRP) and steel cord reinforced grout (SCRG) are new materials (Huang *et al.* 2004) that have found their applications in structural strengthening. SCRP and SCRG are made of thin high-strength steel fibers bundled into cords. The cords are then woven into unidirectional sheets with synthetic textiles. These steel fiber reinforced sheets can be glued onto a concrete structure by either a polymer or a grout similar to the operations for CFRPs or GFRPs. SCRP and SCRG combine the advantages of steel plates and CFRP. They are not as heavy as steel plates and they are covered by polymers or grouts in the application so that the corrosion of the steel is nor longer a problem. Furthermore, the material cost is relatively low compared to that of CFRP; and ductility of the materials is higher than that of carbon and glass fibers such that the strengthened structures have a better performance under excessive dynamic loadings.

5 Glass fiber reinforced plastics components for bridge deck replacement

5.1 Introduction

The deterioration of reinforced concrete bridge decks is a major problem in cold regions where salt is commonly employed for de-icing the road surface. The penetration of chloride ions depassivates the steel reinforcement and initiates corrosion. Subsequent water absorption and expansion of the rust lead to cracking and surface spalling of the concrete deck. Moreover, with salt on the road surface, an osmotic pressure will be set up to draw water towards the top of the deck. Once the upper surface reaches a high level of saturation (above 91 percent), the increase in volume associated with freezing water will also lead to the formation of cracks in the concrete. With repeated freezing and thawing, the deck surface can be severely damaged.

With the use of conventional repair methods, such as patching and placement of additional steel reinforcements, the functionality and load capacity of the deck can be recovered but the degradation problem is not eliminated. In recent years, a new approach to solving the problem has been developed. When a concrete deck is found to be significantly deteriorated, it is replaced by glass fiber reinforced plastics (GFRP) components, which are prefabricated in the factory and transported to the site for assembly. Since both the glass fiber and the polymeric matrix employed for making GFRP components are non-corroding in the presence of salt, and GFRP is not vulnerable to freezing and thawing, the durability of the GFRP deck is expected to be greatly improved over its concrete counterpart. Moreover, as the strength to weight ratio of GFRP is much higher than that of reinforced concrete, the total weight of the GFRP deck is significantly lower. As a result, the dead load acting on the supporting structure (including the beam girders and columns) is reduced. Such a reduction in dead load is particularly advantageous for composite bridges with steel girders and a concrete deck that have been in service for many years. After years of repeated loading, the load-carrying capacities of the steel members, as well as that of the connections, are likely to be reduced. By decreasing the total load acting on the supporting members, their serviceable lifetime under further traffic loading can be extended. It should be noted that the relatively low weight of GFRP members also facilitates the ease of construction. Therefore, while GFRP components are currently used for the replacement of concrete deck in existing bridges, they are also applicable to new bridges under severe environmental conditions.

The field application of FRP bridge decks started in 1997. Since then, this technology has been adopted in many projects. An excellent summary of FRP deck applications can be found in a report by Keller (2001).

5.2 Materials

Various kinds of glass fibers are available commercially for different performance requirements. Due to cost considerations, E-glass is commonly used to make GFRP components for construction applications. For the polymer, polyester and vinylester are commonly employed. Their advantages include low viscosity at uncured stage and short curing time. Their major disadvantage is the high volumetric shrinkage on curing, which may induce significant residual stresses. Epoxy resin shrinks less during curing, but is not used due to its high cost and long curing time (up to several hours).

The general properties of the fiber and matrix materials are summarized in Table 5.1. From Table 5.1, it is clear that the glass fiber has much higher tensile strength and modulus than both polyester and vinylester. Also, in a typical GFRP, the fiber volume fraction ranges from 30 to 60 percent. As a result, the strength and stiffness along the fiber direction are governed by the fibers. The compression strength of glass fiber is not given, due to the fact that single fibers (or fiber bundles) buckle easily under compressive load. When embedded inside a polymer, the composite can carry significant compression and failure occurs through the formation of a shear band by collaborative fiber buckling rather than crushing of the glass fiber. The compressive strength of the glass fiber itself is therefore not a relevant material

Table 5.1 Properties of the fiber and matrix materials for the fabrication of GFRP

	Glass fiber	Polyester	Vinylester
Fiber diameter (μm)	8–14	–	–
Density (kg/m^3)	2,560	1,200–1,500	1,030–1,170
Young's modulus (GPa)	76	2–4.5	3.1–3.7
Tensile strength (MPa)	1,400–2,500	40–90	64–89
Strain at tensile failure (%)	1.8–3.2	2	3–6
Compressive strength	–	90–250	116–295
Coefficient of expansion (°C)	4.9×10^{-6}	$1–2 \times 10^{-4}$	$3.2–3.8 \times 10^{-5}$
Thermal conductivity (W °C/m)	1.04	0.2	0.25
Shrinkage on curing (%)	–	4–8	1–8
Glass transition temperature (°C)	–	70–120	60–195
Water absorption 24 hrs at 20 °C (%)	–	0.1–0.3	0.1–0.3
Cost (U.S.$/kg)	2–3	2–3	~4

parameter to be measured. From Table 5.1, one can also note the large difference in thermal expansion coefficient between the fiber and matrix. However, with the low modulus of polymers relative to glass, the thermally induced stress within the composite is not high. Also, with proper coupling agents applied on the glass surface, good bonding with polymer can be achieved, so debonding will not occur at the end of fibers where high shear stresses are required for strain compatibility to be achieved.

In the fabrication of GFRP, both polyester and vinylester are commonly employed. Vinylester is often chosen over polyester for its better chemical and thermal resistance. It also has low viscosity and short curing time but the volumetric shrinkage can be high.

One important consideration in the use of GFRP is the reduction of load carrying capacity under sustained loading or cyclic fatigue. For glass fiber alone, the phenomenon of stress rupture (i.e. drop in static strength with the duration of loading) is well known. This effect is attributed to the action of water moisture at the tip of surface cracks. Interaction between water and glass molecules in highly stressed regions results in bond breakage and propagation of cracks. When the crack size increases under sustained load-ing to reach the critical value, fracture will occur. According to this mechan-ism, static fatigue should be less of a problem in a dry environment, and this has indeed been experimentally verified. In other words, by limiting the contact of glass with moisture, the static fatigue problem can be alleviated.

In GFRP, fibers are embedded inside a polymeric matrix that forms a barrier to water moisture penetration. However, according to many results in the literature, the static fatigue phenomenon is still present in GFRP com-ponents. This is an indication that the polymer can slow down the penetra-tion of water but cannot stop it completely. To account for the stress rupture of GFRP members, the design strength is often taken to be much lower (e.g. at about 20 percent) than the value measured in the laboratory.

Under cyclic loading conditions, due to the low stiffness of the matrix, the glass fibers carry most of the loading. The fatigue phenomenon observed for single fiber is hence also expected to be found for the GFRP. To design GFRP components under cyclic loading, relevant strength reduction factors should be adopted.

5.3 Fabrication process and example systems

Various processes have been developed for the fabrication of GFRP com-ponents. In this section, we'll describe four processes that are particularly suitable for making bridge deck components.

5.3.1 Pultrusion

Pultrusion is most suitable for the mass production of components with uniform cross section. Figure 5.1 shows the process with dry fibers as the

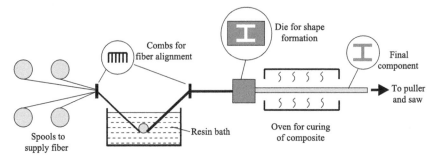

Figure 5.1 The pultrusion process for composite fabrication.

starting material. Fibers from spools are first pulled through a resin bath so they are soaked by the polymeric resin to be employed as a matrix. A pair of combs is often located before and after the bath to ensure proper fiber alignment. The wet fibers then pass through a die (which is often heated to facilitate the flow of resin) to form the required shape of the component. The formed component is pulled slowly through an oven for the curing process to be completed. Finally, the fabricated component is cut into specific lengths to be stored and/or transported.

Besides continuous fiber strands, weaved fiber cloth or fiber mat (with random chopped fibers embedded inside a resin) can also be pultruded to form composite members. Actually, various combinations of fiber strand, weaved cloth and mat can be employed to produce components with different performance requirements.

Various kinds of pultruded GFRP bridge deck components have been commercially produced and applied in the field. Examples are given in Figure 5.2. The Asset system manufactured by Fiberline, Denmark (Figure 5.2(a)) is the result of a research project funded by the European Community. Components are produced in the form of a parallelogram stiffened by a diagonal plate (which divides the parallelogram into two triangles). The individual members are glued together on site to form the bridge deck. To facilitate member connections, extrusions and recesses are provided on the top, bottom and sides. The Superdeck system by Creative Pultrusion, Inc., U.S.A. (Figure 5.2(b)) and the Duraspan system by Martin Marietta and Creative Pultrusion, Inc., U.S.A. (Figure 5.2(c)) are each composed of two matching components. Each kind of component in the Superdeck system is produced by pultrusion alone. For the Duraspan system, the rectangular part (which is divided into two trapezoids by the inclined stiffener) is pultruded, while the upper plate is bonded on afterwards. In the field, the matching components are glued together to form the complete deck. For the above three components, the total depth is 225 mm or below, making them unsuitable for spanning over long distances. In practice, these GFRP members are used to construct a laterally spanning deck system which

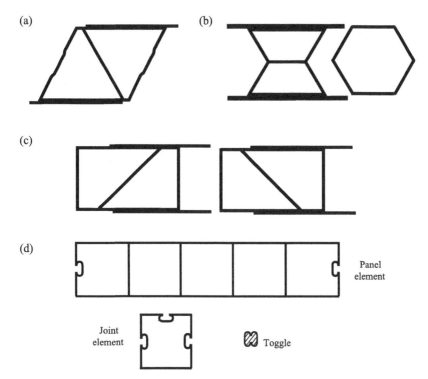

Figure 5.2 Examples of pultruded GFRP components: (a) Asset system; (b) Super-deck system; (c) Duraspan system; (d) ACCS system.

has to be supported by longitudinal girders. Field application of these systems as replacement of the concrete bridge deck can be found in the US and Europe.

The ACCS (Advanced Composite Construction System) by Maunsell Structural Plastics, UK (Figure 5.2(d)) offers the flexibility to produce box sections of various sizes that can be placed directly between two sets of piers. In other words, both the longitudinal girder and the laterally spanning deck can be replaced by this system. The system consists of three kinds of pul-truded elements, including (1) the panel element, formed by multiple squares (each of 75 × 75 mm in size), with special details at the ends for connection; (2) the joint element, which is a single square to be placed at the connection of horizontal and vertical panels; and (3) the toggle element, which is a solid component for connection purposes. With the proper combination of joint elements (which can be either two-way or three-way) and horizontal/vertical panel elements, single or multiple box sections of various sizes can be pro-duced. Using this system, pedestrian bridges and road bridges have been constructed in the UK.

5.3.2 Filament winding

Filament winding is a common process for making composite members in tubular forms. One variation of this process is illustrated in Figure 5.3. The key component of the process is a mandrel or internal form, which is fixed to a rotary fixture. The mandrel can be removed from the composite after the fabrication process is complete, or it can stay with the final product. An example of the latter case is a protruded tubular mandrel with fibers along the pultruding direction alone. With fibers wound around its surfaces, the shear and torsional capacity can be significantly improved. As in pultrusion, fibers are first passed through a resin bath and the ends of the wet fibers are fixed on one side of the mandrel. By setting the mandrel in rotation and moving the fibers (together with the whole resin bath) in a direction along the mandrel's axis of rotation, the fibers are wound onto the mandrel to form a composite member. As a variation of the above process, it is also possible to wind plain fibers onto the mandrel and then add the resin into the fiber assembly through an impregnation process. In either case, after the composite is formed, curing is required for the polymer resin to gain stiffness and strength.

5.3.3 Hand lay-up

Hand lay-up is a very versatile process for the fabrication of composites with different fiber arrangements. It relies on the availability of fiber pre-pregs, which are fiber sheets with aligned fibers already wet with uncured or partially cured resins. These pre-pregs are made by passing fiber strands through a resin bath and then wrapping them onto a big roller to form sheets with fiber aligned along the same direction. During the wrapping process, waxed paper is wrapped onto the roller together with the sheet to avoid individual fiber sheets from sticking together. The pre-pregs prepared in this

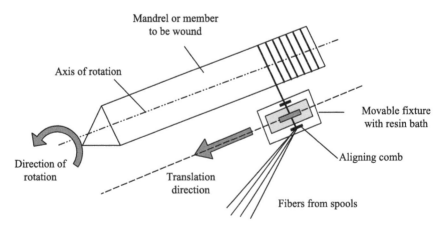

Figure 5.3 The filament winding process.

manner are then sold as a roll of composite sheet to the user. The typical thickness of glass fiber pre-preg sheet is between 0.25 to 0.50 mm. To make a composite member, the sheet is cut into the required size with the proper fiber orientation. The individual sheets are then stacked on top of one another, until the required thickness is achieved with the specified fiber arrangement. The process is illustrated in Figure 5.4. One can easily see that hand lay-up is suitable for making parts in the form of a plate. It offers excellent design flexibility as the engineer can specify different fiber orientations in different sheets to achieve the required structural performance. In most structural applications, plates formed by hand lay-up are used together with other components to form a highly effective system. For example, in the production of aircraft wings, composite plates are employed as the surface skins separated by metallic stiffeners to take advantage of the high in-plane stiffness and strength of composite materials. For composite bridge decks, the Manitoba system, developed in Canada, adopted a similar approach. The system, which is illustrated in Figure 5.5, makes use of all the composite fabrication techniques discussed so far. Hand-laid plates are placed on the top and bottom, separated by filament wound triangular tubes in the middle. In the wrapping process, to avoid excessively large curvature of the fibers, the corners of the triangular sections are rounded. Pultruded solid rods are added at the corner locations to prevent the formation of voids and

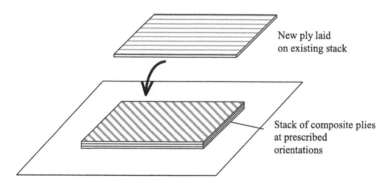

New ply laid on existing stack

Stack of composite plies at prescribed orientations

Figure 5.4 The hand lay-up process.

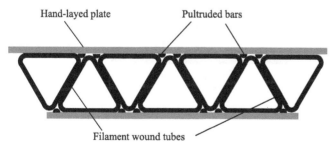

Hand-layed plate   Pultruded bars

Filament wound tubes

Figure 5.5 The Manitoba deck system.

to further strengthen the member. The Manitoba system is again designed as a laterally spanning deck and needs to be supported by longitudinal girders.

5.3.4 *Vacuum-assisted resin transfer molding (VARTM)*

Vacuum-assisted resin transfer molding (VARTM) is a relatively new composite fabrication method. As illustrated in Figure 5.6, it involves the use of a mold in which plain fibers are placed in a prescribed arrangement. On one side of the mold is an opening connecting to a resin container, while an opening on the other side is connected to a vacuum pump. By drawing a vacuum, resin will flow slowly into the mold and impregnate the fiber. The major advantage of this technique is the shape flexibility, as components of complex shapes can be produced.

A modified VARTM method has been used by Hardcore Composites USA to fabricate GFRP components for bridge deck replacement. The deck, as illustrated in Figure 5.7, consists of GFRP plates on the top and bottom, separated by foam material in between. The foam core of the deck is made with hard foam blocks that are wrapped with fiber composites. In the fabrication of the deck, hard foam blocks are placed side by side in between the upper and lower layers of plain glass fibers. After the whole deck is assembled, it is sealed within a bag, and vacuum is drawn to facilitate resin impregnation. In this method, members over 9 m in span have been produced for use as longitudinally spanning deck without the need for additional girders.

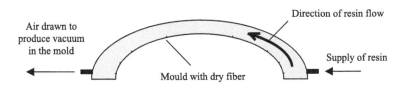

Figure 5.6 Illustration of the VARTM.

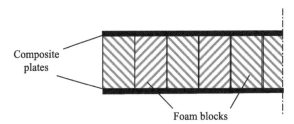

Figure 5.7 The Hardcore Composite system.

5.3.5 *Connections for FRP deck systems*

In the examples given in Figure 5.2 and Figure 5.5, the connection of adjacent FRP members is illustrated. In practice, the FRP deck system has to be properly secured to the bridge girders or piers. Also, safety rails have to be constructed on the sides of the deck. Typical connection methods and details are provided by the deck manufacturers. However, since typical designs may not cover all site conditions, it is highly advisable for the structural engineer to work together with the manufacturer to come up with the most effective connections that are also easy to implement.

5.4 Analysis of FRP bridge deck members

As shown in the last section, most practical FRP bridge deck members are in the form of a multiple tube spanning along one direction or composite plates separated by a foam core. In the former case, as each plate element constituting the tube member has a thickness much smaller than its width, the FRP tube can be treated as a thin-walled structural component, the analysis of which is covered in texts on mechanics of materials. In the latter case, the FRP deck can be analyzed as a sandwich construction with stiff plate elements separated by a soft core. Compared to conventional members made with homogeneous materials, the additional complexity is that each plate element may be made up of many fiber layers with different fiber orientations (Figure 5.8). In other words, the properties of the wall vary over its thickness according to the local fiber arrangement. To analyze the behavior of the FRP member, the equivalent properties of each plate element needs to be obtained first. The theory for finding the equivalent stiffness of a layered composite, as well as the stresses in each layer when loading is applied, is referred to as the classical lamination plate theory. In the following sections, a summary of this theory will be given. Readers interested in further details should refer to the classical text-books by Halpin (1984) and Jones (1998).

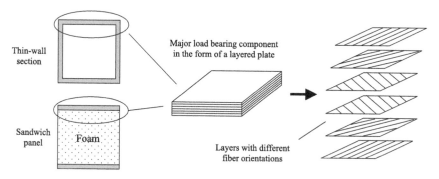

Figure 5.8 The composition of an FRP structural member.

Before describing the theory, a number of terminologies and notations need to be introduced. First, a layer of material with fibers aligning along the same direction is called a lamina. A composite made up of a number of bonded lamina is called a laminated composite or simply a laminate. Conventionally, laminated composites are often constructed with pre-pregs using the hand lay-up process. Each layer of pre-preg in the composite (which is technically a lamina) is referred to as a ply. The notation of the composition of a laminated composite follows the fiber orientation in each ply. For example, [0/0/30/–30/90/90/–30/30/0/0] represents a composite built with 10 plies of a particular lamina, with fibers in each layer oriented at the stated angles. As one can see, such a notation can get very lengthy. To simplify the notation, the following is commonly performed: (1) n subsequent plies with the same fiber orientation θ are denoted as θ_n; (2) two subsequent plies with opposite fiber orientations ($+\theta$ and $-\theta$) are grouped together as $\pm\theta$; (3) symmetrical layer arrangements (which are commonly found in practice for a reason to be explained later) are written as [...]$_s$. With these simplifying approaches, the above composite is denoted as [0$_2$/±30/90]$_s$.

5.4.1 Elastic behavior of composite lamina

The composite lamina comprises aligned fibers in a polymeric matrix. As it is in the form of a thin sheet, we are mainly interested in its in-plane properties. Following the axis notation in Figure 5.9, the in-plane strain and stress components are related by:

$$\begin{bmatrix} \varepsilon_1 \\ \varepsilon_2 \\ \gamma_{12} \end{bmatrix} = \begin{bmatrix} \dfrac{1}{E_1} & \dfrac{-v_{21}}{E_2} & 0 \\ \dfrac{-v_{12}}{E_1} & \dfrac{1}{E_2} & 0 \\ 0 & 0 & \dfrac{1}{G_{12}} \end{bmatrix} \begin{bmatrix} \sigma_1 \\ \sigma_2 \\ \tau_{12} \end{bmatrix} \tag{5.1}$$

where $v_{ij} = -\varepsilon_j/\varepsilon_i$ when there is direct stress along the i-direction alone. Also, due to symmetry of the elastic matrix, $v_{21}/E_2 = v_{12}/E_1$.

The elastic constants in Eq (5.1) have similar physical meaning to Young's modulus and Poisson's ratio in isotropic materials. E_i and v_{ij} can be obtained by applying loading along the i-direction alone, and measuring the strain along both the i and j directions. To find G_{12}, one can either perform the in-plane shear test, or the torsion test on a thin cylinder made of the lamina, with the fibers aligning along the axis of the cylinder.

In the practical analysis of a component under loading, the strain components often need to be calculated first (see example on p. 259). It is hence necessary to express the stress components in terms of the strain components as:

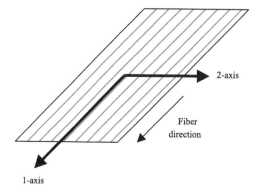

Figure 5.9 Local (1–2) coordinates for the lamina.

$$\begin{bmatrix} \sigma_1 \\ \sigma_2 \\ \tau_{12} \end{bmatrix} = \begin{bmatrix} Q_{11} & Q_{12} & 0 \\ Q_{12} & Q_{22} & 0 \\ 0 & 0 & Q_{66} \end{bmatrix} \begin{bmatrix} \varepsilon_1 \\ \varepsilon_2 \\ \gamma_{12} \end{bmatrix} = [Q] \begin{bmatrix} \varepsilon_1 \\ \varepsilon_2 \\ \gamma_{12} \end{bmatrix} \qquad (5.2)$$

where:

$$Q_{11} = \frac{E_1}{(1 - v_{12}v_{21})}; \qquad Q_{12} = \frac{v_{12}E_2}{(1 - v_{12}v_{21})} = \frac{v_{21}E_1}{(1 - v_{12}v_{21})};$$

$$Q_{22} = \frac{E_2}{(1 - v_{12}v_{21})}; \qquad Q_{66} = G_{12}$$

It should be noted that the subscript "66" for the shear stiffness arises from the condensation of the 6 × 6 stiffness matrix in three dimensions to the 3 × 3 matrix in two dimensions. We retain this subscript to maintain consistency with other texts on composite laminate theory.

5.4.1.1 Stress–strain relation for a lamina at arbitrary orientation

In a composite laminate, the individual layers of lamina are oriented at different directions to achieve the required properties for a particular application. To obtain the "averaged" elastic behavior of the laminate in a given coordinate system, it is necessary to know the transformed properties of each lamina in that system. In Figure 5.10, the 1–2 coordinates correspond to the directions parallel and perpendicular to the fibers, while the *x–y* axis denotes the "global" coordinates of the composite laminate. For example, for a [0/±30]$_s$ laminate, the fibers in the 0-degree lamina is along the *x*-direction.

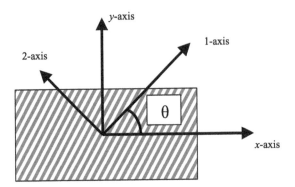

Figure 5.10 Definition of local (1–2) and global (x–y) coordinates.

The transformed elastic properties can be derived from stress and strain transformation rules as follows. The stress components in the *x–y* coordinates are related to those in the 1–2 coordinates by:

$$\begin{bmatrix} \sigma_x \\ \sigma_y \\ \tau_{xy} \end{bmatrix} = \begin{bmatrix} \cos^2\theta & \sin^2\theta & -2\sin\theta\cos\theta \\ \sin^2\theta & \cos^2\theta & 2\sin\theta\cos\theta \\ \sin\theta\cos\theta & -\sin\theta\cos\theta & \cos^2\theta - \sin^2\theta \end{bmatrix} \begin{bmatrix} \sigma_1 \\ \sigma_2 \\ \tau_{12} \end{bmatrix} = [T]^{-1} \begin{bmatrix} \sigma_1 \\ \sigma_2 \\ \tau_{12} \end{bmatrix} \quad (5.3a)$$

Similarly,

$$\begin{bmatrix} \varepsilon_x \\ \varepsilon_y \\ \gamma_{xy}/2 \end{bmatrix} = [T]^{-1} \begin{bmatrix} \varepsilon_1 \\ \varepsilon_2 \\ \gamma_{12}/2 \end{bmatrix} \quad (5.3b)$$

Defining $[R] = \begin{bmatrix} 1 & 0 & 0 \\ 0 & 1 & 0 \\ 0 & 0 & 2 \end{bmatrix}$, we have $\begin{bmatrix} \varepsilon_1 \\ \varepsilon_2 \\ \gamma_{12} \end{bmatrix} = [R] \begin{bmatrix} \varepsilon_1 \\ \varepsilon_2 \\ \gamma_{12}/2 \end{bmatrix} \quad (5.4a)$

and $\begin{bmatrix} \varepsilon_x \\ \varepsilon_y \\ \gamma_{xy} \end{bmatrix} = [R] \begin{bmatrix} \varepsilon_x \\ \varepsilon_y \\ \gamma_{xy}/2 \end{bmatrix}. \quad (5.4b)$

From Eq. (5.2) and the transformation of strain components:

$$\begin{bmatrix} \sigma_1 \\ \sigma_2 \\ \tau_{12} \end{bmatrix} = [Q] \begin{bmatrix} \varepsilon_1 \\ \varepsilon_2 \\ \gamma_{12} \end{bmatrix} = [Q][R] \begin{bmatrix} \varepsilon_1 \\ \varepsilon_2 \\ \gamma_{12}/2 \end{bmatrix} = [Q][R][T] \begin{bmatrix} \varepsilon_x \\ \varepsilon_y \\ \gamma_{xy}/2 \end{bmatrix}$$

$$= [Q][R][T][R]^{-1} \begin{bmatrix} \varepsilon_x \\ \varepsilon_x \\ \gamma_{xy} \end{bmatrix} \quad (5.5)$$

From stress transformation,

$$
\begin{bmatrix} \sigma_x \\ \sigma_y \\ \tau_{xy} \end{bmatrix} = [T]^{-1}[Q][R][T][R]^{-1} \begin{bmatrix} \varepsilon_x \\ \varepsilon_y \\ \gamma_{xy} \end{bmatrix} = \begin{bmatrix} \bar{Q}_{11} & \bar{Q}_{12} & \bar{Q}_{16} \\ \bar{Q}_{21} & \bar{Q}_{22} & \bar{Q}_{26} \\ \bar{Q}_{61} & \bar{Q}_{62} & \bar{Q}_{66} \end{bmatrix} \begin{bmatrix} \varepsilon_x \\ \varepsilon_y \\ \gamma_{xy} \end{bmatrix} \qquad (5.6)
$$

where:

$$
\bar{Q}_{11} = Q_{11} \cos^4\theta + 2(Q_{12} + 2Q_{66}) \sin^2\theta \cos^2\theta + Q_{22} \sin^4\theta
$$
$$
\bar{Q}_{12} = \bar{Q}_{21} = (Q_{11} + Q_{22} - 4Q_{66}) \sin^2\theta \cos^2\theta + Q_{12}(\sin^4\theta + \cos^4\theta)
$$
$$
\bar{Q}_{22} = Q_{11} \sin^4\theta + 2(Q_{12} + 2Q_{66}) \sin^2\theta \cos^2\theta + Q_{22} \cos^4\theta \qquad (5.7\ \text{a–f})
$$
$$
\bar{Q}_{16} = \bar{Q}_{61} = (Q_{11} - Q_{12} - 2Q_{66}) \sin\theta \cos^3\theta + (Q_{12} - Q_{22} + 2Q_{66}) \sin^3\theta \cos\theta
$$
$$
\bar{Q}_{26} = \bar{Q}_{62} = (Q_{11} - Q_{12} - 2Q_{66}) \sin^3\theta \cos\theta + (Q_{12} - Q_{22} + 2Q_{66}) \sin\theta \cos^3\theta
$$
$$
\bar{Q}_{66} = (Q_{11} + Q_{22} - 2Q_{12} - 2Q_{66}) \sin^2\theta \cos^2\theta + Q_{66}(\sin^4\theta + \cos^4\theta)
$$

Alternatively, the components of the $[\bar{Q}]$ matrix can be expressed in terms of the invariants U_is of a given lamina as:

$$
\bar{Q}_{11} = U_1 + U_2 \cos 2\theta + U_3 \cos 4\theta
$$
$$
\bar{Q}_{12} = \bar{Q}_{21} = U_4 - U_3 \cos 4\theta
$$
$$
\bar{Q}_{22} = U_1 - U_2 \cos 2\theta + U_3 \cos 4\theta
$$
$$
\bar{Q}_{16} = \bar{Q}_{61} = \tfrac{1}{2}U_2 \sin 2\theta + U_3 \sin 4\theta \qquad (5.8\ \text{a–f})
$$
$$
\bar{Q}_{26} = \bar{Q}_{62} = \tfrac{1}{2}U_2 \sin 2\theta - U_3 \sin 4\theta
$$
$$
\bar{Q}_{66} = U_5 - U_3 \cos 4\theta
$$

with

$$
U_1 = \tfrac{1}{8}(3Q_{11} + 3Q_{22} + 2Q_{12} + 4Q_{66})
$$
$$
U_2 = \tfrac{1}{2}(Q_{11} - Q_{22})
$$
$$
U_3 = \tfrac{1}{8}(Q_{11} + Q_{22} - 2Q_{12} - 4Q_{66}) \qquad (5.9\ \text{a–e})
$$
$$
U_4 = \tfrac{1}{8}(Q_{11} + Q_{22} + 6Q_{12} - 4Q_{66})
$$
$$
U_5 = \tfrac{1}{8}(Q_{11} + Q_{22} - 2Q_{12} + 4Q_{66})
$$

5.4.2 Classical lamination theory

The classical lamination theory covers the analysis of composite laminates under combined effects of bending, torsion and in-plane forces. It is based on Kirchhoff's assumption of "plane section remaining plane", and thus is applicable only to plates with a large ratio of span/thickness. Due to the small thickness of common composite laminates in practice, this condition is normally satisfied. With Kirchhoff's kinematic assumption, the basic unknowns in the problems are the in-plane displacements (which can be related to the in-plane strains) and the curvatures of the laminate. Let us define u_o, v_o an w_o to be the displacements at the neutral plane of the laminate along the x, y and z directions respectively. From Figure 5.11, the displacement of the laminate along the x-direction (u), at any distance z from the neutral plane, is given by:

$$u = u_o - z \frac{\partial w_o}{\partial x} \tag{5.10a}$$

Similarly, the displacement v along the y-direction is:

$$v = v_o - z \frac{\partial w_o}{\partial y} \tag{5.10b}$$

The in-plane strain components are then:

$$
\begin{bmatrix} \varepsilon_x \\ \varepsilon_y \\ \gamma_{xy} \end{bmatrix} =
\begin{bmatrix} \partial u/\partial x \\ \partial v/\partial y \\ \partial u/\partial y + \partial v/\partial x \end{bmatrix} =
\begin{bmatrix} \partial u_o/\partial x \\ \partial v_o/\partial y \\ \partial u_o/\partial y + \partial v_o/\partial x \end{bmatrix} + z
\begin{bmatrix} -\partial^2 w_o/\partial^2 x \\ -\partial^2 w_o/\partial^2 x \\ -\partial^2 w_o/\partial x \partial y \end{bmatrix}
$$

$$
= \begin{bmatrix} \varepsilon_x^o \\ \varepsilon_y^o \\ \gamma_{xy}^o \end{bmatrix} + z
\begin{bmatrix} \kappa_x \\ \kappa_x \\ \kappa_{xy} \end{bmatrix}
\tag{5.11}
$$

where κ_x, κ_y and κ_{xy} represent the curvatures of the plate.

To find the governing equations for the laminate, the notations in Figure 5.12 will be followed. In Figure 5.12, an element of the laminate is shown.

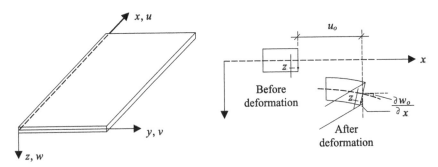

Figure 5.11 Illustration of the plane-section-remains-plane assumption.

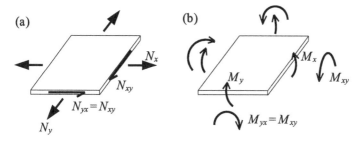

Figure 5.12 Definition of forces and bending/torsional moments for the plate.

N_x, N_y and N_{xy} are the in-plane normal and shear forces per unit length along the sides of the element, M_x, M_y and M_{xy} represent the bending and torsional moments per unit length.

From equilibrium, the forces and moments can be expressed in terms of stress components by the following integrals over the total thickness (t) of the laminate:

$$
\begin{bmatrix} N_x \\ N_y \\ N_{xy} \end{bmatrix} = \int_{-t/2}^{t/2} \begin{bmatrix} \sigma_x \\ \sigma_y \\ \tau_{xy} \end{bmatrix} dz \tag{5.12a}
$$

$$
\begin{bmatrix} M_x \\ M_y \\ M_{xy} \end{bmatrix} = \int_{-t/2}^{t/2} \begin{bmatrix} \sigma_x \\ \sigma_y \\ \tau_{xy} \end{bmatrix} z\,dz \tag{5.12b}
$$

For a laminate comprising N plies of lamina, each integral in Eq. (5.12) can be written as the summation of N separate integrals, each evaluated over a specific layer. Eq. (5.12a) and (5.12b) can then be combined with Eq. (5.6) and Eq. (5.11) to give:

$$
\begin{bmatrix} N_x \\ N_y \\ N_{xy} \end{bmatrix} = \sum_{k=1}^{N} \int_{z_{k-1}}^{z_k} \begin{bmatrix} \sigma_x \\ \sigma_y \\ \tau_{xy} \end{bmatrix}_k dz = \sum_{k=1}^{N} \begin{bmatrix} \bar{Q}_{11} & \bar{Q}_{12} & \bar{Q}_{16} \\ \bar{Q}_{21} & \bar{Q}_{22} & \bar{Q}_{26} \\ \bar{Q}_{61} & \bar{Q}_{62} & \bar{Q}_{66} \end{bmatrix}_k \int_{z_{k-1}}^{z_k} \left(\begin{bmatrix} \varepsilon_x^o \\ \varepsilon_y^o \\ \varepsilon_y^o \end{bmatrix} + \begin{bmatrix} \kappa_x \\ \kappa_y \\ \kappa_{xy} \end{bmatrix} z \right) dz
$$

$$
\tag{5.13a}
$$

$$
= \begin{bmatrix} A_{11} & A_{12} & A_{16} \\ A_{21} & A_{22} & A_{26} \\ A_{61} & A_{62} & A_{66} \end{bmatrix} \begin{bmatrix} \varepsilon_x^o \\ \varepsilon_y^o \\ \gamma_{xy}^o \end{bmatrix} + \begin{bmatrix} B_{11} & B_{12} & B_{16} \\ B_{21} & B_{22} & B_{26} \\ B_{61} & B_{62} & B_{66} \end{bmatrix} \begin{bmatrix} \kappa_x \\ \kappa_y \\ \kappa_{xy} \end{bmatrix}
$$

$$
\begin{bmatrix} M_x \\ M_y \\ M_{xy} \end{bmatrix} = \sum_{k=1}^{N} \int_{z_{k-1}}^{z_k} \begin{bmatrix} \sigma_x \\ \sigma_y \\ \tau_{xy} \end{bmatrix}_k z\,dz = \sum_{k=1}^{N} \begin{bmatrix} \bar{Q}_{11} & \bar{Q}_{12} & \bar{Q}_{16} \\ \bar{Q}_{21} & \bar{Q}_{22} & \bar{Q}_{26} \\ \bar{Q}_{61} & \bar{Q}_{62} & \bar{Q}_{66} \end{bmatrix}_k \int_{z_{k-1}}^{z_k} \left(\begin{bmatrix} \varepsilon_x^o \\ \varepsilon_y^o \\ \gamma_{xy}^o \end{bmatrix} z + \begin{bmatrix} \kappa_x \\ \kappa_y \\ \kappa_{xy} \end{bmatrix} z^2 \right) dz
$$

$$
= \begin{bmatrix} B_{11} & B_{12} & B_{16} \\ B_{21} & B_{22} & B_{26} \\ B_{61} & B_{62} & B_{66} \end{bmatrix} \begin{bmatrix} \varepsilon_x^o \\ \varepsilon_y^o \\ \gamma_{xy}^o \end{bmatrix} + \begin{bmatrix} D_{11} & D_{12} & D_{16} \\ D_{21} & D_{22} & D_{26} \\ D_{61} & D_{62} & D_{66} \end{bmatrix} \begin{bmatrix} \kappa_x \\ \kappa_y \\ \kappa_{xy} \end{bmatrix} \qquad (5.13b)
$$

where:

$$
A_{ij} = \sum_{k=1}^{N} (\bar{Q}_{ij})_k \, (z_k - z_{k-1});
$$

$$
B_{ij} = \frac{1}{2} \sum_{k=1}^{N} (\bar{Q}_{ij})_k \, (z_k^2 - z_{k-1}^2);
$$

$$
D_{ij} = \frac{1}{3} \sum_{k=1}^{N} (\bar{Q}_{ij})_k \, (z_k^3 - z_{k-1}^3)
$$

Eq. (5.13a) and Eq. (5.13b) can be combined together to form a 6 × 6 matrix:

$$
\begin{bmatrix} \tilde{N} \\ \tilde{M} \end{bmatrix} = \begin{bmatrix} A & B \\ B & D \end{bmatrix} \begin{bmatrix} \tilde{\varepsilon}_o \\ \tilde{\kappa} \end{bmatrix} \qquad (5.14)
$$

with:

$\tilde{N}$ being a 3 × 1 vector representing the forces/unit length;
$\tilde{M}$ being a 3 × 1 vector representing the moments/unit length;
$\tilde{\varepsilon}_o$ being a 3 × 1 vector representing the in-plane strain components along the mid-plane of the laminate;
$\tilde{\kappa}$ being a 3 × 1 vector representing the curvatures of the laminate.

Also, [A], [B] and [D] are the 3 × 3 matrices in Eqs (5.13a) and (5.13b).

Eq. (5.14) is the governing equation to find the strains and curvatures associated with a given set of applied forces and moments. Due to the presence of the sub-matrix [B] in the equation, there is plausible coupling between the in-plane deformation and the bending/torsional moments as well as the curvatures and the in-plane forces. In other words, if pure bending is applied to the laminate, there may be elongation, shortening or shearing of the mid-plane. Similarly, under the presence of in-plane forces alone, the laminate may undergo bending or twisting. From the design engineer's point of view, this kind of coupling is not desirable, as any restraint to the induced deformation will induce secondary stresses into the member. It is therefore common practice to arrange the lamina layers in such a way as to make all the components of the $[B]$ matrix zero. A simple approach is to use a symmetric laminate. For each layer at a given orientation above the mid-plane, there is a corresponding layer of the same orientation at the same distance below the mid-plane. The contribution of these two layers to any component of the $[B]$ matrix will then sum to zero (the proof is left to the reader as an exercise). With zero contribution from each pair of lamina above and below the mid-plane, the $[B]$ matrix becomes the null matrix and the coupling effects described above disappear. In the remaining part of the present chapter, we will only consider symmetric laminates which are commonly employed in practice.

For symmetric laminates, there is still the possibility of coupling between the direct stress and shear strain through the A_{16} and A_{26} components. In other words, when direct tension (or compression) is applied, shear deformation is also resulted. To eliminate this kind of coupling, the laminate should be designed with each angled lamina accompanied by a lamina at opposite angle (i.e. for any number of $+\theta$ lamina, there should be an equal number of $-\theta$ lamina). Also, through the D_{16} and D_{26} components, a bending moment will result in the twisting of the member. Since it is impossible to design a laminate with D_{16}, D_{26}, and all the components of the $[B]$ matrix to be zero at the same time, the bending/twisting coupling is allowed to remain in the practical design of composite laminates.

With Eq. (5.14), the strain components $\tilde{\varepsilon}_o$ at the middle of the composite laminate as well as the curvatures $\tilde{\kappa}$ can be calculated from the applied force and moments/torsion. Knowing $\tilde{\varepsilon}_o$ and $\tilde{\kappa}$, the in-plane strain components in each lamina layer (which vary linearly over the thickness of the lamina) can be obtained. Then, using the corresponding $[\bar{Q}]$ matrix for each lamina, the stress components are calculated. To illustrate the calculation procedures, an example is given below.

Example 5.1

A $[0_2/\pm30]_s$ laminate is constructed with GFRP lamina with the following properties:

$$E_1 = 53.8 \text{ GPa}, E_2 = 17.9 \text{ GPa}, v_{12} = 0.25,$$

$$G_{12} = 8.6 \text{ GPa}, \text{Ply thickness} = 0.25 \text{ mm}$$

Find the in-plane stress components in each ply under:

1   in-plane forces $N_x =$ 20,000 MN/m;
2   Moment $M_x =$ 200 Nm/m.

The first step in solving this problem is to derive the $[A]$ and $[D]$ matrices. From the given elastic properties,

$$v_{21} = \frac{v_{12}E_2}{E_1} = 0.08318$$

$$Q_{11} = \frac{E_1}{(1 - v_{12}v_{21})} = 54.94 \text{ GPa}$$

$$Q_{12} = \frac{v_{12}E_2}{(1 - v_{12}v_{21})} = \frac{v_{21}E_1}{(1 - v_{12}v_{21})} = 4.57 \text{ GPa}$$

$$Q_{22} = \frac{E_2}{(1 - v_{12}v_{21})} = 18.28 \text{ GPa}$$

$$Q_{66} = G_{12} = 8.6 \text{ GPa}$$

The invariants are given by:

$$U_1 = \tfrac{1}{8}(3Q_{11} + 3Q_{22} + 2Q_{12} + 4Q_{66}) = 32.90 \text{ GPa}$$
$$U_2 = \tfrac{1}{2}(Q_{11} - Q_{22}) = 18.33 \text{ GPa}$$
$$U_3 = \tfrac{1}{8}(Q_{11} + Q_{22} - 2Q_{12} - 4Q_{66}) = 3.71 \text{ GPa}$$
$$U_4 = \tfrac{1}{8}(Q_{11} + Q_{22} + 6Q_{12} - 4Q_{66}) = 8.28 \text{ GPa}$$
$$U_5 = \tfrac{1}{8}(Q_{11} + Q_{22} - 2Q_{12} + 4Q_{66}) = 12.31 \text{ GPa}$$

Each component of the $[A]$ matrix can be calculated from:

$$A_{ij} = \sum_{k=1}^{N} (\bar{Q}_{ij})_k (z_k - z_{k-1})$$

with $(z_k - z_{k-1})$ being the thickness of each ply which is 0.25 mm.

According to Eq. (5.8), the above summation involves the summation of $\cos 2\theta$, $\cos 4\theta$, $\cos 2\theta$ and $\cos 2\theta$ over all the plies. This can be performed systematically in a table form (Table 5.2).

$$A_{11} = t_{ply} \sum_{k=1}^{8} [U_1 + U_2 \cos 2\theta + U_3 \cos 4\theta]_k$$

$$= t_{ply} \left[8U_1 + \left(\sum_{k=1}^{8} \cos 2\theta \right) U_2 + \left(\sum_{k=1}^{8} \cos 4\theta \right) U_3 \right]$$

$$= t_{ply} [8U_1 + 6U_2 + 2U_3] = 95.15 \text{ MN/m}$$

Table 5.2 Summation over all the plies

Ply	Angle	cos 2θ	cos 4θ	sin 2θ	sin 4θ
1	0	1	1	0	0
2	0	1	1	0	0
3	+30	0.5	−0.5	0.866	0.866
4	−30	0.5	−0.5	−0.866	−0.866
5	−30	0.5	−0.5	−0.866	−0.866
6	+30	0.5	−0.5	0.866	0.866
7	0	1	1	0	0
8	0	1	1	0	0
Sum		6	2	0	0

With similar calculations, the other components of the [A] matrix are found to be:

$A_{12} = 14.71$ MN/m

$A_{22} = 40.16$ MN/m

$A_{66} = 22.77$ MN/m

$$A_{16} = A_{26} = 0 \left(\because \sum_{k=1}^{8} \sin 2\theta = 0 \text{ and } \sum_{k=1}^{8} \sin 4\theta = 0 \right)$$

To find the [D] matrix, it is necessary to determine the $\bar{Q}_{ij}$ component for *each ply* first.

- *For the 0° plies:* The components of the $[\bar{Q}]$ matrix is the same as that of the [Q] matrix.
- *For the +30° ply:*

$$\bar{Q}_{11}^{[+30]} = U_1 + U_2 \cos(2 \times 30°) + U_3 \cos(4 \times 30°) = 40.21 \text{ GPa}$$

$$\bar{Q}_{12}^{[+30]} = U_4 - U_3 \cos(4 \times 30°) = 10.14 \text{ GPa}$$

$$\bar{Q}_{22}^{[+30]} = U_1 - U_2 \cos(2 \times 30°) + U_3 \cos(4 \times 30°) = 21.88 \text{ GPa}$$

$$\bar{Q}_{16}^{[+30]} = \tfrac{1}{2} U_2 \sin(2 \times 30°) + U_3 \sin(4 \times 30°) = 11.15 \text{ GPa}$$

$$\bar{Q}_{26}^{[+30]} = \tfrac{1}{2} U_2 \sin(2 \times 30°) - U_3 \sin(4 \times 30°) = 4.72 \text{ GPa}$$

$$\bar{Q}_{66}^{[+30]} = U_5 - U_3 \cos(4 \times 30°) = 14.17 \text{ GPa}$$

- *For the −30° ply:* Since $\cos(-a) = \cos(a)$ and $\sin(-a) = -\sin(a)$,

$$\bar{Q}_{11}^{[-30]} = \bar{Q}_{11}^{[+30]} = 40.21 \text{ GPa}$$

$$\bar{Q}_{12}^{[-30]} = \bar{Q}_{12}^{[+30]} = 10.14 \text{ GPa}$$

$$\bar{Q}_{22}^{[-30]} = \bar{Q}_{22}^{[+30]} = 21.88 \text{ GPa}$$

$$\bar{Q}_{66}^{[-30]} = \bar{Q}_{66}^{[+30]} = 14.17 \text{ GPa}$$

$$\bar{Q}_{16}^{[-30]} = -\bar{Q}_{16}^{[+30]} = -11.15 \text{ GPa}$$

$$\bar{Q}_{26}^{[-30]} = -\bar{Q}_{26}^{[+30]} = -4.72 \text{ GPa}$$

The components of the $[D]$ matrix are then given by:

$$D_{11} = \frac{1}{3} \sum_{k=1}^{8} (\bar{Q}_{11})_k (z_k^3 - z_{k-1}^3)$$

$$= 2 \times \tfrac{1}{3} [(64 - 27)(54.94) + (27 - 8)(54.94) + (8 - 1)(40.21)$$

$$+ (1 - 0)(40.21)] \, t_{\text{ply}}^3$$

$$= 35.40 \text{ Nm}$$

$$D_{12} = 2 \times \tfrac{1}{3} [(64 - 8)(4.57) + (8 - 1)(10.14) + (1 - 0)(10.14)] t_{\text{ply}}^3 = 3.51 \text{ Nm}$$

$$D_{22} = 2 \times \tfrac{1}{3} [(64 - 8)(18.28) + (8 - 1)(21.88) + (1 - 0)(21.88)] t_{\text{ply}}^3$$

$$= 12.49 \text{ Nm}$$

$$D_{16} = 2 \times \tfrac{1}{3} [(64 - 8)(0) + (8 - 1)(11.15) + (1 - 0)(-11.15)] t_{\text{ply}}^3 = 0.697 \text{ Nm}$$

$$D_{26} = 2 \times \tfrac{1}{3} [(64 - 8)(0) + (8 - 1)(4.72) + (1 - 0)(-4.72)] t_{\text{ply}}^3 = 0.295 \text{ Nm}$$

$$D_{66} = 2 \times \tfrac{1}{3} [(64 - 8)(8.6) + (8 - 1)(14.17) + (1 - 0)(14.17)] t_{\text{ply}}^3 = 6.20 \text{ Nm}$$

(i) When $N_x = 20,000$ N/m is applied, the in-plane strains at the mid-plane are given by:

$$\begin{bmatrix} 95.15 & 14.71 & 0 \\ 14.71 & 40.16 & 0 \\ 0 & 0 & 22.77 \end{bmatrix} \begin{bmatrix} \varepsilon_x^o \\ \varepsilon_y^o \\ \gamma_{xy}^o \end{bmatrix} = \begin{bmatrix} 20,000 \\ 0 \\ 0 \end{bmatrix} \times 10^{-6}$$

Note that the factor 10^{-6} appears in the above matrix as the A_{ij} components are in MN/m while the applied force is in N/m.

On solving,

$$\begin{bmatrix} \varepsilon_x^o \\ \varepsilon_y^o \\ \gamma_{xy}^o \end{bmatrix} = \begin{bmatrix} 222.80 \\ -81.61 \\ 0 \end{bmatrix} \times 10^{-6}$$

The stress components in each ply can then be calculated.

- *For the 0° plies:*

$$\begin{bmatrix} \sigma_x \\ \sigma_y \\ \tau_{xy} \end{bmatrix} = \begin{bmatrix} 54.94 & 4.57 & 0 \\ 4.57 & 18.28 & 0 \\ 0 & 0 & 8.6 \end{bmatrix} (\times 10^3 \text{MPa}) \begin{bmatrix} 22.80 \\ -81.61 \\ 0 \end{bmatrix} \times 10^{-6}$$

$$= \begin{bmatrix} 11.87 \\ -0.474 \\ 0 \end{bmatrix} (\text{MPa})$$

- *For the +30° plies:*

$$\begin{bmatrix} \sigma_x \\ \sigma_y \\ \tau_{xy} \end{bmatrix} = \begin{bmatrix} 40.21 & 10.14 & 11.15 \\ 10.14 & 21.88 & 4.72 \\ 11.15 & 4.72 & 14.17 \end{bmatrix} \times 10^3 \begin{bmatrix} 222.80 \\ -81.61 \\ 0 \end{bmatrix} \times 10^{-6} = \begin{bmatrix} 8.13 \\ -0.474 \\ 2.10 \end{bmatrix} (\text{MPa})$$

- *For the −30° plies:*

$$\begin{bmatrix} \sigma_x \\ \sigma_y \\ \tau_{xy} \end{bmatrix} = \begin{bmatrix} 40.21 & 10.14 & -11.15 \\ 10.14 & 21.88 & -4.72 \\ -11.15 & -4.72 & 14.17 \end{bmatrix} \times 10^3 \begin{bmatrix} 222.80 \\ -81.61 \\ 0 \end{bmatrix} \times 10^{-6}$$

$$= \begin{bmatrix} 8.13 \\ 0.474 \\ -2.10 \end{bmatrix} (\text{MPa})$$

The stress distribution along the upper 4 plies of the laminate is shown in Figure 5.13(a). Due to symmetry, the stresses in the lower 4 plies are mirror images of those in the upper plies. One can take a closer look at the calculated results to see if they are consistent with physical intuition. First of all, the stresses must satisfy global equilibrium. For example,

$$\sum_{k=1}^{N} t_{ply}\sigma_x = 0.25 \times 10^{-3}[(11.87 + 11.87 + 8.13 + 8.13) \times 10^6] \times 2$$
$$= 20{,}000 \text{ N/m}$$

Along the x-direction, the 0° ply is stiffer than the 30° ply. Under the same value of ε_x, the stress is higher for the 0° ply. Also, as both the +30° and −30° plies exhibit the same stiffness, they carry the same σ_x.

With fibers at 30° to the loading direction, rotation of fibers under the applied loading leads to increased lateral deformation. The Poisson's effect is hence more significant than the 0° ply. To achieve compatibility, lateral compressive stresses are induced in the 0° ply while the 30° ply is under lateral tension. This explains the presence of σ_y in all the plies, as well as the opposite sign of the stresses in the 0° and 30° plies.

The application of N_x on the laminate does not produce any shear strain. For the 0° ply, the shear stress is also zero. For the +30° and –30° plies, shear stresses of the same magnitude but opposite signs are induced.

(ii) When M_x = 200 Nm/m is applied on the laminate, the curvatures are given by:

$$
\begin{bmatrix} 35.40 & 3.51 & 0.697 \\ 3.51 & 12.49 & 0.295 \\ 0.697 & 0.295 & 6.20 \end{bmatrix} \begin{bmatrix} \kappa_x \\ \kappa_y \\ \kappa_{xy} \end{bmatrix} = \begin{bmatrix} 200 \\ 0 \\ 0 \end{bmatrix}
$$

On solving, $\begin{bmatrix} \kappa_x \\ \kappa_y \\ \kappa_{xy} \end{bmatrix} = \begin{bmatrix} 5.82 \\ -1.62 \\ -0.58 \end{bmatrix}$ /m

According to Kirchhoff's assumption for a plate under bending and/or torsion, the strain is linearly distributed along the depth of the plate. The stresses in each ply also vary linearly over its thickness, with a magnitude directly proportional to distance from the mid-plane. In the following, the stresses at the top of the laminate as well as the boundaries between plies of different properties are derived.

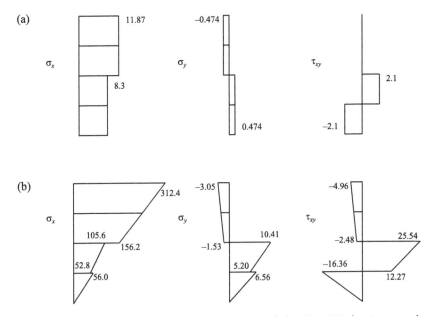

Figure 5.13 Stress distribution in the upper half of the [0₂/±30]ₛ laminate under (a) N_x = 20,000 MN/m, and (b) M_x = 200 Nm/m.

At the top of the laminate

$$
\begin{bmatrix} \sigma_x \\ \sigma_y \\ \tau_{xy} \end{bmatrix} = \begin{bmatrix} 54.94 & 4.57 & 0 \\ 4.57 & 18.28 & 0 \\ 0 & 0 & 8.6 \end{bmatrix} \times 10^3 \begin{bmatrix} 5.82 \\ -1.62 \\ -0.58 \end{bmatrix} \times 4 \times 0.25 \times 10^{-3}
$$

$$
= \begin{bmatrix} 312.44 \\ -3.05 \\ -4.96 \end{bmatrix} \ (MPa)
$$

At the boundary between the 0° and +30° plies
Right above the boundary, within the 0° ply,

$$
\begin{bmatrix} \sigma_x \\ \sigma_y \\ \tau_{xy} \end{bmatrix} = \begin{bmatrix} 312.35 \\ -3.02 \\ 4.558 \end{bmatrix} \times \tfrac{2}{4} = \begin{bmatrix} 156.22 \\ -1.53 \\ -2.48 \end{bmatrix} \ (MPa)
$$

Right below the boundary, within the +30° ply,

$$
\begin{bmatrix} \sigma_x \\ \sigma_y \\ \tau_{xy} \end{bmatrix} = \begin{bmatrix} 40.21 & 10.14 & 11.15 \\ 10.14 & 21.88 & 4.72 \\ 11.15 & 4.72 & 14.17 \end{bmatrix} \times 10^3 \begin{bmatrix} 5.82 \\ -1.62 \\ -0.58 \end{bmatrix} \times 2 \times 0.25 \times 10^{-3}
$$

$$
= \begin{bmatrix} 105.61 \\ 10.41 \\ 24.54 \end{bmatrix} \ (MPa)
$$

At the boundary between the +30° and −30° plies
Right above the boundary, within the +30° ply,

$$
\begin{bmatrix} \sigma_x \\ \sigma_y \\ \tau_{xy} \end{bmatrix} = \begin{bmatrix} 105.84 \\ 10.53 \\ 24.87 \end{bmatrix} \times \tfrac{1}{2} = \begin{bmatrix} 52.81 \\ 5.20 \\ 12.27 \end{bmatrix} \ (MPa)
$$

Right below the boundary, within the −30° ply,

$$
\begin{bmatrix} \sigma_x \\ \sigma_y \\ \tau_{xy} \end{bmatrix} = \begin{bmatrix} 40.21 & 10.14 & -11.15 \\ 10.14 & 21.88 & -4.72 \\ -11.15 & -4.72 & 14.17 \end{bmatrix} \times 10^3 \begin{bmatrix} 5.82 \\ -1.62 \\ -0.58 \end{bmatrix} \times 1 \times 0.25 \times 10^{-3}
$$

$$
= \begin{bmatrix} 56.02 \\ 6.56 \\ -16.36 \end{bmatrix} \ (MPa)
$$

The distribution of stresses in the upper 4 plies is shown in Figure 5.13(b). As expected, σ_x is higher in the outer layers because the strains are higher and the 0° plies possess higher stiffness along the x-direction than the 30° plies. The discontinuity in stress between the 0° and +30° plies is mainly due to the stiffness difference. Interestingly, the stress is also discontinuous at the interface between the +30° and −30° plies. This is due to the presence of shear strains arising from the torsional curvature κ_{xy}, which have different effects on plies of opposite angles. As an exercise, the reader can show that the sum of moments from stresses in each ply is in equilibrium with the applied moment. For σ_y and τ_{xy}, one can observe sign changes in the stress values. Also, the stresses tend to be smaller for the plies away from the mid-plane. These observations are consistent with physical intuition because the total moments produced by these stresses must sum to zero over all the plies.

In this example, the effect of axial force and bending are considered separately. In practice, both may be present simultaneously, together with shear force and torsion. However, as long as the laminate is symmetric (which is the case in most practical situations), so the matrix [B] is zero, the effects of in-plane forces and bending/torsional moments can be considered separately to determine $\tilde{\varepsilon}_o$ and $\tilde{\kappa}$. The total strain (or stress) in each ply is then obtained by superposition.

5.4.3 Equivalent forces and moments arising from thermal and hygroscopic effects

In the above, we have focused on the effect of externally applied forces and moments on the laminate. However, for a composite laminate with different fiber orientations in different plies, the thermal expansion coefficient along any direction will also be different for each ply. As a result, when temperature changes occur, stresses will be induced in the individual plies to maintain compatibility of the laminated composite. This is of particular relevance for laminates made with resins that require a high curing temperature. After the material hardens at the curing temperature, cooling to room temperature will introduce high residual stresses in the plies.

Besides temperature changes, variation in moisture content of the composite (due to water absorption by the polymer) also results in dimensional changes. This is known as the hygroscopic effect. The dimensional change is less significant along the fiber direction than along the direction perpendicular to the fiber. Again, with fibers at different orientations in different plies, the hygroscopic effect will result in stresses in each ply.

To quantify the thermal and hygroscopic effects (which are often combined together and called the hygrothermal effects), the superposition principle can be employed. One can assume that external forces and moments have been applied to the laminate to resist the deformation so the laminate stays un-deformed. Since these external effects are fictitious and not present in reality, opposite forces and moments need to be applied to cancel them out. The strains and curvatures arising from the latter set of forces and moments represent the effect of temperature and moisture change. Therefore, once the equivalent forces and moments corresponding to the hygroscopic effects are obtained, the same approach as in the last section can be employed to find the mid-plane strains and curvatures.

To find the equivalent forces and moments, it is noted that the strain in each ply induced by temperature and moisture changes can be written as:

$$\begin{bmatrix} \varepsilon_x^H \\ \varepsilon_y^H \\ \gamma_{xy}^H \end{bmatrix} = \begin{bmatrix} a_x \Delta T + \beta_x \Delta m \\ a_y \Delta T + \beta_y \Delta m \\ a_{xy} \Delta T + \beta_{xy} \Delta m \end{bmatrix} \tag{5.15}$$

where

> ΔT and Δm are the change in temperature and moisture content respectively;
> a_x, a_y and a_{xy} are the transformed thermal expansion coefficients in the x–y coordinate system;
> β_x, β_y and β_{xy} are the transformed hygroscopic expansion coefficients.

Also, by defining a_1 and a_2 as the thermal expansion coefficients of the lamina along and perpendicular to the fiber direction, and β_1 and β_2 as the corresponding hygrosopic expansion coefficients, it can be shown by simple strain transformations that:

$a_x = a_1 \cos^2\theta + a_2 \sin^2\theta;\ \beta_x = \beta_1 \cos^2\theta + \beta_2 \sin^2\theta$

$a_y = a_1 \sin^2\theta + a_2 \cos^2\theta;\ \beta_y = \beta_1 \sin^2\theta + \beta_2 \cos^2\theta$

$a_{xy} = (a_1 - a_2) \sin\theta \cos\theta;\ \beta_{xy} = (\beta_1 - \beta_2) \sin\theta \cos\theta$

To prevent the laminate from deformation, fictitious forces and moments are required to produce strains that are equal in magnitude but opposite in sign to the thermal and hygroscopic strains give in Eq. (5.15). In each ply, the required stress is given by:

$$\begin{bmatrix} \sigma_x^H \\ \sigma_y^H \\ \tau_{xy}^H \end{bmatrix} = \begin{bmatrix} \bar{Q}_{11} & \bar{Q}_{12} & \bar{Q}_{16} \\ \bar{Q}_{21} & \bar{Q}_{22} & \bar{Q}_{26} \\ \bar{Q}_{61} & \bar{Q}_{62} & \bar{Q}_{66} \end{bmatrix}_k \left(\begin{bmatrix} a_x \\ a_y \\ a_{xy} \end{bmatrix}_k \Delta T - \begin{bmatrix} \beta_x \\ \beta_y \\ \beta_{xy} \end{bmatrix}_k \Delta m \right) \tag{5.16}$$

$$
\begin{bmatrix} N_x^H \\ N_y^H \\ N_{xy}^H \end{bmatrix} = \sum_{k=1}^{N} \int_{z_{k-1}}^{z_k} \begin{bmatrix} \sigma_x^H \\ \sigma_y^H \\ \tau_{xy}^H \end{bmatrix}_k dz = \sum_{k=1}^{N} \begin{bmatrix} \bar{Q}_{11} & \bar{Q}_{12} & \bar{Q}_{16} \\ \bar{Q}_{21} & \bar{Q}_{22} & \bar{Q}_{26} \\ \bar{Q}_{61} & \bar{Q}_{62} & \bar{Q}_{66} \end{bmatrix}_k
$$

(5.17a)

$$
\int_{z_{k-1}}^{z_k} \left(- \begin{bmatrix} a_x \\ a_y \\ a_{xy} \end{bmatrix}_k \Delta T - \begin{bmatrix} \beta_x \\ \beta_y \\ \beta_{xy} \end{bmatrix}_k \Delta m \right) dz
$$

$$
\begin{bmatrix} M_x^H \\ M_y^H \\ M_{xy}^H \end{bmatrix} = \sum_{k=1}^{N} \int_{z_{k-1}}^{z_k} \begin{bmatrix} \sigma_x^H \\ \sigma_y^H \\ \tau_{xy}^H \end{bmatrix}_k z\,dz = \sum_{k=1}^{N} \begin{bmatrix} \bar{Q}_{11} & \bar{Q}_{12} & \bar{Q}_{16} \\ \bar{Q}_{21} & \bar{Q}_{22} & \bar{Q}_{26} \\ \bar{Q}_{61} & \bar{Q}_{62} & \bar{Q}_{66} \end{bmatrix}_k
$$

(5.17b)

$$
\int_{z_{k-1}}^{z_k} \left(- \begin{bmatrix} a_x \\ a_y \\ a_{xy} \end{bmatrix}_k \Delta T - \begin{bmatrix} \beta_x \\ \beta_y \\ \beta_{xy} \end{bmatrix}_k \Delta m \right) z\,dz
$$

The equivalent forces and moments are opposite to the fictitious actions applied to maintain zero deformation, and can be written as:

$$
\begin{bmatrix} N_x^{eq} \\ N_y^{eq} \\ N_{xy}^{eq} \end{bmatrix} = - \begin{bmatrix} N_x^H \\ N_y^H \\ N_{xy}^H \end{bmatrix} = \sum_{k=1}^{N} \begin{bmatrix} \bar{Q}_{11} & \bar{Q}_{12} & \bar{Q}_{16} \\ \bar{Q}_{21} & \bar{Q}_{22} & \bar{Q}_{26} \\ \bar{Q}_{61} & \bar{Q}_{62} & \bar{Q}_{66} \end{bmatrix}_k \left(\begin{bmatrix} a_x \\ a_y \\ a_{xy} \end{bmatrix}_k \right.
$$

(5.18a)

$$
\left. \int_{z_{k-1}}^{z_k} \Delta T dz + \begin{bmatrix} \beta_x \\ \beta_y \\ \beta_{xy} \end{bmatrix}_k \int_{z_{k-1}}^{z_k} \Delta m\,dz \right)
$$

$$
\begin{bmatrix} M_x^{eq} \\ M_y^{eq} \\ M_{xy}^{eq} \end{bmatrix} = -\begin{bmatrix} M_x^H \\ M_y^H \\ M_{xy}^H \end{bmatrix} = \sum_{k=1}^{N} \begin{bmatrix} \bar{Q}_{11} & \bar{Q}_{12} & \bar{Q}_{16} \\ \bar{Q}_{21} & \bar{Q}_{22} & \bar{Q}_{26} \\ \bar{Q}_{61} & \bar{Q}_{62} & \bar{Q}_{66} \end{bmatrix}_k \left(\begin{bmatrix} a_x \\ a_y \\ a_{xy} \end{bmatrix}_k \right.
$$

$$
\left. \int_{z_{k-1}}^{z_k} \Delta T z \, dz + \begin{bmatrix} \beta_x \\ \beta_y \\ \beta_{xy} \end{bmatrix}_k \int_{z_{k-1}}^{z_k} \Delta m z \, dz \right)
$$

(5.18b)

Generally speaking, a thermal analysis or moisture diffusion analysis needs to be performed first to obtain the temperature or moisture variation over the thickness of the laminate. Knowing the variation of ΔT and Δm with z (the position along the thickness direction), the above equations can be employed to calculate the equivalent forces and moments. These can then be superposed to the externally applied forces/moments to calculate $\tilde{\varepsilon}_o$ and $\tilde{\kappa}$ for the laminate.

In the presence of hygrothermal effects, an important point should be noted about the calculation of stresses. For each ply, the stress is given by:

$$
\begin{bmatrix} \sigma_x \\ \sigma_y \\ \tau_{xy} \end{bmatrix}_k = \begin{bmatrix} \bar{Q}_{11} & \bar{Q}_{12} & \bar{Q}_{16} \\ \bar{Q}_{21} & \bar{Q}_{22} & \bar{Q}_{26} \\ \bar{Q}_{61} & \bar{Q}_{62} & \bar{Q}_{66} \end{bmatrix}_k \left(\begin{bmatrix} \varepsilon_x^o \\ \varepsilon_y^o \\ \gamma_{xy}^o \end{bmatrix} + \begin{bmatrix} \kappa_x \\ \kappa_y \\ \kappa_{xy} \end{bmatrix} z - \begin{bmatrix} a_x \\ a_y \\ a_{xy} \end{bmatrix}_k \Delta T - \begin{bmatrix} \beta_x \\ \beta_y \\ \beta_{xy} \end{bmatrix}_k \Delta m \right)
$$

(5.19)

The physical meaning of the equation is that the stresses are arising from the difference in final strain and the relative dimensional changes caused by hygrothermal effects alone.

5.4.4 Failure of composite laminates

In the above, a framework is presented for the derivation of strains and stresses in a composite laminate under both external loading and hygro-thermal effects. After knowing the stresses in the individual plies, one would like to know if failure would occur in any of the layers. To do so, a failure criterion for the lamina is required. In the literature, various failure criteria have been proposed, and several major ones are described below.

5.4.4.1 Maximum stress theory

According to this theory, the composite lamina will NOT fail as long as:

$$X_C < \sigma_1 < X_T$$
$$Y_C < \sigma_2 < Y_T \qquad\qquad (5.20)$$
$$|\tau_{12}| < S$$

where:

σ_1, σ_2 and τ_{12} are the in-plane stress components with respect to the principal directions of the lamina (i.e. the directions parallel and perpendicular to the fibers);

X_C and X_T are the compressive and tensile strength along the fiber direction respectively. (Note: compressive stress is taken to be negative here.)

Y_C and Y_T are the compressive and tensile strength perpendicular to the fiber direction;

S is the shear strength of the laminate.

In the analysis of composite laminates, the strain and stress components are first obtained with respect to the global coordinates. For layers with fibers not aligning with one of the global coordinate axes, it is necessary to transform the stresses back to the principal axes of the lamina first, using the transformation matrix $[T]$ in Eq. (5.3).

This failure criterion is very simple and does not consider any interaction between direct and shear stresses, or that between σ_1 and σ_2. The use of this criterion may therefore over-estimate the failure load of the lamina.

5.4.4.2 Maximum strain theory

According to this theory, the composite lamina will NOT fail as long as:

$$\varepsilon_1^C < \varepsilon_1 = \frac{\sigma_1 - v_{12}\sigma_2}{E_1} < \varepsilon_1^T$$

$$\varepsilon_2^C < \varepsilon_1 = \frac{\sigma_2 - v_{21}\sigma_1}{E_2} < \varepsilon_2^T \qquad\qquad (5.21)$$

$$|\gamma_{12}| = \left| \frac{\tau_{12}}{G_{12}} \right| < \Gamma_{12}$$

where:

ε_1, ε_2 and γ_{12} are the in-plane strain components with respect to the principal directions of the lamina;

ε_1^C and ε_1^T are the compressive and tensile failure strain along the fiber direction respectively. (Note: compressive strain is taken to be negative here.)

ε_2^C and ε_2^T are the compressive and tensile failure strain perpendicular to the fiber direction;

Γ_{12} is the shear failure strain of the laminate.

By considering strain components in the failure criterion, the interaction between stresses along and perpendicular to the fibers is somewhat accounted for. However, the interaction between the direct and shear stresses is still not considered.

5.4.4.3 The Tsai-Hill theory

According to this theory, failure of the lamina occurs when:

$$\frac{\sigma_1^2}{X^2} - \frac{\sigma_1\sigma_2}{X^2} + \frac{\sigma_2^2}{Y^2} + \frac{\tau_{12}^2}{S^2} = 1 \tag{5.22}$$

σ_1, σ_2, τ_{12} and S are defined as above. X is taken to be X_T if σ_1 is in tension, and X_C if σ_1 is in compression. Similarly Y is taken to be Y_T or Y_C depending on the sign of σ_2. In this failure criterion, the interaction of the various stress components is considered. Experimental results on the failure of lamina with loading applied at various degrees to the fiber direction show that the Tsai-Hill criterion can predict the failure load better than either the maximum stress or maximum strain criterion.

One disadvantage of the Tsai-Hill criterion is that the user has to determine which strength to use (e.g. X_T or X_C) according to the sign of the stresses, which will result in a certain degree of inconvenience in practice. Moreover, if the criterion is plotted in stress space, the surface is not smooth at locations where σ_1 or σ_2 changes in sign. The presence of discontinuities in the gradient can impose difficulties in numerical analysis. Such a disadvantage is alleviated by the next criterion to be discussed.

5.4.4.4 The tensor polynomial failure criterion

According to this criterion, failure occurs when:

$$\left(\frac{1}{X_T} + \frac{1}{X_C}\right)\sigma_1 + \left(\frac{1}{Y_T} + \frac{1}{Y_C}\right)\sigma_2 - \frac{\sigma_1^2}{X_TX_C} - \frac{\sigma_2^2}{Y_TY_C} + \frac{\tau_{12}^2}{S^2} = 1 \tag{5.23}$$

Under many practical situations, predictions from this criterion and the Tsai-Hill criterion are very similar. As this criterion gives a smooth surface in the stress space, it is commonly employed when numerical analysis is to be performed.

5.4.4.5 The first-ply failure and post-failure analysis

After knowing the stress components in each ply of a composite laminate, application of a lamina failure criterion allows us to tell if failure has occurred in any of the plies. The simplest approach for failure prediction of laminates is to assume ultimate failure to occur at the instant of first-ply failure. In other words, if any of the plies in a laminate fails, the composite is considered to have reached its full load capacity. It is quite obvious that such an approach can be very conservative in many cases. For example, one can consider the situation with tension applied to a laminate consisting of ten 0°-plies and two 90°-plies, along the 0° direction. As the strength of the lamina is much lower at a direction perpendicular to the fibers, the 90°-plies will fail at a rather low load level. However, after the 90°-plies fail, the 0°-plies can continue to support a much higher loading. The first-ply failure approach will then significantly underestimate the load capacity of the composite.

To avoid over-conservativeness in some cases, behavior after first-ply failure can be analyzed with updated [A] and [D] matrices calculated with reduced stiffness of the failed ply. If failure in a ply is due to fiber rupture (i.e. dominated by σ_1), all the stiffness components of the ply are taken to be zero. In other words, that particular ply is completely removed from the analysis. If ply failure occurs in the matrix instead (i.e. failure dominated by σ_2 or τ_{12}), only the stiffness along the fiber is taken to remain. That is, only Q_{11} is non-zero for that particular ply. Analysis of the composite can be continued with updated matrices after the failure of each ply until the load can no longer be increased. Ultimate failure is then reached.

5.4.4.6 Interlaminar failure

In practice, the failure of laminates may not initiate within a particular ply. Instead, adjacent plies in the laminate may separate from one another due to the presence of shear and normal tensile stresses between the plies. A typical interlaminar failure is illustrated in Figure 5.13, which shows two plies with opposite angles delaminating from one another, inducing final failure by shearing along the fibers in each ply. The presence of interlaminar stresses can be explained with a simple example, where direct tension is applied along the x-direction of a laminate (Figure 5.14). Due to different fiber

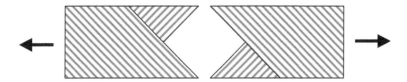

Figure 5.14 Interlaminar failure between opposite angle plies.

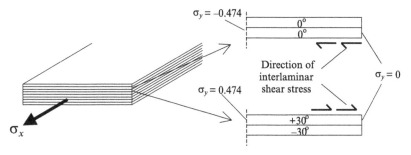

Figure 5.15 Physical explanation for the presence of interlaminar shear stresses.

orientations in each ply, the Poisson's ratio of a particular ply (along the loading direction) can be very different to that of an adjacent ply. As shown in Example 1, such a mismatch in Poisson's ratio can induce in-plane stresses in the y-direction, which are required for displacement compatibility to be maintained. However, as shown in Figure 5.15, σ_y must be zero along the lateral surface of the laminate. In order for non-zero σ_y to appear in a particular ply, force equilibrium requires the presence of shear stresses along one or both interfaces between the ply and its adjacent plies. In practice, σ_y builds up very quickly and reaches the required value in a certain ply at a small distance from the lateral side. The shear stresses are hence very high near the side, but decreases rapidly to zero. The presence of normal interlaminar stresses is related to the shear stresses. As one can imagine, the interlaminar shear stresses on the top and bottom of a certain ply may not produce the same resultant force. As a result, the moment produced by the corresponding shear forces are not balanced, and will cause the ply to bend (in the y–z plane). In general, different plies in the laminate may bend to different curvatures under the shear stresses, and normal stresses will be induced (also near the lateral sides) to maintain curvature compatibility of the plate. When the normal stress is tensile, interlaminar failure is facilitated.

The analysis of interlaminar failure is an advanced topic beyond the scope of this book. For the interested reader, a more thorough discussion of composite damage and failure, together with various analytical approaches, can be found in Herakovich (1998).

5.5 Analysis of composite sections

5.5.1 Equivalent elastic properties for the symmetric laminate

In this section, we will deal with the analysis of composite sections made of symmetric laminates. The first step is to obtain the equivalent elastic properties. When the composite laminate is symmetric in its ply arrangements, A_{16} and A_{26} are both equal to zero. For such a case,

$$[A] = \begin{bmatrix} A_{11} & A_{12} & 0 \\ A_{21} & A_{22} & 0 \\ 0 & 0 & A_{66} \end{bmatrix} \text{ and } \begin{bmatrix} N_x \\ N_y \\ N_{xy} \end{bmatrix} = \begin{bmatrix} A_{11} & A_{12} & 0 \\ A_{21} & A_{22} & 0 \\ 0 & 0 & A_{66} \end{bmatrix} \begin{bmatrix} \varepsilon_x^o \\ \varepsilon_y^o \\ \gamma_{xy}^o \end{bmatrix} \quad (5.24)$$

For a laminate of total thickness t, the above can be re-written as:

$$\begin{bmatrix} \varepsilon_x^o \\ \varepsilon_y^o \\ \gamma_{xy}^o \end{bmatrix} = \begin{bmatrix} a_{11} & a_{12} & 0 \\ a_{21} & a_{22} & 0 \\ 0 & 0 & a_{66} \end{bmatrix} \begin{bmatrix} N_x \\ N_y \\ N_{xy} \end{bmatrix} = \begin{bmatrix} a_{11}t & a_{12}t & 0 \\ a_{21}t & a_{22}t & 0 \\ 0 & 0 & a_{66}t \end{bmatrix} \begin{bmatrix} N_x/t \\ N_y/t \\ N_{xy}/t \end{bmatrix}$$

$$= \begin{bmatrix} a_{11}t & a_{12}t & 0 \\ a_{21}t & a_{22}t & 0 \\ 0 & 0 & a_{66}t \end{bmatrix} \begin{bmatrix} \sigma_x \\ \sigma_y \\ \tau_{xy} \end{bmatrix}_{av} \quad (5.25)$$

The [a] matrix is the inverse of the [A] matrix, with its component given by:

$a_{11} = A_{22}A_{66}/\det[A]$
$a_{12} = a_{21} - A_{12}A_{66}/\det[A]$
$a_{22} = A_{11}A_{66}/\det[A]$
$a_{66} = 1/A_{66}$

where $\det[A] = [A_{11}A_{22} - (A_{12})^2]A_{66}$ is the determinant of the matrix.

Assuming the laminate to be a homogeneous material, the constitutive relation can be represented by equivalent moduli and Poisson's ratios as:

$$\begin{bmatrix} \varepsilon_y^o \\ \varepsilon_y^o \\ \gamma_{xy}^o \end{bmatrix} = \begin{bmatrix} \dfrac{1}{E_x} & \dfrac{-v_{yx}}{E_y} & 0 \\ \dfrac{-v_{xy}}{E_x} & \dfrac{1}{E_y} & 0 \\ 0 & 0 & \dfrac{1}{G_{xy}} \end{bmatrix} \begin{bmatrix} \sigma_x \\ \sigma_y \\ \tau_{xy} \end{bmatrix}_{av} \quad (5.26)$$

Comparing Eqs (5.25) and (5.26), a direct relation between the equivalent elastic properties and the components of the [a] matrix can be obtained.

$(E_x)_{eq} = 1/(a_{11}t)$ (5.27a)

$(E_y)_{eq} = 1/(a_{22}t)$ (5.27b)

$(G_{xy})_{eq} = 1/(a_{66}t)$ (5.27c)

$(v_{xy})_{eq} = -a_{21}/a_{11}$ (5.27d)

$(v_{yx})_{eq} = -a_{12}/a_{22}$ (5.27e)

Similarly, if we define [d] as the inverse matrix of [D], the equivalent bending stiffness $(EI)_x$ per unit width for plate bending in the x–z plane is given by $(1/d_{11})$, and stiffness $(EI)_y$ per unit width in the y–z plane is given by $(1/d_{22})$.

5.5.2 Stress and deformation analysis of composite members

Using the equivalent elastic properties derived above, the stiffness of thin-walled composite sections can be derived for the calculation of deformation under applied loading. Based on the deformations, the force and/or moment per unit width on each part of the member (e.g. a web or a flange) can be obtained. The stresses in the individual plies forming the particular part are then calculated with the approach described in Section 5.4. In the following, the analysis is illustrated with two examples of composites sections, the *I*-section and the box section.

5.5.2.1 The I-section

An *I*-section under a combination of axial load (N_x^T), bending moment (M_x^T) and torsion (T_x^T) is shown in Figure 5.16. Since we will focus on elastic analysis alone, the effects of various kinds of loading can be considered separately and then superposed.

Under axial loading, the strain ε_x is related to N_x^T by:

$$N_x^T = [2bt_f E_{x,f} + ht_w E_{x,w}]\varepsilon_x, \text{ or}$$
$$N_x^T = (EA)_{eq}\varepsilon_x = [2b/(a_{11})_f + h/(a_{11})_w]\varepsilon_x \tag{5.28}$$

where the subscripts $()_f$ and $()_w$ represent the web and flange respectively.

After finding ε_x, the force/width on the flange and web, $N_{x,f}$ and $N_{x,w}$, are obtained as:

$$N_{x,f} = 1/(a_{11})_f \varepsilon_x \tag{5.29a}$$
$$N_{x,w} = 1/(a_{11})_w \varepsilon_x \tag{5.29b}$$

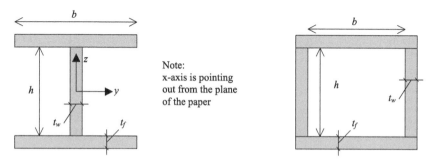

Note:
x-axis is pointing
out from the plane
of the paper

Figure 5.16 Definition of dimensions for the *I*-section and box-section.

Knowing $N_{x,f}$ and $N_{x,w}$, the stresses in each ply within the flange and web are calculated according to the classical laminate plate theory.

Under applied moment M_x^T, the curvature κ_x is given by:

$$M_x^T = (EI)_{eq} \kappa_x = \{2[b/(d_{11})_f + b(h/2)^2/(a_{11})_f] + h^3/(12a_{11})_w\} \kappa_x \qquad (5.30)$$

In the above expression, the first term involving $(d_{11})_f$ represents the moment of inertia of the flange about its own mid-plane. This is often small compared to the second term which gives the stiffness about the centroid of the I-section, and can hence be ignored in practice.

Assuming the plane section remains plane, the strain $\varepsilon(z)$ along the x-direction at any distance z from the centroid of the section is given by $(\kappa_x z)$. Depending on the location of a particular ply, the lateral strain can be deduced by multiplying $(\kappa_x z)$ with the equivalent Poisson's ratio of the flange or web accordingly. Once the strains in both directions are known, the stress components are calculated by multiplying the strains with the stiffness matrix of the ply. Note that for a flange restrained by a stiff slab above, the lateral strain can be taken to be zero.

With torsion T_x^T applied to the section, the rate of twist β $(= d\theta/dx)$ is:

$$T_x^T = (GJ)_{eq} \beta = [2bt_f^2/(3a_{66})_f + ht_w^2/(3a_{66})_w]\beta \qquad (5.31)$$

The torsion and width acting on the flange and web are then given by:

$$T_f = [t_f^2/(3a_{66})_f]\beta \qquad (5.31a)$$

$$T_w = [t_w^2/(3a_{66})_w]\beta \qquad (5.31b)$$

With T_f and T_w, the stresses in each ply are calculated from classical laminated plate theory.

In the above, we have focused on the calculated of stresses. The equivalent stiffness $(EA)_{eq}$, $(EI)_{eq}$ and $(GJ)_{eq}$ can be used in the calculation of member deflection using conventional approaches. In the calculation of deflection under bending, it should be noted that the ratio of shear to flexural stiffness of composite laminates is usually much smaller than that for homogeneous materials. It is therefore advisable to consider shear deformation in deflection calculations. For the I-section, one can assume all the shear force to be carried by the web, and obtain the shear stiffness $(GA)_{eq}$ as:

$$(GA)_{eq} = h/(a_{66})_w \qquad (5.32)$$

To calculate the shear deformation, a shear factor $k = 1.2$ should be applied. For details, the reader should refer to textbooks on mechanics of materials.

5.5.2.2 The box section

For the box section, the effects of axial load and bending load are analyzed in exactly the same way as the *I*-section, using the following equations for the equivalent stiffness:

$$(EA)_{eq} = 2[b/(a_{11})_f + h/(a_{11})_w] \tag{5.33}$$

$$(EI)_{eq} = 2\{[b/(d_{11})_f + b(h/2)^2/(a_{11})_f] + h^3/(12a_{11})_w\} \tag{5.34}$$

$$(GA)_{eq} = 2h/(a_{66})_w \tag{5.35}$$

Under torsion, the behavior of a closed section (e.g. a box) and an open section (e.g. an *I*-section) is very different. For the closed section, the shear per unit width (N_s) is uniform along the wall. The following equation then arises from the equilibrium of torsional moment:

$$T_x^T = (N_s b)h + (N_s h)b = 2(bh)N_s \tag{5.36}$$

With N_s, the stresses in each layer within the flange or web can be calculated.
   The torsional stiffness is given by:

$$(GJ)_{eq} = 2(bh)^2/[b(a_{66})_f + h(a_{66})_w] \tag{5.37}$$

As shown in Figure 5.16, the width b of the box is taken to be the total flange width minus the width of the web. This is different from the case of the *I*-section where b is taken to be the total width of the flange. The reason is due to the uncertainly of fiber arrangements at the corners of the box section, especially when wrapping has been performed as part of the fabrication process. By ignoring the material in the joints, a conservative estimate of the torsional stiffness is obtained. With knowledge of the fabrication process and the actual fiber arrangements, the engineer should be able to make an appropriate judgment on the value of b and h to be used for a particular composite section.

5.6 Summary

The replacement of concrete bridge decks with GFRP components is a viable technique to extend the lifetime of bridges. In this chapter, the fabrication method of GFRP bridge deck components first is introduced. Then, a theoretical approach to the analysis and design of laminated fiber composites as well as fiber reinforced plastic structural components is provided. This chapter can hopefully serve as a starting point for civil engineers who are not familiar with composite manufacturing and/or analysis, but are interested in investigating the use of GFRP bridge deck components in real-world projects.

6 Strengthening of reinforced concrete structures with fiber reinforced polymers

6.1 Introduction

A significant portion of the world's civil infrastructure, including highways, bridges, buildings, hydraulic structures and dams, is constructed with reinforced concrete. After years of service, many of these structures are deteriorating. To recover the original factor of safety, strengthening of the structure needs to be performed. Also, for many structures, the current load demand is significantly beyond the value adopted in the original design many years ago. As a notable example, there are 40,000 bridges in the United Kingdom falling short of the new requirements of European highways to carry 44-tonne vehicles. Such bridges have to be strengthened to satisfy current needs.

Conventionally, a concrete structure can be strengthened through the enlargement of its members by casting additional concrete, together with the incorporation of additional steel reinforcements. This approach, while simple in principle, has several disadvantages that can make the retrofitting process a real nuisance in practical situations. First of all, to cast new concrete, formwork has to be placed around an existing member and a falsework system set up beneath the formwork is normally required to provide the support. The process will therefore take up significant space under the member to be strengthened. If the member is a beam or a slab in the ceiling of a room, the room has to be vacated for several weeks, to allow sufficient time for the preparation/execution of structural repair, and for the concrete to cure and harden. In other words, the room will be unusable for an extended period of time. To strengthen the girders of a flyover bridge above an underlying highway, one or more lanes of the highway may have to be blocked to accommodate the falsework system. The efficiency of transportation is affected, and this may lead to severe traffic jams during the rush hours. Second, when strengthening needs to be performed in a relatively confined space, transportation of wood for formwork construction, steel members for the falsework system as well as the steel reinforcements and concrete for strengthening can be difficult and labor-intensive. The cost of the process will then be greatly increased. Third, the strengthened member is larger than the original one, implying reduction in headroom and space between members. Fourth, the additional concrete increases the dead-weight of the member, which increases the inertial force acting on the structure when earthquakes strike.

Alternatively, concrete structures can be strengthened with the use of steel members. For examples, a steel *I*-beam can be placed alongside a degraded concrete beam to support the loading from the slab. To strengthen a concrete column, one can either employ steel struts to share some of its loading, or confine it with steel jackets. The use of steel members, however, has its own disadvantages. Similar to the casting of new concrete, the procedures involve in the addition of steel members will take up significant space around the concrete member for an extended period of time. Also, significant weight is added to the structure while space/headroom is reduced. Considering the weight of steel, lifting it into the right position requires the use of powerful machinery. Also, the in-situ processes to ensure the proper connection of the steel member to the existing concrete structure can be rather complicated. All of these will translate into high costs. Moreover, exposed steel members are prone to rusting, so corrosion prevention is an additional issue to be considered.

Besides the use of steel structural sections, concrete beams can also be strengthened with the use of bonded or bolted steel plates, on or near the tensile surface (for flexural strengthening) or on the sides (for shear strengthening). Steel plates are lighter than steel I-sections but still very heavy. Powerful machinery is required to move them into place. Also, if adhesive bonding is performed, a temporary supporting system is required, and this will again take up a lot of space. To ensure durability of the retrofit, corrosion protection has to be performed as well.

Due to the limitations of conventional strengthening methods, a new approach to strengthen concrete members through the adhesive bonding of fiber reinforced polymers (FRP) was developed in the 1980s and has been applied all over the world. Design recommendations for this strengthening method have been developed by ACI in the United States, FIB in Europe and JSCE in Japan. In the following, the principle of FRP strengthening is first introduced, together with a description of common FRP materials and adhesives. The strengthening procedures are then described. Then, the behavior of concrete beams strengthened in flexure and shear, as well as concrete columns strengthened in compression, will be discussed in detail. For each kind of strengthened member, a design approach will also be introduced.

6.2 Structural strengthening with bonded fiber reinforced polymer (FRP)

6.2.1 *Materials for FRP fabrication*

FRP for the strengthening of concrete members is available in different forms, including pre-fabricated plates or shells as well as fiber sheets, fabrics and tows (note: tows are continuous fiber bundles that are well aligned). Pre-fabricated plates are produced by the pultrusion process. Shell elements can be made by hand lay-up or wrapping around a mandrel. Elements formed by wrapping can be cut to form open sections for fitting onto structural members. Alternatively, it is possible to produce a single cut along the longitudinal direction, which allows the shell to "open up" and then be placed around a concrete column. To bond pre-fabricated plates or shells onto concrete members, adhesives are used.

When fiber sheets, fabrics, and tows are used, the FRP is often formed directly on the member by applying alternate layers of resin and fiber until the required thickness is obtained. Alternatively, the FRP can be prepared at the site first with the use of fiber and resin. While the resin is still wet, the FRP can be bonded to the concrete member. Some material suppliers pre-impregnate their fiber products with resin. Depending on the product, the use of additional resin during in-situ installation may or may not be necessary. For systems requiring additional resin, the hardener is placed in the additional resin alone, and the FRP hardens at ambient temperature. For pre-impregnated fiber products that do not require additional resin, curing needs to be performed at a higher temperature, so a heating source is required.

Three kinds of fiber, made of glass, carbon and aramid, are commonly used in the fabrication of FRP for structural strengthening. Each has its own advantages and limitations. For glass fiber, advantages include its high strength (>2 GPa), high ultimate strain (>3 percent) and relatively low cost. The specific gravity is about 2.6. The major disadvantages include low modulus ($E = 72$ GPa), sensitivity to abrasion and alkaline environment, as well as low resistance to moisture, sustained loads and cyclic loads. The most commonly used fiber is called E-glass, which is for general purpose. S-glass has better mechanical properties but is more susceptible to alkaline attack. AR-glass (i.e. alkaline-resistant glass) has improved durability in an alkaline environment, but the degradation problem is not completely eliminated. Although glass fibers in FRP are embedded inside a polymer matrix, alkaline attack by ions penetrating the matrix is still a concern.

Carbon fibers possess much higher strength and stiffness than glass fibers but are far more expensive. The strength of carbon fibers ranges from 2.2–5 GPa, its modulus from 800–250 GPa, and the ultimate strain from 0.3–1.8 percent. The specific gravity ranges from 2.2 for high modulus fiber to 1.8 for low modulus fiber. It should be noted that the stronger fibers are associated with lower moduli and higher ultimate strain. Besides high strength and stiffness, carbon fiber also possesses excellent resistance to moisture and chemicals, and is insensitive to fatigue. The weaknesses of carbon fiber are its low impact resistance and low ultimate strain.

Aramid fibers have very good properties when put in direct tension. The strength ranges from 2.8–4.1 GPa, the modulus from 80–190 GPa and the failure strain can be up to 4 percent. The specific gravity is about 1.4–1.5. However, flexural and compressive properties are relatively poor. Also, aramid fibers creep significantly when exposed to moisture and degrade under UV radiation. The cost of the fiber is in the same range as carbon fibers.

For each of the above fibers, the stress–strain relation stays linear up to failure. Final failure of the fiber is due to fracturing with no signs of ductile yielding. With fibers dominating their behavior, common FRP plates and sheets for strengthening also exhibit a linear stress–strain relation and a brittle rupture failure. The relation of this behavior to structural design will be discussed later.

Generally speaking, FRP can be prepared with various kinds of thermoplastics and thermosets as matrix material. In practice, all commonly available FRP systems for the strengthening of concrete members employ epoxy as the matrix material. Although more expensive than other thermosetting polymers (such as polyester and vinylester), epoxy has high strength, good creep resistance, strong resistance to chemicals and solvents as well as low shrinkage and volatile emission. For the bonding of prefabricated FRP to a concrete member, epoxy adhesive is also commonly used.

The properties of FRP depend on the kind of fiber employed and the percentage of fiber in the composite. Based on values reported in the literature, the typical ranges of mechanical properties for carbon fiber reinforced polymers (CFRP), aramid fiber reinforced polymers (AFRP) and glass fiber reinforced polymers (GFRP), as well as qualitative information regarding long-term performance under various conditions (fatigue, chloride environment, etc.) are listed in Table 6.1. Corresponding properties of steel are also included for comparison. Note that the values in the table are for general reference only. When a certain FRP system is to be employed, specific properties should be obtained from the material supplier.

Table 6.1 Typical properties of fiber reinforced polymers and steel

	Carbon FRP	*Aramid FRP*	*Glass FRP*	*Steel*
Density (kg/m³)	1600–900	1050–1250	1600–2000	7800
Elastic modulus (GPa)	56–300	11–125	15–70	190–210
Tensile strength (MPa)	630–4200	230–2700	500–3000	250–500
Fatigue	Excellent	Good	Adequate	Good
Sustained loading	Very good	Adequate	Adequate	Very good
Alkaline environment	Excellent	Good	A concern	Excellent
Acid/chloride exposure	Excellent	Very good	Very good	Poor

6.2.2 Strengthening principles and configurations

The flexural strengthening of concrete beams with bonded FRP is illustrated in Figure 6.1. With loading acting on the beam, the FRP at the bottom is subjected to tension. The tensile force in the FRP will generate an additional moment about the neutral axis of the member to increase the total bending capacity of the beam. To maximize the effectiveness of strengthening, uni-directional fiber sheets, pultruded fiber plates or fabric with fibers running predominantly along the longitudinal direction should be used. The bonded FRP should cover the region that needs to be strengthened, with additional anchorage lengths on both sides to ensure effective stress transfer from the concrete member to the FRP. Under normal conditions, FRP bonding is performed without unloading of the concrete member (e.g. by jacking it to an un-deformed configuration). In other words, the FRP only contributes to carrying the live load of the member, as the concrete member itself is already carrying the full dead load when FRP is applied. To increase the proportion of loading carried by the FRP, the FRP can be pre-stressed before it is bonded to the beam. With uniform pre-stress applied on the plate before bonding, the bonded plate may peel off easily at the ends due to the presence of high stress concentrations. To resolve this problem, one can apply the full pre-stressing to the middle part of the FRP, and gradually reduce the pre-stressing force towards the plate end. A mechanical system to perform such kind of pre-stressing operation is commercially available.

Reinforced concrete beams are always designed to fail in bending, so the inherent shear capacity is higher than the flexural capacity. However, if the beam is strengthened in flexure, shear strengthening may be necessary to maintain the more ductile flexural failure mode. Figure 6.2(a) illustrates the shear strengthening of beams with the use of individual FRP strips that are either prefabricated plates or fiber sheets/fabrics. The function of the strips is similar to that of steel stirrups. Once inclined shear cracks form along the span, the strips act as ties between the concrete on the two sides of each crack. The tensile forces in the strips will then add to the forces in the

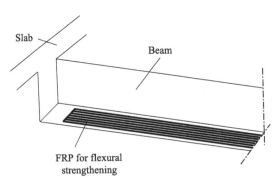

Figure 6.1 Flexural strengthening of concrete beam with bonded FRP.

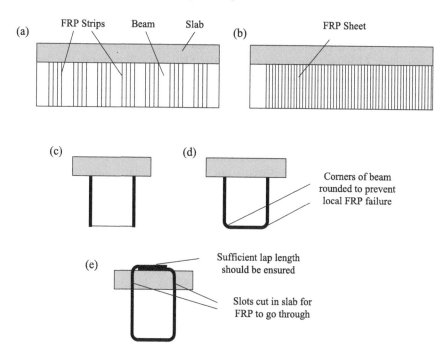

Figure 6.2 Shear strengthening of concrete beam with (a) FRP strips; (b) FRP sheet; (c) side-bonded FRP, (d) U-jacketing; (e) full wrapping.

stirrups to increase the total shear force that can be carried. In this case, to maximize effectiveness, the fibers should preferentially be oriented along the length of the strip. In shear strengthening, the use of strips provides greater flexibility in practical design. By varying the strip thickness and/or spacing, different degrees of strengthening can be achieved. However, the procedures for preparing and bonding a large number of individual strips are very labor-intensive. An alternative approach, which applies a single FRP sheet over the whole region to be shear-strengthened, is illustrated in Figure 6.2(b). With this approach, more material is likely to be used, but the labor cost is reduced. Moreover, by covering up the whole region, further penetration of water and other corrosive agents into the structure is prevented.

As illustrated in Figures 6.2(c), 6.2(d) and 6.2(e), shear strengthening can be performed with three different configurations, namely side-bonding, U-jacketing and full wrapping. For side bonding, FRP is applied only on the two vertical sides of the member. This is the only possible approach if the beam is connected to a wall underneath, and cutting of a horizontal slot at the beam/wall junction is not allowed. This configuration is the simplest of the three and involves the least labor. However, the strengthening effectiveness is also the lowest since the FRP can debond easily from the beam when shear cracks are formed. (Note: more detailed discussions on this aspect will be provided in a later section.) With the U-jacketing configuration, FRP debonding at the lower side of the vertical beam surface is controlled. Although the upper part of the FRP can still debond, the degree of strengthening is improved over that of the side-bonding configuration. To further enhance the strengthening, mechanical anchors can be applied at the upper part of the vertical strip. When U-jacketing is performed, it is necessary to round off the corners of the beam to a specified radius of curvature, in order to avoid FRP rupture due to high stresses at sharp corners. If a large increase in shear capacity is required, the most effective approach is to perform full wrapping. Vertical slots are cut through the slab on the two sides of the beam, for a strip of FRP sheet to go through. A thin strip can be wrapped around the beam section (which includes part of the slab) several times until the right thickness is reached. Alternatively, a wet strip of the designed thickness can be prepared first and then inserted through the slots. In the latter case, it is necessary to provide sufficient overlapping on the strip to avoid failure at the joint (see Figure 6.2(e)). While full wrapping is very effective, it is highly labor-intensive. Moreover, in many cases, cutting through the slab may not be allowed.

When U-jacketing or full wrapping is performed, there is an additional benefit. The bottom strip, which runs in the transverse direction, can resist the separation of the longitudinal FRP from the beam and hence increase the effectiveness of flexural strengthening.

Research on the use of inclined strips for shear strengthening has been carried out. With the strips making a higher angle to the crack, the effectiveness is improved. However, the use of angled strips is only possible if there is no load reversal. If the beam is under dynamic loading that can create shear cracks in two inclined directions, the approach is ineffective unless two sets of inclined strips at opposite angles are employed.

Compressive strengthening with FRP is based on the enhancement of concrete strength through lateral confinement. When a concrete column is loaded, internal damage leads to an increase in Poisson's ratio, and a consequent increase in lateral displacement. If a circular column is surrounded by FRP, the tensile stretching of FRP will provide inward pressure on the column. This confining pressure resists the lateral deformation of the concrete and significantly enhances its strength along the vertical direction. The compressive load capacity is therefore improved. For maximum effectiveness, the fibers in the FRP should be running along the hoop direction of the column. Various configurations for column strengthening are illustrated in Figure 6.3. In Figure 6.3(a), the column is wrapped by a single continuous FRP sheet, but individual strips or a continuous helical strip can also be employed. In addition, prefabricated FRP elements in the form of half-shells or shell with a vertical slit can be used. In the former case (Figure 6.3(b)), there should be sufficient lap length between the half-shells to avoid failure at the joint. In the latter case (Figure 6.3(c)), the shell element should be thin

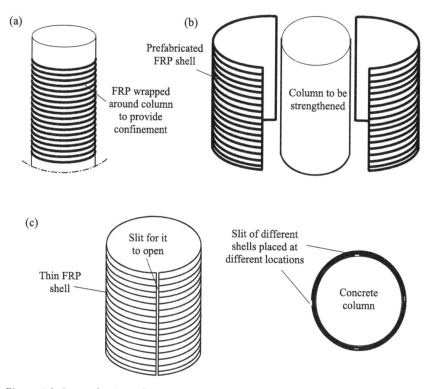

Figure 6.3 Strengthening of concrete column with (a) FRP wrapping; (b) prefabricated FRP shell; (c) prefabricated thin FRP shell with a slit.

enough for it to be easily opened at the slit and wrapped around the column. To come up with the required total thickness, several shell elements are required, and the slit in each element should be at different locations around the member circumference. In design, the thickness of an N-layer shell is taken to be $(N-1)$ times the thickness of each layer.

According to the strengthening principle, FRP wrapping is most effective for circular columns, and is applicable to elliptical columns as well. For rectangular columns, the tension in the FRP only produces significant inward pressure at the corners. The strengthening is only effective if the corners are properly rounded. In this chapter, we will only focus on the FRP strengthening of circular columns. The strengthening of rectangular columns will only be briefly discussed.

6.2.3 The strengthening process

Various commercial FRP systems are available for the strengthening of concrete structures. Normally, each material supplier will provide clear guidelines related to the use of their system, and these should be closely followed in practice. While the detailed procedure may vary for different systems, the basic principle is the same. In the following, the general strengthening process is presented with explanations of major steps.

Before the FRP system is applied to a concrete member, damage or cracks on the member should first be properly repaired (with the approaches discussed in Chapter 4). If there is plaster or other soft material on the concrete surface, it will have to be removed. Bonding of FRP onto a weak substrate will greatly compromise the effectiveness of strengthening as failure would occur within the substrate well before the FRP's full contribution is attained. After ensuring that the substrate is sound and strong, surface preparation is the next step. Specifically, a thin layer of mortar should be removed from the surface to expose the underlying aggregates. By removing the weak mortar layer and increasing the surface roughness, the FRP/concrete bond can be improved. Common surface preparation methods include blasting with sand, grid or water jet, and surface grinding using mechanical tools. The choice of method depends on the FRP system as well as the kind of strengthening to be performed. For example, blasting is more efficient than grinding but it produces a surface that is more uneven. It is suitable for beams if the FRP system consists of prefabricated strips with significant stiffness and an adhesive that is effective in filling up voids on the surface. However, if dry or uncured fiber sheets that tend to follow the concrete surface profile are employed, the grinding process is preferred.

When FRP has to be bent around the edge of a beam (e.g. in U-jacketing or U-wrapping for shear strengthening) or a rectangular column, the corner needs to be rounded to avoid premature failure due to high stress concentration. Theoretically speaking, the higher the radius of curvature, the lower the reduction in FRP strength due to concentrated stress. In practice, the radius of curvature at any corner should be 25 mm or above.

In the strengthening of columns, the concrete surface should be kept convex everywhere for the member to be effectively confined at all locations. In this situation, excessive unevenness on the surface is also undesirable. Even when grinding is performed, the prepared surface may still be overly uneven. In this case, a sticky polymeric paste that can bond well with concrete, called a putty, can be applied to smoothen the surface first. For most commercial FRP strengthening systems, a putty compatible with the adhesive or resin is available from the supplier.

After the concrete substrate is repaired (if necessary) and the surface properly prepared, the FRP can be applied. Before the bonding of FRP, dust, dirt, oil and other contaminations should be completely removed from the concrete surface. In addition, the surface needs to be dry. The adhesive or resin, which is normally supplied in two parts, can then be mixed according to the proportions recommended by the supplier. For prefabricated plates, adhesive is applied on both the concrete surface and one side of the plate. For the adhesive on the plate, it should be applied to form a dome-shaped cross-section with the middle part slightly higher than the side. The plate is then lifted to the right position and pressed onto the concrete surface with a rubber roller. With the adhesive slightly thicker in the middle, the lateral movement of adhesive during the bonding process will push air to the side and significantly reduce the probability of air being trapped within the adhesive. When fiber sheets or fabrics are employed with resin, a primer is sometimes applied first. The primer is usually of lower viscosity than the resin, and penetrates better into the concrete substrate. Then, the first layer of resin is applied, followed by a layer of fiber sheet, and an additional layer of resin. To facilitate the impregnation of resin into the FRP sheet, and to remove trapped air, a roller brush is used to apply pressure on the layers of resin and fiber. After one layer is completed, the resin is left to cure for a short period of time before the second layer is applied. This is to prevent the movement of the first layer of material during the applying of the next layer. The procedure is repeated until the required FRP thickness is reached.

For the repair of columns, a semi-automatic wrapping process has been developed. The concept is illustrated in Figure 6.4. A ring holding fiber spools is first placed around a concrete column. Fiber tows from the various spools are then glued to a number of points around the circumference of a section near the top or bottom of the column. By rotating the ring and moving it vertically at the same time, the fibers are wrapped around the column surface, with the fiber angle depending on the speeds of rotation and vertical movement. In this process, the fiber tows can be passed through a resin bath before they are wrapped around the column. The composite can then be cured at ambient temperature. Alternatively, pre-impregnated tows

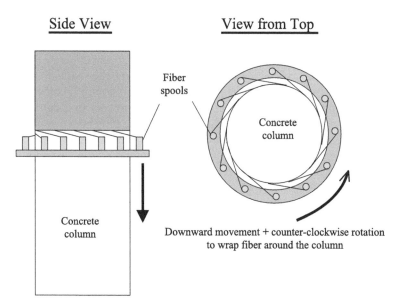

Figure 6.4 Schematics of an automatic FRP wrapping system.

can be used, but then curing needs to be performed with the use of a heating source such as a cylindrical heater around the column.

Due to the light weight of fiber sheets/fabrics and FRP plates, the epoxy resin or adhesive is sufficient to hold them in place under most situations, so temporary supporting systems are not required. In some cases, a simple clamping system providing small compression on the FRP is sufficient to prevent its movement before the epoxy hardens. Epoxy for FRP strengthening systems can undergo curing at room temperature, and gain sufficient strength within a week. After curing, the strength of epoxy is higher than that of the underlying concrete. Providing the surface has been properly prepared, failure of the bond will occur within the concrete substrate rather than along the epoxy/concrete interface.

6.2.4 Advantages and limitations

In comparison to conventional strengthening methods, the major advantages of FRP strengthening arise from the high strength/weight ratio of FRP and the relative simplicity of the retrofitting process. Take, for example, a typical CFRP with strength of 4200 MPa and relative density of 1.8. Considering the possibility of pre-mature debonding failure and variability of the FRP bonding process, we may assume the CFRP stress at structural failure to be about 2000 MPa. To carry the same tensile force as a steel plate with a yield strength of 500 MPa and relative density of 7.8, the weight of the equivalent CFRP plate is only 6 percent that of the steel member. A significant reduction in material weight can also be shown for other kinds of FRP. In most cases, the weight of the FRP element (plate or shell) is low enough for it to be handled by between one and three workers without the need of heavy machinery. If a large and heavier element is required for strengthening, it can always be produced directly on the concrete member with thin FRP sheets. Using uncured individual sheets, the shape of concrete members (such as the round surface of a cylindrical column) can easily be followed. Bonding is also a simple and efficient process that can be performed during the off-hours of facilities, such as the time between midnight and the early morning. The disturbance to users of the facilities is hence minimal. After the FRP is bonded, it does not require temporary support. In common strengthening applications, the required thickness of FRP seldom exceeds several millimeters. The reduction in headroom/space and increase in weight of structure are both negligible.

The FRP bonding technique, while being efficient and effective, is not without its limitations. Fire resistance is always a concern. At high temperature, the load-carrying capacity of FRP is greatly reduced by the softening of resin or adhesive. It is necessary to ensure that the loss of FRP strengthening will not lead to sudden collapse. According to the ACI recommendations (2002), the non-strengthened structure should be able to carry the unfactored dead and live loads. This seems to be a reasonable requirement that can be adopted in general. Also, as polymeric materials can actually burn, any adhesive or resin used in the FRP should not increase the severity of the fire. In some cases, the load-carrying capacity of the FRP has to be maintained in a fire. An example is a bridge that has been widened, so the columns need to be strengthened to carry additional dead and live loads. Under this situation, a fire protection system should be installed. An external polymer layer that can foam up during heating to insulate the FRP is effective in protecting the FRP from fire.

Another common concern about the use of FRP is the possibility of attack by ultraviolet (UV) light from the sun. This will be a problem for FRP around an exposed column or FRP on the lateral sides of an exposed beam. UV light tends to break down molecular chains in a polymer, and hence will degrade the mechanical performance of the polymer and the FRP. With temperature change and moisture penetration, the effect may become more severe. To protect the FRP against UV light, the simplest way is to apply a light-colored paint that will reflect most of the energy away.

While FRP can easily be bonded to a concrete structure, it can also easily be removed. As a result, FRP strengthened structures are vulnerable to vandalism. For critical locations, access to strengthened parts of the structure may have to be restricted.

As FRP bonding is a new technique that is still under research and development, the lack of a universally accepted approach for failure analysis limits its practical application. While various design guidelines have been developed separately in the USA, Europe, Japan and other countries, the design equations in various guidelines may be very different. As a result, the predicted load capacity of a strengthened member may vary significantly among different guidelines. With continuous research and development activities on FRP strengthening of concrete structures, the situation is likely to improve in the coming years. A widely accepted approach for the analysis and design of strengthened member will eventually evolve and lead to the convergence of various guidelines.

In the following sections, the mechanics of FRP strengthening will be discussed separately for beams under flexure and shear, and columns under compression. The focus of discussion is on failure mechanisms and factors affecting the failure load. For each kind of FRP strengthening, a design framework is presented together with design equations to predict the contribution of FRP to the load-carrying capacity. While these equations are likely to be modified or replaced in the future, the overall framework should still be applicable for practical design.

6.3 Flexural strengthening of beams

6.3.1 Failure modes

For a FRP-strengthened beam under flexural loading, failure can occur in many different modes. Generally speaking, one can distinguish between the situation with full composite action and that with a loss of composite action. Full composite action takes place when there is perfect bonding between the FRP and concrete, or limited debonding that has little effect on the FRP force. However, in many cases, loading results in significant debonding, so the FRP is no longer effectively coupled to the concrete member. In other words, the composite action is lost so the FRP force and the level of strengthening are greatly reduced. In the following, the various failure modes are discussed.

6.3.1.1 Failure modes under full composite action

When there is full composite action, the failure modes of strengthened beams are similar to those for normal reinforced concrete members, with the additional mechanism of FRP rupture when its ultimate strength is reached. Plausible failure modes include: (1) steel yielding followed by concrete crushing; (2) steel yielding followed by FRP rupture; (3) concrete crushing before steel yielding; and (4) shear failure. The occurrence of mode (1), (2) or (3) depends on the area fraction of steel and FRP at the critical cross-section. Before strengthening, a properly designed beam should fail with steel yielding followed by concrete crushing. If only a small amount of FRP is added, the same failure mode (Mode (1)) should be maintained. However, if a large volume of FRP is bonded to the bottom of the member, the neutral axis will exhibit a significant downward shift that increases the concrete strain at the top and reduces the strain in the tensile steel. Crushing of concrete (Mode (3)) may hence occur before steel yielding, which is undesirable. Also, depending on the failure strain of the FRP, it may rupture after steel yielding before the concrete crushes (Mode (2)). Theoretically speaking, FRP may also rupture before steel yielding. If this is the case, the strengthening is not effective at all, because the contribution from the steel reinforcements is not yet fully activated when the FRP fails. For properly designed FRP strengthened beams, this mode of failure should be eliminated.

Mode (4) occurs when the increase in load capacity due to flexural strengthening exceeds the original shear capacity of the member. In this case, strengthening in shear is also required, and this will be discussed in Section 6.4.

6.3.1.2 Failure modes with loss of composite action

Loss of composite action is caused by stress concentrations along the FRP/concrete interface. The locations along the beam with stress concentrations are illustrated in Figure 6.5. The FRP stress is zero at the section where the plate is terminated (or cut off). For tensile force to develop in the FRP plate, interfacial shear stress must exist near the plate end for equilibrium to be satisfied. Detailed analysis, first performed by Roberts (1989), reveals the presence of a concentrated shear stress at the cut-off point. Also, to have compatible curvatures for the concrete beam and FRP plate, concentrated tensile stress normal to the interface is also generated. Under shear and tensile stress concentrations, local failure initiates at the plate end with the formation of an inclined crack that propagates upwards (Figure 6.6(a)). Once the crack grows beyond the steel reinforcements, load transfer to the concrete cover will occur mainly through shear and normal stresses at the level of the reinforcements (Figure 6.6(b)). Under increased loading, a horizontal crack forms at the level of the steel reinforcement and its propagation will cause the concrete cover near the plate end to debond together with the FRP. In the literature, this failure mode is referred to as "plate end debonding", "cover separation" or "peel-off". It should be pointed out that initial cracking at the cut-off point can occur at a load level significantly below the ultimate load of the strengthened member, even if the plate is terminated close to the support (as in the specimen shown in Figure 6.6(a)). As a result, while the interfacial shear stress concentrations are important for the initiation of debonding at the plate end, they are not applicable to the prediction of ultimate failure load (Leung 2007).

Stress concentration is also present at the bottom of flexural or flexural/shear cracks along the span of the concrete beam. Under loading, these cracks tend to open but their openings are restrained by the FRP plate. In return, concentrated shear stresses are induced along the interface on the two sides of the opening crack (Leung 2001). Under this situation, debonding initiated at the vicinity of the concrete crack, and propagates towards the end of the plate. The mode of failure is referred to as crack-induced debonding. For this failure mode, material separation occurs within the concrete at a small distance from the concrete/adhesive interface. This

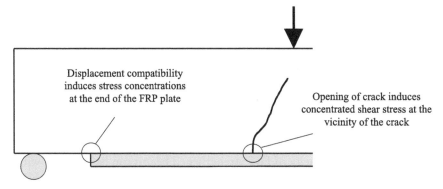

Figure 6.5 Locations with stress concentrations along the FRP/concrete interface.

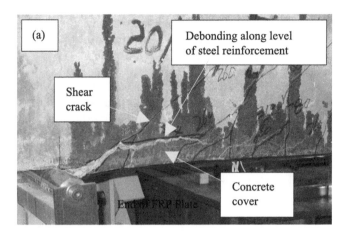

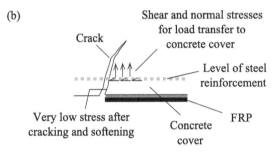

Figure 6.6 Debonding failure at the plate end.

is evidenced from the thin layer of concrete (of about several mms) attached to the FRP plate after debonding failure occurs (Figure 6.7(a)). The occurrence of debonding within the concrete member at a distance from the concrete/adhesive interface can be explained as follows. First, penetration of adhesive into the concrete may increase the strength of a thin layer of material right next to the interface. Second, high shear stresses acting along the concrete/adhesive interface will produce micro-cracks that tend to propagate away from the interface at a certain angle along the principal compression direction (Figure 6.7(b)). The interaction and coalescence of these inclined cracks will produce the final debonding surface inside the concrete substrate.

6.3.2 Factors affecting flexural load capacity when there is loss of composite action

The flexural load capacity of FRP strengthened beams can be obtained with the same approach for conventional reinforced concrete, as long as the FRP force at failure is known. When there is full composite action, the FRP

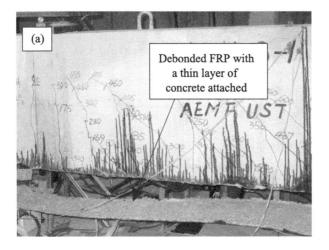

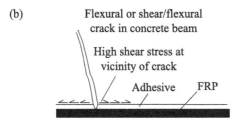

Figure 6.7 Crack-induced FRP debonding.

behaves as a linear elastic material up to its rupture strength (or, till concrete crushing occurs), and its failure stress is easy to find. When plate-end debonding or crack-induced debonding occurs, various models have been proposed to obtain the ultimate failure load directly, or to calculate the FRP stain at the instant of failure. Unfortunately, a universally accepted model is yet to be developed, so different approaches for calculating the debonding failure load are adopted in different design recommendations. Detailed discussion of the existing models is beyond the scope of this book. The interested reader should refer to Teng *et al.* (2002), where an excellent summary of various modeling approaches can be found. In what follows, we will first discuss the effect of various parameters on the debonding failure load. Then, an empirical formula for predicting the FRP failure strain, based on the fitting of a large number of test data, is presented. To facilitate the discussions, the following terms are defined (see also Figure 6.8):

S = shear span of the beam
h = total depth of the beam
d = effective depth of the tensile reinforcement
b = width of the beam

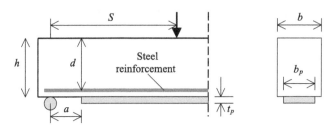

Figure 6.8 Definition of geometrical parameters for the strengthened beam.

b_p = width of the FRP plate
t_p = thickness of the FRP plate
a = cut-off distance, which is the distance between the plate end and the closer support
E_p = Young's modulus of FRP
f_c' = compressive cylinder strength of concrete.

6.3.2.1 Effect of cut-off distance

The variation of FRP debonding load with cut-off distance is illustrated by a set of test data given in Table 6.2. When the cut-off distance is sufficiently small (i.e. the plate end is sufficiently close to the support), the failure load and maximum FRP strain are essentially independent of cut-off distance. However, as the cut-off distance increases, significant reduction in strengthening effectiveness can be observed, with the failure mode changing from crack-induced debonding to plate end failure. The trend of experimental results can be explained with the help of Figure 6.9, which shows schematically the variation of crack-induced debonding load and plate end failure load with cut-off distance. For crack-induced debonding, the maximum FRP force depends on the distance of the plate end from the major crack initiating the debonding process. As the length of plate beyond the major crack increases (i.e. the cut-off distance decreases), both experimental bond tests and theoretical models indicate that the FRP force would approach an

Table 6.2 Effect of cut-off distance on strengthening effectiveness

Cut-off distance a (mm)	a/S	Failure Load (kN)	Max. FRP strain at failure ($\times 10^{-6}$)	Failure mode
25	0.004167	263.4	7057	Crack-induced
300	0.25	265.1	7280	Crack-induced
450	0.375	234.8	4943	End Failure
600	0.50	204.1	3072	End Failure

Notes: For all beams, $b = b_p = 150$ mm, $h = 400$ mm, $d = 370$ mm, $S = 1.2$ m, $t_p = 0.44$ mm, $E_p = 235$ GPa and $f_c' = 33$ MPa. Failure Load of Control Beam (with no FRP) is 197.8 kN. Tests are performed at the Hong Kong University of Science and Technology.

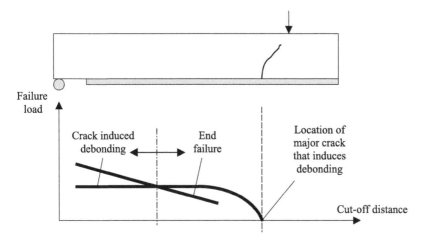

Figure 6.9 Transition of failure mode with increasing cut-off distance.

asymptotic value. As a result, the failure load also approaches a constant value as shown in Figure 6.9. For plate end failure, the failure load decreases continuously with cut-off distance, due to increasing stresses in the concrete adjacent to the plate end. Since the failure loads for the two modes do not decrease at the same rate with the cut-off distance, a transition of failure mode may occur.

6.3.2.2 Effect of FRP stiffness

Table 6.3 shows the test results on similar beams strengthened with the same FRP of different thickness. In these tests, the width of FRP is kept constant. The stiffness of the FRP, which is the product of its Young's modulus and area, can hence be represented by $E_p t_p$ (as in the ACI Design recommendation). According to the results in Table 6.3, increasing FRP stiffness at

Table 6.3 Effect of plate stiffness on strengthening effectiveness

Plate thickness t_p (mm)	$E_p t_p$ ($\times 10^3$) N/mm)	Failure load (kN)	Max. FRP strain at failure ($\times 10^{-6}$)	Failure mode
0.22	51.7	216.2	7254	Crack-induced
0.44	103.4	239.1	6475	Crack-induced
0.66	155.1	255.2	5655	Crack-induced
0.88	206.8	275.9	4934	Crack-induced
1.10	258.5	276.8	4421	End failure
1.32	310.2	251.8	3097	End failure

Notes: For all beams, $b = b_p = 150$ mm, $h = 400$ mm, $d = 370$ mm, $S = 1.2$ m, $a = 15$ mm, $E_p = 235$ GPa and $f'_c = 29$ MPa. Failure Load of Control Beam (with no FRP) is 180.8 kN.
Tests are performed at the Hong Kong University of Science and Technology.

constant width initially increases the failure load and decreases the maximum FRP strain, but eventually leads to reduction in both. The reduction in strengthening effectiveness is associated with a change of failure mode from crack-induced debonding to plate end failure. Based on fracture mechanics arguments, the FRP force for crack-induced debonding to occur should increase with plate thickness. However, as the plate becomes stiffer, the plate end stress concentration also becomes more severe. Once the inclined crack initiated at the plate end propagates beyond the steel reinforcements, higher tensile stresses will be induced at the level of the steel reinforcement by a stiffer cover (which consists of both the concrete and the FRP plate). Cover separation is then easier to take place. Since the debonding failure loads for the two modes vary in opposite trends with the plate stiffness, a transition can be expected.

A very interesting observation here is the existence of an optimal FRP thickness (or stiffness) for flexural strengthening. Excessive increase in thickness will lead to a transition to plate end failure and associated reduction in ultimate load. It should be noted that the cut-off distance for the tested beams is very small. According to the discussions in the former sub-section, when the cut-off distance is larger, plate end failure is easier to occur, so the optimal FRP thickness is expected to decrease.

From Table 6.3, the maximum FRP strain decreases continuously with plate stiffness. The rate of decrease is greatly increased after the transition of failure mode. When crack-induced debonding occurs, a four-fold increase in stiffness (as t_p increases from 0.22 mm to 0.88 mm) leads to about one-third reduction in maximum strain. If plate end failure occurs, 50 percent increase in plate stiffness (due to increase of t_p from 0.88 mm to 1.32 mm) results in a reduction of nearly 40 percent. This result raises an issue for practical applications. Real structural members are much larger in size than laboratory specimens. To provide sufficient force for strengthening, thick FRP plates are required, and the significant reduction in strengthening effectiveness is a matter of concern. To address this issue, the effect of member size on flexural strengthening should be studied.

6.3.2.3 Effect of member size

Table 6.4 shows two sets of tests on the effect of member size on failure load and maximum failure strain. In each series, the geometry of members and steel reinforcement ratios are kept the same. From the results, two interesting observations can be made. First, while the results for beams with $d = 400$ mm (Table 6.3) indicate failure mode transition when t_p is between 0.88 mm and 1.1 mm, the failure mode for beams with $d = 800$ mm remains the same for t_p up to 1.76 mm. Second, comparing the maximum FRP strain for the same t_p but different member size d, the FRP plate is able to carry a higher strain when it is bonded to a larger beam. These observations are encouraging as they show that the reduction in strengthening effectiveness

Table 6.4 Effect of member size on strengthening effectiveness

Beam depth h (mm)	t_p (mm)	Cut-off Distance a (mm)	f_c (MPa)	Load capacity (kN)	Max. FRP strain at failure ($\times 10^{-6}$)	Failure mode
200	0.22	200	47	75.7	9737	Crack-induced
400	0.44	400	47	273.5	7364	Crack-induced
800	0.88	800	47	1025.3	6548	Crack-induced
200	0.44	15	29	74.4	5904	Crack-induced
400	0.88	30	29	275.9	4934	Crack-induced
800	1.76	60	29	1097	4816	Crack-induced

Notes: For all beams, $b = b_p = 0.375\ d$, $S = 3\ h$, $E_p = 235$ GPa.
Tests are performed at the Hong Kong University of Science and Technology.

for thick plates on large members employed in practice is not as drastic as test results on laboratory specimens may indicate.

Summarizing the results in Tables 6.1 to 6.3, one may postulate that the failure mode transition is dependent on a/S and $E_p t_p/d$. The second parameter considers the effect of member size as well as plate stiffness on the failure mode. Either the total depth (h) or effective depth (d) can be used as the dimension parameter. As most equations for conventional concrete design involved d rather than h, d is also employed here. In Figure 6.10, the reported failure modes for 111 tests are summarized in a plot with a/S and

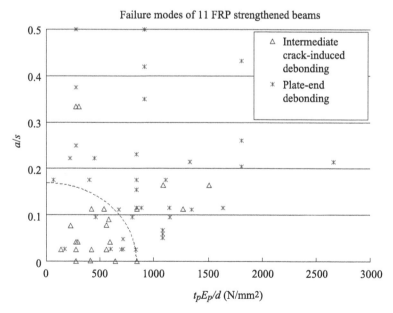

Figure 6.10 Effect of $t_p E_p/d$ and a/S on the failure mode.

$E_p t_p/d$ being the two axes. For most of the tests, crack-induced debonding occurs when both a/S and $E_p t_p/d$ are relatively small. When one or both of them becomes large, plate end failure occurs. The dotted curve in the figure seems to provide a reasonable boundary between the two failure modes.

While $E_p t_p/d$ affects the debonding failure mode, members with the same $E_p t_p/d$ but different sizes (d) do not exhibit the same maximum FRP strain. This is illustrated by the test results in Table 6.3 for geometrically similar members of different sizes, strengthened with the same kind of FRP. In the ACI design recommendation (2001), an empirical equation for the maximum FRP strain is proposed in terms of the parameter $E_p t_p$ alone. However, for a more accurate model, both $E_p t_p/d$ and d should be employed as independent parameters.

6.3.2.4 Effect of FRP width ratio

Table 6.5 shows the result on beams strengthened with FRP of the same thickness but different width to study the effect of the parameter b_p/b. With reduced b_p/b, the maximum FRP strain increases to compensate for the reduction in FRP area. When the FRP width is 150 mm, the load of the strengthened beam is 33 percent beyond the control member. With the FRP width reduced to 50 mm (i.e. the total FRP area is reduced to one-third), 17 percent increase in load can still be achieved. The enhanced effectiveness with reduced b_p/b is attributed to the presence of concrete material beyond the two sides of the bonded plate, which helps to resist debonding failure. In the flexural design of strengthened beams, this factor should not be overlooked.

6.3.2.5 Effect of concrete cover thickness

The concrete cover thickness is related to both the total beam depth (h) and the effective depth (d). According to Raoof and Hassanen (2000), a higher cover thickness increases the likelihood of failure at the plate end. This can

Table 6.5 Effect of plate width on strengthening effectiveness

Plate width b_p (mm)	b_p/b	Failure load (kN)	Max. FRP strain at failure	Failure mode
150	1.0	263.4	7057	Crack-induced
100	0.667	247.9	8401	Crack-induced
75	0.50	239.8	9475	Crack-induced
50	0.333	232.2	11441	Crack-induced

Notes: For all beams, b = 150 mm, h = 400 mm, d = 370 mm, S = 1.2 m, a = 25 mm, E_p = 235 GPa and f_c' = 33 MPa. Failure Load of Control Beam (with no FRP) is 197.8 kN
Tests are performed at the Hong Kong University of Science and Technology.

be explained by the higher stiffness of a thicker cover, which induces higher stresses at the level of the steel reinforcement, making it easier for cover separation to occur.

6.3.2.6 Effect of shear span to effective beam depth

As pointed out by Sebastian (2003), FRP debonding is sensitive to the ratio between maximum moment and maximum shear (M_{max}/V_{max}). When (M_{max}/V_{max}) increases, the failure mode may go from plate end debonding to crack-induced debonding. For beams tested under four-point loading, which is the most common configuration adopted in the literature, the effect of (M_{max}/V_{max}) can be represented by the factor (S/d).

6.3.2.7 Effect of concrete properties

Debonding always occurs within the concrete, at a small distance from the interface for crack-induced debonding, and at the level of the steel reinforcement for plate end failure. In the literature, the debonding load has been expressed in terms of various material parameters of concrete, including its compressive strength, tensile strength, surface pull-off strength and interfacial fracture energy. Of all these parameters, the concrete compressive strength is the only one consistently reported in experimental studies. Equations to obtain the other parameters in terms of the compressive strength have also been proposed. While the compressive strength may not be the parameter governing the actual failure, it is perhaps the only one that can be employed for developing an empirical model for design.

6.3.3 An empirical model for determining the maximum FRP strain at failure

Summarizing the above discussions, the maximum FRP strain at debonding failure can be related to the following parameters: (1) $E_p t_p/d$, (2) d, (3), b_p/b, (4) h/d, (5) S/d, (6) a/S and (7) f_c'. It should be noted that the adhesive thickness and stiffness have not been included, due to controversial findings regarding the effect of the adhesive on FRP debonding. While some researchers find it beneficial to reduce the shear stiffness of the adhesive layer, others discover no effect. Also, the adhesive thickness is hard to measure and seldom reported. Data related to the adhesive are hence insufficient for the development of an empirical model.

In Leung (2006), an empirical approach is proposed to determine the ultimate FRP strain from the seven parameters presented above. With a comprehensive experimental database of 143 tests, a neural network relating the ultimate FRP strain to the various parameters is trained and validated. Using the validated network, an empirical curve for the maximum FRP strain together with several correction equations are generated to provide a simple means to obtain the FRP debonding strain for practical design. For details, the readers should refer to the above paper. Here, only the final results are shown. The design curve given in Figure 6.11 is for the determination of $(\varepsilon_{pu})_{ref}$ in terms of $E_p t_p/d$. The maximum FRP strain (ε_{pu}) is then obtained from:

$$\varepsilon_{pu} = (\varepsilon_{pu})_{ref} * C_{bp/b} * C_d * C_{h/d} * C_{S/d} * C_{a/S} * C_{fc'}/RF \tag{6.1a}$$

RF = reduction factor with recommended value of 1.6:

$$C_{bp/b} = -0.274 \, ((b_p/b)/0.8 - 1) + 1 \tag{6.1b}$$

$$C_d = 0.0051 \, (d/200 - 1)^2 - 0.1453 \, (d/200 - 1) + 1 \tag{6.1c}$$

$$C_{h/d} = 0.987 \, ((h/d)/1.2 - 1) + 1 \tag{6.1d}$$

$$C_{S/d} = 0.1967 \, ((S/d)/4 - 1) + 1 \tag{6.1e}$$

$$C_{a/S} = 0.0008 \, ((a/S)/0.1 - 1)^2 - 0.0627 \, ((a/S)/0.1 - 1) + 1 \tag{6.1f}$$

$$C_{fc'} = 0.2862 \, (f_c'/40 - 1) + 1 = 0.2862 \, (f_{cu}/50 - 1) + 1 \tag{6.1g}$$

When $b_p/b = 0.8$, $d = 200$ mm, $h/d = 1.2$, $S/d = 4$, $a/S = 0.1$, $f_c' = 40$ MPa, all the correction factors become 1.0, and the maximum FRP strain is given by $(\varepsilon_{pu})_{ref}/RF$. $(\varepsilon_{pu})_{ref}$ can hence be interpreted as the maximum FRP strain when all the other parameters are at specific reference values. The C_i's are then correction factors for the general case when the various parameters deviate from their reference values. In Eq (6.1g), f_c' is the cylinder strength of

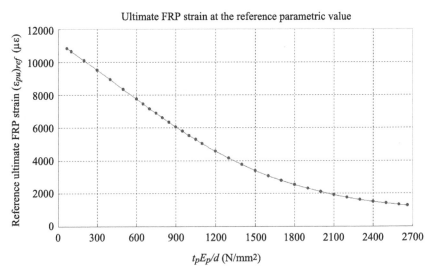

Figure 6.11 Graph for the determination of $(\varepsilon_{pu})_{ref}$ in terms of $E_p t_p/d$.

concrete while f_{cu} is the cube strength which can be taken as 25 percent higher, according to the British code.

In Figure 6.12 the normalized ultimate moments computed with ε_{pu} from Eq. 6.1 is compared with experimental results for 143 tests. The data points for $RF = 1.0$ (shown as crosses) are obtained from the original expressions generated from the neural network. As one can see, they spread around the 45 degree line within a narrow band, showing good fitting of data with the model. In design, we recommend to employ a reduction factor of 1.6. In this case, less than 5 percent of the predicted values will go below the experimental results. Together with the use of additional material safety factors, this will provide a safe design for practical applications.

6.4　Design of beams strengthened in flexure

6.4.1　General assumptions and material behavior

The flexural design of strengthened beams follows closely the conventional method for reinforced concrete design. Specifically, concrete is taken to carry compression only. The "plane section remains plane" assumption is followed and relative slip between FRP and concrete surface is neglected. The latter is justified as the relative slip remains small until FRP debonding becomes unstable at the ultimate load. In the analysis, the following material behaviors for concrete, steel and FRP are assumed.

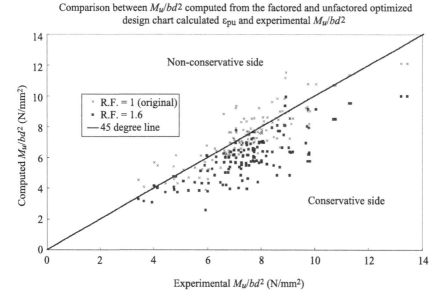

Figure 6.12 Comparison of normalized ultimate moment obtained from empirical approach and experimental results.

6.4.1.1 Concrete

The compressive stress–strain behavior for concrete from BS8110 is employed. It is given by the following form:

$$\sigma_c = (E_c \varepsilon_c - \frac{E_c}{2\varepsilon_0} \varepsilon_c^2) / \gamma_m \text{ for } \varepsilon_c < \varepsilon_0 \tag{6.2a}$$

$$\sigma_c = 0.67 f_{cu} / \gamma_m \text{ for } \varepsilon_0 \leq \varepsilon_c \leq \varepsilon_{cu} \tag{6.2b}$$

where σ_c and ε_c are the stress and strain respectively, f_{cu} (N/mm^2) is the cube compressive strength, E_c ($= 5500\sqrt{f_{cu}}$ N/mm^2) is the initial tangent modulus of concrete, γ_m ($= 1.5$) is the partial safety factor for the strength of concrete, ε_0 ($= 0.00024\sqrt{f_{cu}}$) is the strain at the end of the parabolic part of the stress–strain diagram and ε_{cu} ($= 0.0035$) is the ultimate strain of concrete.

6.4.1.2 Reinforcing steel

Elastic perfectly plastic behavior is assumed for steel reinforcements in both tension and compression. The stress (σ_s) and strain (ε_s) are related by:

$$\sigma_s = E_s \varepsilon_s / \gamma_m \text{ for } \varepsilon_s < \varepsilon_y \tag{6.3a}$$

$$\sigma_s = f_y / \gamma_m \text{ for } \varepsilon_s \geq \varepsilon_y \tag{6.3b}$$

where E_s is the steel elastic modulus, f_y and ε_y are respectively the yield strength and yield strain, and γ_m ($= 1.05$) is the partial factor of safety for steel.

6.4.1.3 FRP plate

The FRP is taken to be linear elastic until failure occurs either by rupture or debonding. Since failure is brittle, the stress drops to zero right after the maximum value is reached. The FRP stress (f_f) is then related to its strain (ε_f) by:

$$f_f = E_f \varepsilon_f / \gamma_m \text{ for } \varepsilon_f < \varepsilon_{fe} \tag{6.4}$$

where E_f and ε_{fe} are respectively the Young's modulus and maximum sustainable strain of the FRP, and γ_m, the partial factor of safety. Depending on the type of FRP, the kind of application and the failure mode, different values of γ_m are given in Table 6.6. The definition of application type in Table 6.6 follows that in the FIB Report (2001). Application Type A refers to

Table 6.6 Partial safety factor for the FRP

Type of FRP	Application type A		Application type B	
	Fiber rupture	Debonding	Fiber rupture	Debonding
CFRP	1.20	1.25	1.35	1.40
AFRP	1.25	1.30	1.45	1.50
GFRP	1.30	1.40	1.50	1.60

pre-cured systems under normal quality control conditions, or wet lay-up systems under conditions with high degree of quality control on both the application conditions and application procedures. Application Type B refers to wet lay-up systems under normal quality control conditions, or any system under difficult on-site working conditions. The safety factors for fiber rupture are also taken from FIB. For debonding failure, we recommend using slightly higher values as the failure load is likely to exhibit higher variability. It should be pointed out that the application of FRP in strengthening still has a relatively short history, so the proposed safety factors are only tentative. Modified values may be used to account for different environmental exposures (especially when GFRP is used). With the accumulation of additional test data and field experience in the future, the proposed factors will be refined.

6.4.2 Initial situation of the member

In most cases, the bonding of FRP to a concrete member is conducted without unloading of the concrete structure or pre-stressing of the FRP. The FRP will therefore only support additional loading beyond that when it is applied. To find the strain (and stress) in the FRP, it is important to know the initial strain at the concrete substrate (ε_{fo}) during the bonding of FRP. When the concrete strain at the substrate increases to ε_c $(> \varepsilon_{fo})$ on further loading, the FRP strain is given by $(\varepsilon_c - \varepsilon_{fo})$.

Let M_0 be the moment acting on the critical section of the concrete beam during FRP bonding. As M_0 is typically larger than the cracking moment M_{cr}, a cracked concrete section should be employed in the calculation. If M_0 is smaller than M_{cr}, the initial strain ε_{fo} is very small and its influence on the FRP strain can be neglected. Considering the cracked section in Figure 6.13 and assuming linear elastic behavior of concrete in compression, the neutral axis depth (x_o) is obtained from:

$$\tfrac{1}{2}bx_0^2 + \left(\frac{E_s}{E_c} - 1\right) A_s'(x_0 - d') = \frac{E_s}{E_c} A_s(d - x_0) \qquad (6.5)$$

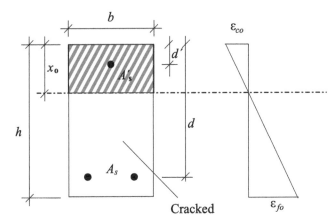

Figure 6.13 Initial condition of the beam.

where A_s and A'_s are the areas of tensile and compressive steel respectively, d and d' the corresponding distances of the tensile and compressive reinforcements from the top of the beam, and b is the beam width.

The concrete strain (ε_{fo}) at the bottom of the concrete beam is then

$$\varepsilon_{f0} = \frac{M_0(h - x_0)}{E_c I_c} \qquad (6.6)$$

where I_c is the moment inertia of the transformed cracked section, given by:

$$I_c = \frac{bx_0^3}{3} + \left(\frac{E_s}{E_c} - 1\right)A'_s(x_0 - d')^2 + \frac{E_s}{E_c}A_s(d - x_0)^2 \qquad (6.7)$$

On further loading after the FRP is bonded, the stress in the composite plate (f_f) is calculated from:

$$f_f = E_f \varepsilon_f = E_f(\varepsilon_c - \varepsilon_{fo}) \qquad (6.8)$$

where ε_c is the strain at the tensile surface of the concrete beam and ε_f is the actual strain in the FRP plate. Theoretically speaking, the change in strain at the level of the FRP is different to that at the concrete substrate. However, as both the FRP and adhesive are very thin, the difference can be neglected in all practical calculations.

6.4.3 A design framework for the strengthened beam

A general design framework should cover both the ultimate state and serviceability state. To find the ultimate moment of the section, a systematic approach proposed by Malek and Patel (2002) is followed. Equations will be derived for the rectangular beam. For the analysis of an *I*-beam (which includes part of the slab above the beam), the interested reader should refer to the original reference. For serviceability considerations as well as avoidance of creep rupture, approaches suggested by ACI (2002) are adopted.

6.4.3.1 Design for ultimate state

DETERMINATION OF FRP STRAIN AT FAILURE

Theoretically speaking, if the FRP fails by rupture, the failure strain should be equal to ε_{fu}, the failure strain measured in a direct tension test. However, many test results indicate the occurrence of FRP rupture in strengthened beams at a lower strain value. For design, we propose to take the rupture strain to be $0.8\varepsilon_{fu}$. Debonding failure will occur when the strain reaches ε_{pu} given by Eq. (6.1). The maximum sustainable strain of the FRP (ε_{fe}) is given by the smaller value between $0.8\varepsilon_{fu}$ and ε_{pu}, divided by the appropriate material safety factor. (Note: as the material safety factors for debonding and rupture are different, $0.8\varepsilon_{fu}$ and ε_{pu} should be directly compared first. The smaller value is then divided by γ_m for the corresponding failure mode.)

6.4.3.2 The balanced plate ratio for simultaneous steel yielding and concrete crushing

In conventional reinforced concrete design, steel yielding should occur before the crushing of compressive concrete. For the strengthened beam, the same should be ensured. The plate ratio ρ_f is given by A_f/bd, where A_f is the cross-sectional area of the composite plate, b the width of the section and d the effective depth to the steel reinforcement. The balanced plate ratio ($\rho_{f,b}$) is defined as the plate ratio at which steel yielding occurs simultaneously with concrete crushing at the ultimate state. Its determination is illustrated in Figure 6.14, where the concrete stress is represented by the equivalent stress block given by BS8110. For the stress block in Figure 6.14, the compressive stress of $0.45f_{cu}$ has already incorporated a material safety factor of 1.5.

In the balanced condition, the tensile steel reinforcement attains its yield strain (ε_y) while the concrete at the compressive surface reaches its ultimate strain (ε_{cu}). With the strain varying linearly along the depth of the beam, the neutral axis position is given by:

$$x_b = \frac{\varepsilon_{cu}}{\varepsilon_{cu} + \varepsilon_y} d \qquad (6.9)$$

From force equilibrium, the balanced plate ratio can be calculated from:

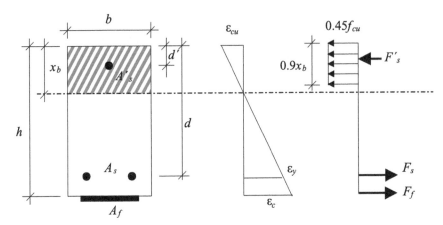

Figure 6.14 Determination of balanced plate ratio for simultaneous steel yielding and concrete crushing.

$$\rho_{f,b} = \frac{0.405 f_{cu} x_b / d + 0.95(\rho' \varepsilon'_s E_s - \rho f_y)}{E_f \varepsilon_f} \tag{6.10}$$

where $\varepsilon_f = \varepsilon_{cu} \dfrac{h - x_b}{x_b} - \varepsilon_{f0} \leq \varepsilon_{fe}, \rho = A_s/bd, \rho' = A'_s/bd$

In design, ρ_f should be less than $\rho_{f,b}$ to ensure the yielding of steel before crushing of concrete. If more ductility is required, ρ_f should be limited to a fraction of $\rho_{f,b}$. In Malek and Patel (2002), a fraction of ¾ is suggested. This can be taken as a general guideline, but the specific value should be determined according to experience and specific ductility requirement of the retrofitted member.

6.4.3.3 The balanced steel ratio for simultaneous plate failure and concrete crushing

After ensuring the occurrence of steel yielding before failure, the next step is to determine the ultimate failure mode, which can be either FRP rupture/debonding or concrete crushing. One can define another balanced plate ratio $\rho_{f,bb}$ for FRP rupture/debonding and concrete crushing to occur simultaneously. For $\rho_f > \rho_{f,bb}$, compressive crushing will be the failure mode. Conversely, FRP rupture/debonding will occur. This balanced condition is illustrated in Figure 6.15, with strain in the composite plate reaching ε_{fe} and the strain at the compressive surface of the beam reaching ε_{cu}. With linear strain variation over the depth of the member, the position of the neutral axis (x_{bb}) is given by:

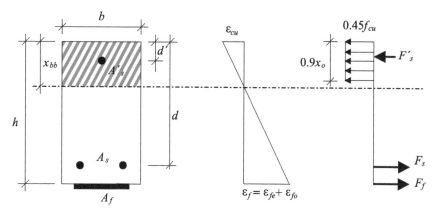

Figure 6.15 Determination of balanced plate ratio for simultaneous FRP rupture/debonding and concrete crushing.

$$\frac{\varepsilon_{cu}}{x_{bb}} = \frac{\varepsilon_{fe} + \varepsilon_{f0}}{h - x_{bb}} \Rightarrow x_{bb} = \frac{\varepsilon_{cu}b}{\varepsilon_{cu} + \varepsilon_{fe} + \varepsilon_{f0}} \tag{6.11}$$

The balanced plate ratio for simultaneous FRP failure and concrete crushing is calculated from:

$$\rho_{f,bb} = \frac{0.405 f_{cu} x_{bb}/d + 0.95(\rho'\varepsilon_s' E_s - \rho f_y)}{\varepsilon_{fe} E_f} \tag{6.12}$$

where $\varepsilon_s' = \varepsilon_{cu} - \dfrac{d'}{h}(\varepsilon_{cu} + \varepsilon_{fe} + \varepsilon_{f0})$ is assumed to be less than ε_y. If the compression steel has yielded, the term $\rho'\varepsilon_s' E_s$ in Eq. (6.12) should be replaced by $\rho\varepsilon_y$.

6.4.3.4 Checking for yielding of compression steel at ultimate state

To calculate the moment at ultimate failure, it is necessary to know if the compression steel has yielded or not. As the plate ratio increases, the neutral axis moves downward and the strain in the compression steel increases. Therefore, compressive yielding of the steel will occur as long as the plate ratio goes beyond a certain critical value ($\rho_{f,cy}$). $\rho_{f,cy}$ depends on the specific failure mode after steel yielding, and is derived below separately for (1) FRP rupture or debonding and (2) concrete crushing.

(A) FRP RUPTURE OR DEBONDING

As illustrated in Figure 6.16, the depth of the neutral axis at this condition is given by:

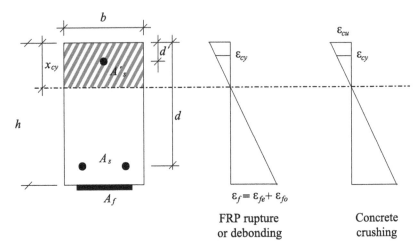

Figure 6.16 Determination of $\rho_{f,cy}$ under two different situations.

$$\frac{\varepsilon_y}{x_{cy} - d'} = \frac{\varepsilon_{fe} + \varepsilon_{f0}}{h - x_{cy}} \Rightarrow x_{cy} = \frac{\varepsilon_y h + (\varepsilon_{fe} + \varepsilon_{f0})d'}{\varepsilon_y + \varepsilon_{fe} + \varepsilon_{f0}} \qquad (6.13)$$

The critical plate ratio, beyond which compression steel yields at or before beam failure is obtained from equilibrium as:

$$\rho_{f,cy} = \frac{0.405 \, f_{cu} x_{cy} / d + 0.95(\rho' - \rho)f_y}{\varepsilon_{fe} E_f} \qquad (6.14)$$

(B) CONCRETE CRUSHING

As illustrated in Figure 6.16, the depth of the neutral axis at this condition is given by:

$$\frac{\varepsilon_y}{x_{cy} - d'} = \frac{\varepsilon_{cu}}{x_{cy}} \Rightarrow x_{cy} = \frac{\varepsilon_{cu} d'}{\varepsilon_y + \varepsilon_{cu}} \qquad (6.15)$$

The critical plate ratio can then be derived from:

$$\rho_{f,cy} = \frac{0.405 \, f_{cu} x_{cy} / d + 0.95(\rho' - \rho)f_y}{\varepsilon_f E_f} \qquad (6.16)$$

where

$$\varepsilon_f = \varepsilon_{cu} \frac{h - x_{cy}}{x_{cy}} - \varepsilon_{f0} \leq \varepsilon_{fe}$$

6.4.3.5 *Calculation of the ultimate moment*

The maximum moment that can be carried by a strengthened beam section depends on the mode of failure. In the following, equations for calculating the ultimate moment are provided separately for FRP rupture/debonding and concrete crushing failure.

(A) FRP RUPTURE OR DEBONDING

(i) Compression steel yields at ultimate load ($\rho_f \geq \rho_{f,cy}$)

$$M_n = 0.95 A_s f_y (d - 0.45x) + A_f \varepsilon_{fe} E_f (h - 0.45x)$$
$$+ 0.95 A_s' f_y (0.45x - d') \tag{6.17}$$

where:

$$x = \frac{0.95 A_s f_y + A_f \varepsilon_{fe} E_f - 0.95\, A_s' f_y}{0.405\, f_{cu} b}$$

(ii) Compression steel does not yield at ultimate load ($\rho_f < \rho_{f,cy}$)

$$M_n = 0.95 A_s f_y (d - 0.45x) + A_f \varepsilon_{fe} E_f (h - 0.45x)$$
$$+ 0.95 A_s' \left(\frac{x - d'}{h - x} \right) (\varepsilon_{fe} + \varepsilon_{f0}) E_s (0.45x - d') \tag{6.18}$$

In this case, the depth of the neutral axis x is calculated using the following equation:

$$\bar{A} x^2 + \bar{B} x + \bar{C} = 0 \tag{6.19}$$

where:

$$\bar{A} = 0.405\, f_{cu} b$$
$$\bar{B} = -0.405\, f_{cu} b h - 0.95(\varepsilon_{fe} + \varepsilon_{f0}) A_s' E_s - 0.95 A_s f_y - A_f \varepsilon_{fe} E_f$$
$$\bar{C} = 0.95(\varepsilon_{fe} + \varepsilon_{f0}) E_s A_s' d' + (0.95 A_s f_y + A_f \varepsilon_{fe} E_f) h$$

(B) CONCRETE CRUSHING

(i) Compression steel yields at ultimate load ($\rho_f \geq \rho_{f,cy}$)

$$M_n = 0.95 A_s f_y (d - 0.45x) + A_f \left(\frac{h - x}{x} \varepsilon_{cu} - \varepsilon_{f0} \right) E_f (h - 0.45x)$$
$$+ 0.95\, A_s' f_y (0.45x - d') \tag{6.20}$$

where x is obtained from Eq. (6.19) using the following parameters:

$$\bar{A} = 0.405 f_{cu} b$$

$$\bar{B} = 0.95(A'_s - A_s)f_y + (\varepsilon_{cu} + \varepsilon_{f0})E_f A_f$$

$$\bar{C} = -\varepsilon_{cu} h A_f E_f$$

(ii) Compression steel does not yield at ultimate load $(\rho_f < \rho_{f,cy})$

$$M_n = 0.95 A_s f_y(d - 0.45x) + A_f\left(\frac{h-x}{x}\varepsilon_{cu} - \varepsilon_{f0}\right)E_f(h - 0.45x)$$

$$+ 0.95 A'_s\left(\varepsilon_{cu}\frac{x-d'}{x}\right)E_s(0.45x - d') \tag{6.21}$$

where x is calculated from Eq. (6.19) with the following parameters:

$$\bar{A} = 0.405 f_{cu} b$$

$$\bar{B} = 0.95 E_s \varepsilon_{cu} A'_s - 0.95 A_s f_y + (\varepsilon_{cu} + \varepsilon_{f0})E_f A_f$$

$$\bar{C} = -\varepsilon_{cu} h A_f E_f - 0.95\varepsilon_{cu} d' A'_s E_s$$

Example calculation

A reinforced concrete beam 400 mm in depth and 270 mm wide is strengthened by the wet lay-up of FRP plate, 0.334 mm in thickness and 250 mm in width, terminated on each side at 100 mm from the supports. The member is loaded under four-point bending with shear span of 1.3 m. The tension steel area is 900 mm² at an effective depth of 340 mm, while the compression steel is 142 mm² in area at a depth of 40 mm from the compressive surface. The concrete cylinder strength is 29.3 MPa. The steel Young's modulus and yield strength are 200 GPa and 484 MPa respectively. The FRP Young's modulus is 230 GPa and its rupture strain is 0.0148. Neglecting the initial strain when FRP is bonded, the ultimate moment of the strengthened member can be obtained as follows:

(1) Compute the effective laminate strain:

$$t_p E_p / d = 225.9\,N/mm^2,\ b_p / b = 0.926,\ d = 340mm,\ h/d = 1.176$$

$$S / d = 3.82,\ a / S = 0.077,\ f'_c = 29.3\ N/mm^2$$

$$(\varepsilon_{fu})_{chart} = 9940 \times 10^{-6}$$

$$C_{bp/b} = -0.274 \cdot (0.926/0.8 - 1) + 1 = 0.957$$

$$C_d = 0.0051 \cdot (340/200 - 1)^2 - 0.1453 \cdot (340/200 - 1) + 1 = 0.901$$

$C_{h/d} = 0.987 \cdot (1.176/1.2 - 1) + 1 = 0.981$

$C_{S/d} = 0.1967 \cdot (3.82/4 - 1) + 1 = 0.991$

$C_{a/s} = 0.0008 \cdot (0.077/0.1 - 1)^2 - 0.0627 \cdot (0.077/0.1 - 1) + 1 = 1.015$

$C_{fc'} = 0.2862 \cdot (29.3/40 - 1) + 1 = 0.923$

$\varepsilon_{pu} = (9940 \times 10^{-6} \cdot 0.957 \cdot 0.901 \cdot 0.981 \cdot 0.991 \cdot 1.015 \cdot 0.923)/1.6$

$\quad = 4878 \times 10^{-6} < 0.8 \cdot 14800 \times 10^{-6} = 8457 \times 10^{-6}$

$\varepsilon_{fe} = 4878 \times 10^{-6}/1.4 = 3484 \times 10^{-6}$

(2) Compute $\rho_{f,b}$

$$x_b = \frac{\varepsilon_{cu}}{\varepsilon_{cu} + \varepsilon_y} \cdot d$$

$$\quad = \frac{3500}{3500 + 2420} \cdot 340$$

$$\quad = 201 \text{ mm}$$

In the calculations, take $f_{cu} = f_c'/0.8 = 36.6$ MPa

Strain in compressive steel $(\varepsilon_s') = \dfrac{201 - 40}{201} 0.0035 = 0.0028 > \varepsilon_y$

$$\quad = 0.00242$$

Stress in compressive steel $= f_y$

$$\rho_{f,b} = \frac{0.405 \cdot f_{cu} \cdot \dfrac{x_b}{d} + 0.95 \cdot (\rho' f_y - \rho f_y)}{E_f \varepsilon_f}$$

$\varepsilon_f = \varepsilon_{cu} \cdot \dfrac{h - x_b}{x_b} = 0.0035 \cdot \dfrac{400 - 201}{201} = 0.003465 < \varepsilon_{fe} = 0.003484$

$\rho = A_s / bd = 900/(270 \cdot 340) = 0.0098$

$\rho' = As' / bd = 142/(270 \cdot 340) = 0.0015$

$$\rho_{f,b} = \frac{0.405 \cdot 36.6 \cdot \dfrac{201}{340} + 0.95 \cdot (0.0015 - 0.0098) \cdot 484}{230000 \cdot 0.003465} = 0.0062$$

$$\rho_f = \frac{A_f}{bd} = \frac{0.334 \cdot 250}{270 \cdot 340} = 0.00091$$

$\rho_f < \rho_{f,b}$ to ensure the steel will yield

(3) Compute $p_{f,bb}$

$$x_{bb} = \frac{\varepsilon_{cu}}{\varepsilon_{cu} + \varepsilon_{fe} + \varepsilon_{fo}} \cdot h = \frac{3500}{3500 + 3484} \cdot 400 = 200 \text{ mm}$$

$$\varepsilon'_s = \varepsilon_{cu} - \frac{d'}{h}(\varepsilon_{cu} + \varepsilon_{fe} + \varepsilon_{fo})$$

$$= 3500 \times 10^{-6} - \frac{40}{400} \cdot (3500 + 3484) \times 10^{-6}$$

$$= 2802 \times 10^{-6} > \varepsilon_y \quad \varepsilon'_s = \varepsilon_y$$

$$p_{f,bb} = \frac{0.405 \cdot f_{cu}\frac{x_{bb}}{d} + 0.95 \cdot (\rho'\varepsilon'_s E_s - \rho f_y)}{E_f \cdot \varepsilon_{fe}}$$

$$= \frac{0.405 \cdot 36.6 \cdot \frac{200}{340} + 0.95 \cdot (0.0015 - 0.0098) \cdot 484}{230000 \cdot 0.003484}$$

$$= 0.00612$$

$$p_f < p_{f,bb}$$

Therefore, debonding is the predominated failure mode.

(4) Compute $p_{f,cy}$

$$x_{cy} = \frac{\varepsilon_y h + (\varepsilon_{fe} + \varepsilon_{fo}) \cdot d'}{\varepsilon_y + \varepsilon_{fe} + \varepsilon_{fo}}$$

$$= \frac{2420 \cdot 400 + (3484 + 0) \cdot 40}{2420 + 3484} = 187.6$$

$$p_{f,cy} = \frac{0.405 \cdot f_{cu} \cdot \frac{x_{cy}}{d} + 0.95 \cdot (\rho' - \rho) \cdot f_y}{E_f \varepsilon_{fe}}$$

$$= \frac{0.405 \cdot 36.6 \cdot \frac{187.6}{340} + 0.95 \cdot (0.0015 - 0.0098) \cdot 484}{230000 \cdot 0.003484}$$

$$= 0.0054$$

$$p_f < p_{f,cy}$$

Using Eq. (6.19) to compute:

$$\bar{A} = 0.405 \cdot f_{cu} \cdot b = 0.405 \cdot 36.6 \cdot 270 = 4002.2$$

$$\bar{B} = -0.405 \cdot f_{cu} \cdot bh - 0.95 \cdot \varepsilon_{fe} \cdot As'Es - 0.95 \cdot Asf_y - A_f\varepsilon_{fe}E_f$$

$$= -0.405 \cdot 36.6 \cdot 270 \cdot 400 - 0.95 \cdot 0.003484 \cdot 200000 \cdot 142$$
$$- 0.95 \cdot 900 \cdot 484$$

$$- 0.003484 \cdot 230000 \cdot 83.5$$

$$= -2.18 \times 10^6$$

$$\bar{C} = 0.95 \cdot (\varepsilon_{fe} + \varepsilon_{fo}) \cdot E_s A_s' d' + (0.95 \cdot A_s f_y + A_f\varepsilon_{fe}E_f) \cdot h$$

$$= 0.95 \cdot 0.003484 \cdot 200000 \cdot 142 \cdot 40$$
$$+ (0.95 \cdot 900 \cdot 484 + 83.5 \cdot 0.003484 \cdot 230000) \cdot 400$$

$$= 1.96 \times 10^8$$

$$x = \frac{-\bar{B} - \sqrt{\bar{B}^2 - 4\bar{A}\bar{C}}}{2\bar{A}} = 114 \text{ mm}$$

$$M_n = 0.95 \cdot A_s f_y (d - 0.45x) + A_f\varepsilon_{fe}E_f(h - 0.45x)$$
$$+ 0.95 \cdot A_s'\left(\frac{x - d'}{h - x}\right)(\varepsilon_{fe} + \varepsilon_{f0})E_s(0.45x - d')$$

$$= 0.95 \cdot 900 \cdot 484 \cdot (340 - 0.45 \cdot 103.8)$$
$$+ 83.5 \cdot 0.003484 \cdot 230000 \cdot (400 - 0.45 \cdot 103.8)$$

$$+ 0.95 \cdot 142 \cdot \left(\frac{103.8 - 40}{400 - 103.8}\right) \cdot 0.003484 \cdot 200000 \cdot (0.45 \cdot 103.8 - 40)$$

$$= 1.43 \times 10^8 \text{ (N.mm) or 143 kNm}$$

6.4.3.6 Serviceability considerations

To avoid excessive cracking and deformation at the serviceability state, yielding of steel should be avoided. Following the suggestion of ACI(2002), the steel stress under service load should not exceed 80 percent of the yield strength. That is:

$$f_{s,s} \leq 0.80 f_y \qquad\qquad (6.22)$$

To find the steel stress under service load, linear-elastic behavior is assumed for the compressive concrete. Analysis of the cracked section gives the following:

$$f_{s,s} = \frac{[M_s + \varepsilon_{f0}A_fE_f(h - x_e/3)](d - x_e)E_s}{A_sE_s(d - x_e/3)(d - x_e) + A_fE_f(h - x_e/3)(h - x_e) + A'_sE_s(x_e - d')(x_e/3 - d')}$$
$$(6.23)$$

where M_s is the maximum moment associated with the service load, and x_e is the depth of neutral axis obtained from:

$$\tfrac{1}{2}bx_e^2 + \left(\frac{E_s}{E_c} - 1\right)A'_s(x_e - d') = \frac{E_s}{E_c}A_s(d - x_e) + \frac{E_f}{E_c}A_f[h - \left(1 + \frac{\varepsilon_{f0}}{\varepsilon_c}\right)x_e] \quad (6.24)$$

6.4.3.7 *Creep-rupture limits*

Under sustained loading, FRP materials may fail at a loading significantly below its short-term strength. This phenomenon, known as creep rupture, is most severe for GFRP and least for CFRP. Adverse environmental conditions such as high temperature, high alkalinity, wetting/drying and freezing/ thawing cycles may aggravate the reduction in strength. Following ACI (2002), the sustained stress levels in GFRP, AFRP and CFRP are limited to 20 percent, 30 percent and 55 percent of their respective short-term strength. Under continuous cyclic loading, the maximum stress during the cycle is also limited to the same value.

As the sustained loading or the maximum load during continuous cyclic loading is unlikely to be very high, the FRP stress can be obtained in the same way as the steel stress in the above serviceability check. Indeed, once the steel stress is found from Eq. (6.23), the FRP stress can be obtained from:

$$f_{f,s} = f_{s,s}\left(\frac{E_f}{E_s}\right)\frac{h - x_e}{d - x_e} - \varepsilon_{f0}E_f \tag{6.25}$$

6.5 Shearing strengthening of beams

6.5.1 *Failure modes*

When the transverse steel reinforcements in an existing concrete beam are not able to provide sufficient shear capacity, shear strengthening can be performed with bonded FRP on the sides of the beam. When loading is applied, shear failure of the strengthened member starts with the formation of an inclined crack in the shear span. With steel stirrups and FRP bridging the crack and transferring stress back to the concrete, additional inclined cracks will form. The cracking pattern is best observed from a member strengthened with FRP strips, where cracks are revealed on the concrete surface between the strips. The opening of a crack will increase the FRP stress and initiate FRP debonding on its two sides. For concrete beams strengthened through side-bonding or U-jacketing of FRP, three different failure modes are possible. If the FRP strength is reached before debonding progresses to a free edge of the bonded FRP, FRP rupture is the failure mode. Conversely, debonding failure occurs if the FRP debonds completely from the concrete substrate before the occurrence of rupture at any location. Here, complete debonding means that the FRP debonds to such

(a)

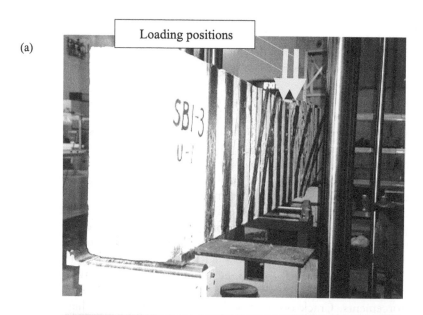

(b)

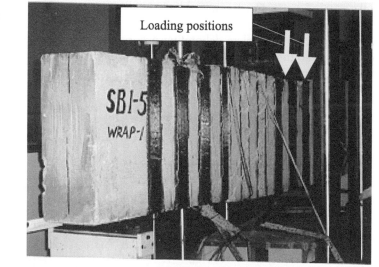

Figure 6.17 (a) Debonding failure for a beam strengthened by U-jacketing;
(b) FRP rupture for a beam strengthened by full wrapping

an extent that it no longer carries any loading. For example, if a major shear crack intersects with a side-bonded FRP strip at a location 100 mm from its bottom edge, complete debonding is considered to occur if the lower 100 mm of the FRP has debonded from the concrete substrate. The third failure mode is a combination of the first two, with rupture of the FRP over part of the bonded area and complete debonding over other parts.

For beams strengthened by full wrapping, final failure always occurs by FRP rupture. Even if the FRP on both sides of the beam has fully debonded from the concrete substrate, the anchorage provided at the top and bottom of the wrap allows the FRP stress to increase continuously until rupture occurs.

Debonding failure and rupture failure are illustrated in Figure 6.17(a) and Figure 6.17(b) respectively. For both failure modes, failure is found to initiate at a location around the middle part of the shear span, rather than near the support or the loading point (for a specimen under four-point loading). Two reasons for this observation can be proposed. First, due to the formation of multiple inclined cracks along the shear span (Figure 6.18), FRP strips near the middle of the span are intersected by several cracks along their lengths, and hence more severely stretched than strips near the support or the loading point. Second, the crack opening, which governs FRP rupture or debonding, is not uniform along a crack. As the tips of inclined cracks are near the loading point, and opening near the crack tip is small, the FRP around the loading point is not heavily stretched. Near the support, the crack opening is controlled by the longitudinal reinforcements. Crack opening is therefore expected to be highest at the middle part of an inclined crack, which is also the location around the middle of the shear span.

When rupture or complete debonding occurs, the FRP stress essentially drops to zero, so the ruptured or fully debonded parts no longer contribute to the load-carrying capacity. The stress in the remaining FRP is increased

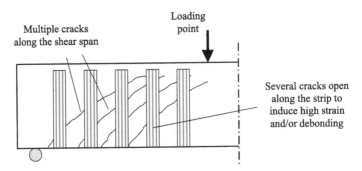

Figure 6.18 Effect of multiple cracks on the straining and debonding of FRP strip around the middle of the shear span.

and they may fail immediately or after additional loading is applied. An important point to note is that the FRP strips at different locations do not reach their ultimate stress (for either full debonding or rupture to occur) at the same time. In practical design, an effective failure strain (ε_{fe}), which represents the average FRP contribution at ultimate loading, should be employed. The determination of ε_{fe} will be discussed in Section 6.5.3.

6.5.2 Factors influencing the shear contribution of FRP reinforcement

6.5.2.1 Effect of FRP stiffness

As discussed in the former section on flexural strengthening, FRP debonding is affected by the stiffness of the bonded sheet or strip. With increasing stiffness, it is easier for debonding to occur. Triantafillou (1998) suggested to represent the FRP stiffness by $\rho_f E_f$ (the product of area fraction and Young's modulus of the FRP), and argued that the effective FRP failure strain (ε_{fe}) should be dependent on this parameter. For failure due to complete debonding, it is well known that the maximum FRP stress (or strain) at failure decreases with stiffness of the FRP. For fiber rupture, failure in most cases will occur after significant (though incomplete) debonding along the interface. ε_{fe}, which is an averaged strain value, should be dependent on the extent of debonding at various parts of the FRP, which affects the stress distribution along the major shear crack. In Triantafillou (1998), ε_{fe} was plotted against $\rho_f E_f$ for a collection of experimental results, and data for both FRP rupture and complete debonding were found to fall along a single curve.

6.5.2.2 Effect of concrete strength

FRP debonding is governed by the surface properties of the concrete, such as its surface tensile strength and interfacial fracture energy. These parameters are often not measured, but they can be related to the concrete compressive strength. In Khalifa et al. (1998) and Triantafillou and Antonopoulos (2000) ε_{fe} is taken to be proportional to $f_c^{2/3}$. The physical basis of this assumption lies in the fact that debonding is affected more by the tensile strength of concrete, which is related to the compressive strength through the 2/3 power relationship. In the ACI and FIB design equations for U-jacketing and side bonding, the parameter $f_c^{2/3}$ appears. Obviously, a higher value of $f_c^{2/3}$ leads to better interfacial bonding and more effective strengthening in such cases. In Triantafillou and Antonopoulos (2000), ε_{fe} was found to be dependent on $(E_f \rho_f / f_c^{2/3})$ for all strengthening configurations. Empirical equations to calculate the effective failure strain from $(E_f \rho_f / f_c^{2/3})$ were therefore proposed. These equations were later incorporated into the FIB design recommendations.

6.5.2.3 Effect of strengthening configuration

The strengthening configuration affects the failure mode. For both side-bonding or U-jacketing, complete debonding of FRP may occur, but the failure load is normally lower for side-bonded FRP, as there is no anchorage at either end of the FRP. For full wrapping, FRP rupture is the failure mode and the effective failure strain is higher than that for complete debonding. Therefore, in terms of strengthening effectiveness, full wrapping is the highest, followed by U-jacketing and then side-bonding. In practice, when U-jacketing and side-bonding are performed, it is desirable to apply anchorage to the free edge(s) of the FRP. The effectiveness of the anchorage should be verified by experiments.

6.5.2.4 Effect of steel shear reinforcement ratio

For strengthened beams failed by complete debonding of side-bonded FRP strips, Pellegrino and Modena (2002) showed that the steel shear reinforcement has a significant effect on the effectiveness of shear strengthening. According to experimental findings, the effectiveness of shear strengthening decreases with an increase in the stiffness ratio between steel shear reinforcements and bonded FRP, given by $\rho_{s,f} = E_s A_{sv} / E_f A_f$.

Physically, the effect is attributed to the change in cracking pattern with increasing amount of steel stirrups in the beam. When there is no shear reinforcement in the beam, only a single major crack will form along the shear span. With more steel reinforcements to transfer stress back into the concrete, more cracks will form. When the FRP is intersected by a large number of opening cracks, complete debonding becomes easier to occur.

6.5.2.5 Effect of member size

In an investigation by Leung *et al.* (2007), geometrically similar concrete beams of three different sizes (with depth of 180 mm, 360 mm and 720 mm) are strengthened with equal area fractions of FRP in both U-jacketing and full wrapping configurations to study the effect of size on the strengthening effectiveness. For full wrapping, the ratio of failure load between strengthened beam and control beam is similar for all specimen sizes, indicating minimum size effect. For U-jacketing, however, the strengthening effectiveness decreases significantly with member size. Further investigations are required to clarify and quantify the effect of size on shear strengthening.

6.5.2.6 Effect of shear span-to-depth ratio

As shear failure of concrete beams occurs in different ways for different span-to-depth ratios, the effectiveness of shear strengthening is also expected to depend on the span-to-depth ratio. In the literature, shear strengthening with FRP bonding is found to be effective for concrete beams with various shear span-to-depth ratios ranging from 1 to 3, but a comprehensive study to quantify the effect of this parameter has yet to be performed. The results from such an investigation will be useful in refining existing design equations in the future.

6.5.3 Design of shear-strengthened concrete beams

To find the shear capacity of a concrete beam strengthened with bonded FRP on the sides, the FRP can be treated in a similar way to steel shear reinforcements, and its contribution (V_f) is added to those from (1) the concrete and longitudinal steel reinforcements (V_c) and (2) the transverse steel reinforcements or bent-up bars (V_s). The total shear capacity (V) is then given by:

$$V = V_c + V_s + V_f \tag{6.26}$$

V_c, which makes up of the shear resistance of concrete in compression, aggregate interlock along the shear crack as well as dowel action of the longitudinal steel reinforcements, is calculated with the same equations for conventional reinforced concrete design. In the calculation of V_s, all transverse steel reinforcements bridging the shear crack are assumed to have yielded. To find V_f, the effective FRP strain at failure (ε_{fe}) needs to be obtained first.

Equations for calculating ε_{fe} are given in design guidelines from ACI, FIB and JSCE. However, only the ACI equations account for the difference between full wrapping, U-jacketing and side bonding. In the FIB guideline, the same equation is proposed for U-jacketing and side bonding. In the JSCE guideline, only a single equation is proposed for all cases. In the following, the ACI equations are adopted.

Treating the FRP sheet or stirrup in a similar way to steel stirrups, V_f is calculated from:

$$V_f = A_f E_f \left(\frac{0.70\varepsilon_{fe}}{\gamma_f}\right) \frac{(\sin a + \cos a)d_f}{s_f} \tag{6.27}$$

The various terms in Eq. (6.27) are illustrated in Figure 6.19. d_f is the effective length of the FRP, which is equal to the FRP length minus the length from the bottom of the beam to the centroid of the tensile steel reinforcements (see Figure 6.19(a)). In Figure 6.19(b), a is the principal fiber orientation for a FRP sheet or the inclination of a FRP strip (with fibers running along the strip). Also, θ is the inclination of shear crack which is taken to be 45 degrees. If FRP strips are used, the center to center spacing is given by s_f (Figure 6.19(c)), and A_f is equal to $2t_f w_f$, where t_f and w_f are respectively the strip thickness and width. For a FRP sheet, (A_f/s_f) in Eq. (6.27) is replaced by $2t_f$.

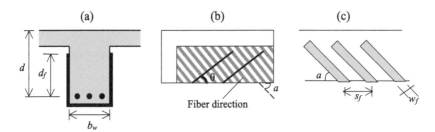

Figure 6.19 Definition of terms in the design equations.

In Eq. (6.27), ε_{fe} is determined from empirical fitting of test data and the factor of 0.70 is incorporated to obtain a value with low probability of failure. Depending on application condition, FRP type and the failure mode, the corresponding material safety factor (γ_f) from Table 6.6 can be used.

To find the mean effective failure strain (ε_{fe}), equations from ACI (2002) are employed. For full wrapping or FRP that is properly anchored, failure occurs by FRP rupture and the effective failure strain is given by:

$$\varepsilon_{fe} = \min(0.75\varepsilon_{fu}, 0.004) \tag{6.28}$$

In Eq. (6.28), ε_{fu} is the rupture strain of the FRP. The effective failure strain is limited to 0.004 to ensure that aggregate interlock is maintained when ultimate failure occurs.

For U-jacketing or side bonding,

$$\varepsilon_{fe} = \min(\kappa_v \varepsilon_{fu}, 0.75\varepsilon_{fu}, 0.004) \tag{6.29}$$

where:

$$\kappa_v = \frac{k_1 k_2 L_e}{11900\varepsilon_{fu}} \tag{6.30}$$

with

$$L_e = \frac{23300}{(t_f E_f)^{0.58}} \tag{6.31}$$

$$k_1 = \left(\frac{f_c}{27}\right)^{2/3} \tag{6.32}$$

$$k_2 = \begin{cases} \dfrac{d_f - L_e}{d_f} & \text{for} \quad \textit{U-Jacketing} \\[2ex] \dfrac{d_f - 2L_e}{d_f} & \text{for} \quad \textit{Side Bonding} \end{cases} \tag{6.33}$$

The design equations proposed above do not take into proper consideration the effects of span depth ratio, member size and steel reinforcement ratio on the FRP failure strain ε_{fe}. With additional experimental results available in the future, and better models developed to quantify the various effects, these equations will be refined.

The calculation of shear capacity from Eqs (6.26) and (6.27) neglects the possibility of crushing of the compressive concrete struts between inclined cracks. To ensure the validity of this assumption, the ultimate shear capacity (V) should be limited to $v_u b_w d$, with $v_u = 0.8 \sqrt{f_{cu}}$ but not exceeding 5 N/mm^2.

6.5.4 Maximum FRP strip spacing

For steel stirrups inside a concrete beam, the maximum FRP spacing is limited by the effective depth to the steel, in order for each inclined crack (assumed to run at 45 degrees to the longitudinal axis) to be intersected by at least one set of stirrups. For FRP strips, a smaller spacing is required to ensure that the crack is not just intersected by at least one FRP strip, but the FRP has to provide significant strengthening at the location of the intersection. For example, if a side-bonded FRP intersects with the crack at a location close to its free edge, complete debonding occurs easily and the effectiveness of the strip is very low. It is much more desirable to have the crack intersecting with the FRP at a location farther away from its edge. To ensure effective strengthening, Teng *et al.* (2002) proposed that there should be at least two FRP strips crossing a crack. The maximum strip spacing is then given by:

$$s_f \leq s_{f,\max} = \frac{d_f(\sin a + \cos a)}{2} = \frac{d_f}{2} \quad \text{if } a = 90° \tag{6.34}$$

where d_f is the effective depth of the FRP strip, as shown in Fig.6.19a.

6.6 Strengthening of concrete columns

6.6.1 Behavior of the FRP strengthened column and failure mode

As discussed in Section 6.2.2, the strengthening of concrete columns relies on the confining effect of FRP on concrete when it cracks and expands laterally under increasing longitudinal compressive strain. To understand the behavior of FRP strengthened columns, it is informative to look at the compressive behavior of concrete under constant lateral confinement first. Here, only circular members with uniform confinement around its circumference are considered. Rectangular members will be discussed in Section 6.6.2.3. Figure 6.20 shows the stress–strain behavior of unconfined concrete, concrete under uniform confinement, and concrete under continuously increasing confinement. As lateral deformation of concrete increases

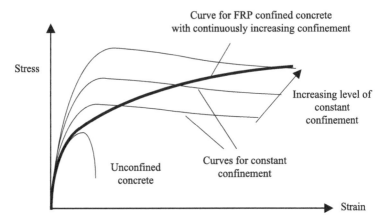

Figure 6.20 The stress–strain behavior of unconfined concrete and concrete under constant and continuously increasing confinement.

with axial strain, FRP-confined concrete belongs to the last category. For unconfined concrete, the stress drops rapidly after the peak value, showing a brittle failure mode. When constant confinement is applied, there is an increase in both the concrete strength and the strain at which the strength is reached. The post-peak load-carrying capacity is greatly improved over that of unconfined concrete, but increasing strain is accompanied by a softening behavior (i.e. decrease in stress). For FRP-confined concrete, the lateral confinement increases continuously with applied axial strain. With increasing strain, the stress value is moving from one constant confinement curve to the next with higher confinement. As illustrated in Figure 6.20, this results in a hardening behavior of the confined column, which is commonly observed in experimental investigations on confined circular columns. In the hardening regime, significant lateral expansion of the member can be observed. Final failure occurs due to FRP rupture (Figure 6.21). Once failure occurs, there is a rapid drop in the load-carrying capacity. Ultimate failure is therefore brittle, but the large deformation before final failure provides sufficient warning and enables effective load redistribution among structural members.

It should be pointed out that the confining effect of FRP is significant only at strain levels approaching or beyond the failure strain (i.e. the strain when the strength is reached) of the unconfined concrete. For concrete under uniaxial compression, splitting cracks tend to form in a direction parallel to the loading direction when the strength is approached. The corresponding increase in lateral strain activates the confining effect of the FRP. Before these cracks are formed, the lateral expansion is too small to induce high confining pressure from the FRP.

Figure 6.21 Failure of an FRP-confined circular concrete specimen by FRP rupture.

6.6.2 Factors influencing the confining effect

6.6.2.1 Effect of FRP strength and thickness

In this sub-section and the next, only circular columns are considered. Here, we further assume that the column is fully wrapped with FRP along its length (as in Figure 6.3(a)). When the column expands laterally, the relation between the confinement pressure and the stress in the FRP can be obtained from force equilibrium over half of the confining FRP, as illustrated in Figure 6.22. With D being the column diameter and t_f the thickness of FRP, the confining pressure f is given by:

$$f = \frac{2\sigma_f t_f}{D} = \frac{p_f \sigma_f}{2} \qquad (6.35)$$

where σ_f is the stress in the FRP and p_f is the area of FRP divided by the area of concrete. The maximum confinement f_l is reached when the stress in the FRP reaches its strength (f_{fu}). The maximum confinement is then given by:

$$f_l = \frac{2f_{fu} t_f}{D} = \frac{p_f f_{fu}}{2} \qquad (6.36)$$

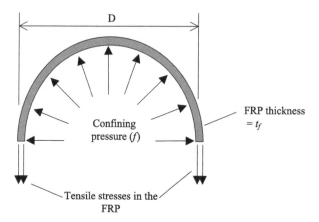

Figure 6.22 Relation between FRP stress and confining pressure.

It is obvious that the performance of the FRP-confined column depends on the maximum confining stress that can be provided. The strength of the FRP and its thickness (or area ratio) are hence important parameters governing the effectiveness of strengthening.

6.6.2.2 Effect of strengthening configuration

Besides full wrapping, it is also possible to strengthen a column with discrete or continuous helical strips. In these cases, some regions of the column surface (specifically, the part between the strips) are not confined. However, even in these regions, confining stresses can still be present at a distance from the surface, due to the spreading of surface compression from the strengthened region towards the interior of the member (Figure 6.23). In this case, failure is governed by the section with the least confinement, which is at the middle of the unconfined region. The effective maximum confining stress in this section is reduced from $2f_{fu}t_f/D$ by two factors, the first one accounting for the stress reduction when the confining stress spreads from the confined regions to cover the whole length of the column, and the second one accounting for the presence of unconfined concrete in the critical section (near the free surface). The first correction factor is given by $b_f/(b_f+s')$, with b_f being the width of the FRP strip and s' the clear spacing between adjacent strips. The second correction factor (k_e) is given by (FIB 2001):

$$k_e = \left(1 - \frac{s'}{2D}\right)^2 \tag{6.37}$$

When helical strips with pitch p are used, an additional correction factor (k_h) is required to account for the increasing radius of curvature of a helix compared to that of the circular section. Following FIB (2001), the factor is:

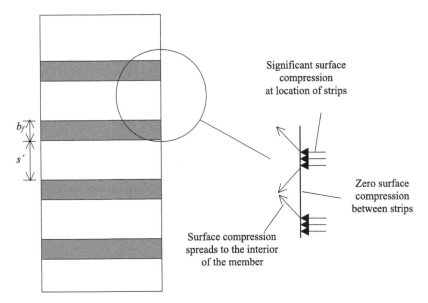

Figure 6.23 Confinement of concrete column by discrete FRP strips.

$$k_b = \left(1 + (\frac{p}{\pi D})^2\right)^{-1}$$ (6.38)

6.6.2.3 *Effect of member section geometry*

While FRP confinement is most effective for circular columns, rectangular columns are commonly employed in structures. When the rectangular column starts to crack and expand laterally, significant pressure only exists at the corners of the member. To strengthen the member, the proper rounding of the corners is crucial. If the corners are not rounded, the high stress concentration will result in pre-mature FRP rupture. Moreover, even if the FRP does not rupture, compressive force is only acting over a very small region of the surface at the vicinity of the corners. In this case, most of the concrete section remains unconfined, and the strengthening is minimal. With a larger radius of curvature at the corners, higher strengthening effectiveness can be achieved. The radius, however, is limited by the presence of reinforcements (especially steel hoops) within the concrete cover. In many cases, even though the strength improvement with FRP wrapping is very small, the post-peak stress softening occurs much slower than the unconfined member. The improved post-peak performance enables stress redistribution to other members and enhances the energy absorption capacity of the member.

To find the failure load of rectangular columns wrapped by FRP, the effective confined area over the concrete section needs to be determined. This is beyond the scope of the present chapter and the interested reader can refer to Teng *et al.* (2002) for a thorough discussion. Also, since this is an ongoing research topic, major journals and conference proceedings in the field can be consulted for the most recent findings.

6.6.3 Design of FRP-strengthened column

As shown in Figure 6.20, the confined circular column shows similar behavior to the unconfined member at the early stage of loading (up to the unconfined concrete strength) and then exhibits a hardening behavior with increasing strain. In design, the stress vs strain relation of the confined concrete can be approximated by a bilinear relation, as illustrated in Figure 6.24. Since the confinement effect is only significant after the concrete is stressed beyond its unconfined strength (f_{co}'), the first branch of the bilinear relation can be taken as the line from zero to ($\varepsilon_{co}', f_{co}'$), where ε_{co}' is the strain corresponding to the unconfined strength. The second branch is then the line from ($\varepsilon_{co}', f_{co}'$) to ($\varepsilon_{cc}', f_{cc}'$), with f_{cc}' and ε_{cc}' being the strength of confined concrete and the strain at ultimate failure. To obtain f_{cc}' and ε_{cc}', both FIB (2001) and ACI (2002) recommend using the empirical equation of Mandar *et al.* (1988), which was originally proposed for concrete under constant confinement, but was found to be applicable to FRP-strengthened columns as well. Using f_l from Eq. (6.36)

$$f_{cc}' = f_{co}'\left[2.25\sqrt{1 + 7.9\frac{f_l}{f_{co}'}} - 2\frac{f_l}{f_{co}'} - 1.25\right] \tag{6.39}$$

$$\varepsilon_{cc}' = \frac{1.71(5\,f_{co}' - 4\,f_{co}')}{E_c} \tag{6.40}$$

As the real stress–strain curve is convex, the use of a bilinear relation will always provide a conservative estimate. Using the bilinear relation, the behavior of a FRP-confined column under combined axial loading and bending can be analyzed with the same approach adopted in conventional reinforced concrete design. If a more accurate description of the stress vs strain relation is required, a procedure proposed by Spoelstra and Monti (1999), described in FIB (2001), can be followed.

In design, partial safety factors have to be applied to the terms in Eq. (6.39). In the term (f_l/f_{co}'), the factor only needs to be applied to f_l to account for the plausible reduction in FRP strength. Then, after f_{cc} is obtained, it should be reduced by the factor of 1.5 as for unconfined concrete.

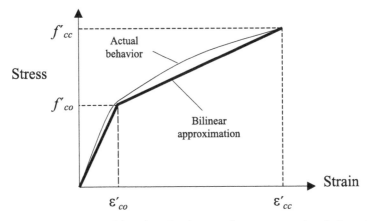

Figure 6.24 A bilinear model to describe the complete stress–strain relation of FRP-confined circular concrete columns.

6.7 Summary

First developed in the 1980s, FRP bonding is now a widely employed strengthening technique for concrete structures. In this chapter, after discussing the need of the technique, the FRP material and repair procedures are described. The applications of bonded FRP to the flexural and shear strengthening of beams as well as the compressive strengthening of columns are then covered. For each application, the failure behavior of the strengthened member and factors affecting the behavior are discussed. Equations for practical design are then given. The reader will notice that the section on flexural strengthening of beams is more thorough than those on the other two applications. This is because the current understanding of flexural strengthening with bonded FRP is beyond that for shear strengthening and compressive strengthening. The effects of various factors such as member size, span/depth ratio and steel reinforcement ratio on the effectiveness of shear strengthening, as well as the modeling of compressive behavior of rectangular members wrapped with FRP, are topics of ongoing research. The interested reader should consult recent issues of major journals and conference proceedings for the most up to-date information.

Bibliography

Abdelrazig, B.E.I., Sharp, J.H. and Jazairi, B.E. (1988) "The Chemical Composition of Mortars Made from Magnesia-Phosphate Cement", *Cement and Concrete Research*, 18: 415–425.

Achenbach, J.D., Komsky, I.N., Lee, Y.C. and Angel, Y.C. (1992) "Self-calibrating Ultrasonic Technique for Crack Depth Measurement", *Journal of Nondestructive Evaluation*, 11: 103–108.

ACI 201.2R-92 (1994a) "Guide to Durable Concrete", in *ACI Manual of Concrete Practice*, Part I: *Materials and General Properties of Concrete*, Detroit, MI: ACI.

ACI 209R-92 (1994b) "Prediction of Creep, Shrinkage, and Temperature Effects in Concrete Structures", in *ACI Manual of Concrete Practice*, Part I: *Materials and General Properties of Concrete*, Detroit, MI: ACI.

ACI 318–95 (1995) *Building Code Requirements for Structural Concrete*, Detroit, MI: ACI.

ACI 201.1R-92 (1997) *Guide for Making a Condition Survey of Concrete in Service*, Detroit, MI: American Concrete Institute.

ACI 305R-99 (1999) "Hot Weather Concreting", in *ACI Manual of Concrete Practice*, Part 2, Detroit, MI: ACI.

ACI 440.2R-02 (2002a) *Guide for the Design and Construction of Externally Bonded FRP Systems for Strengthening Concrete Structures*, Farmington Hills, MI: American Concrete Institute.

ACI-440 (2002b) *Report – Guide for the Design and Construction of Externally Bonded Frp Systems for Strengthening Concrete Structures*. Detroit, MI: ACI.

ACI 437R-03 (2003) *Strength Evaluation of Existing Concrete Buildings*, Detroit, MI: ACI.

ACI 546R-04 (2004) *Concrete Repair Guide*, Detroit, MI: ACI.

ACI (2006) *Vision 2020: A Vision for the Concrete Repair, Protection, and Strengthening Industry*, Farmington Hills, MI: ACI Strategic Development Council.

ACI 216.1–07/TMS-0216–07 (2007a) *Code Requirements for Determining Fire Resistance of Concrete and Masonry Construction Assemblies*, Detroit, MI: ACI.

ACI 224.1R-07 (2007b) *Causes, Evaluation, and Repair of Cracks in Concrete Structures*, Detroit, MI: ACI.

ACI 349.1R-07 (2007c) *Reinforced Concrete Design for Thermal Effects on Nuclear Power Plant Structures*, Detroit, MI: ACI.

ACI 364.1R-07 (2007d) *Guide for Evaluation of Concrete Structures before Rehabilitation*, Detroit, MI: ACI.

ACI Committee 201 (1984) *Guide for Making a Condition Survey of Concrete in Service*, ACI 201.IR-68, Detroit, MI: ACI.

ACI Committee 228 (1998) *Nondestructive Test Methods for Evaluation of Concrete in Structures*, ACI 228.2R-98, Farmington Hills, MI: American Concrete Institute.

ADWR (2006) Arizona Department of Water Resource, http://www.azwater.gov

Ai, H. (2000) "Investigation of the Dimensional Stability in DSP Cement Paste", PhD thesis, University of Illinois at Urbana-Champaign.

Air Force (1994) *Bridge Inspection, Maintenance, and Repair, Chapter 13: Concrete Bridge Maintenance, Repair, and Upgrade, Army TM 5–600*, Air Force AFJPAM 32–1088.

Allen, R.T.L., Edwards, S.C. and Shaw, J.D.N. (1993) *The Repair of Concrete Structures*, 2nd edn, Glasgow: Blackie.

Alonso, C. and Andrade, C. (1994) "Life Time or Rebars in Carbonated Concrete", in J.M. Costa and A.D. Mercer (eds), *Progress in Understanding and Prevention of Corrosion*, London: Institute of Materials.

Alonso, C. Andrade, C. and Gonzalez, J. (1988) "Relation between Resistivity and Corrosion Rate of Reinforcements on Carbonated Mortar Made with Several Cement Types", *Cement and Concrete Research*, 18: 687–698.

Amparano, F.E. and Xi, Y. (1998) "Pumpability of Nonsand Concrete with Anti-Segregative Additives", *ACI Materials Journal*, 95(6): 695–703.

Anonymous (1988) "Concrete Durability: A Multibillion-Dollar Opportunity", *Concrete International*, 10(1): 33–35.

ASCE Standard 7–95 (1995) *Minimum Design Loads for Buildings and Other Structures*, New York: American Society of Civil Engineers.

ASCE (2000) *Guideline for Structural Condition Assessment of Existing Buildings*, SEI/ASCE 11–99, ASCE.

ASTM C 876–91 (1999) "Standard Test Method for Half-Cell Potentials of Uncoated Reinforcing Steel in Concrete", in *Annual Book of ASTM Standards*, West Conshohocken, PA: ASTM.

ASTM C 1383 (2002) "Standard Test Method for Measuring the P-Wave Speed and Thickness of Concrete Plates Using the Impact-Echo Method", in *Annual Book of ASTM Standards*, West Conshohocken, PA: ASTM.

ASTM Test Designation C597–02 (2003a) "Standard Test Method for Pulse Velocity through Concrete", in *Annual Book of ASTM Standards*, West Conshohocken, PA: ASTM.

ASTM E 1316–02a (2003b) "Standard Terminology for Nondestructive Examinations, Section B: Acoustic Emission", in *Annual Book of ASTM Standards*, West Conshohocken, PA: ASTM.

ASTM (2004a) "Standard Specification for Epoxy-Coated Steel Reinforcing Bars, ASTM A775/A775M-04a, 01.04", in *Annual Book of ASTM Standards*, West Conshohocken, PA: ASTM.

ASTM (2004b) "Standard Test Method for Determining the Apparent Chloride Diffusion Coefficient of Cementitious Mixtures by Bulk Diffusion, C1556–04", in *Annual Book of ASTM Standards*, West Conshohocken, PA: ASTM.

ASTM C42 (2004c) "Standard Method of Obtaining and Testing Drilled Cores and Sawn Beams of Concrete", in *Annual Book of ASTM Standards*, West Conshohocken, PA: ASTM.

ASTM C1012–04 "Standard Test Method for Length Change of Hydraulic-Cement Mortars Exposed to a Sulfate Solution", in *Annual Book of ASTM Standards*, West Conshohocken, PA: ASTM.

ASTM C227–03 "Standard Test Method for Potential Alkali Reactivity of Cement-Aggregate Combinations (Mortar-Bar Method)", in *Annual Book of ASTM Standards*, West Conshohocken, PA: ASTM.

ASTM C586–05 "Standard Test Method for Potential Alkali Reactivity of Carbonate Rocks as Concrete Aggregates (Rock-Cylinder Method)", in *Annual Book of ASTM Standards*, West Conshohocken, PA: ASTM.

Attoh-Okine, N. (1995) "Use of Artificial Neural Networks in Ground Penetrating Radar Applications in Pavement Evaluation and Assessment", in G. Schickert, and H. Wiggerhanser (eds), *International Symposium Non-Destructive Testing in Civil Engineering*, Sept. 26–28, Berlin, pp. 93–100.

Austin, S., Robins, P. and Pan, Y. (1999) "Shear Bond Testing of Concrete Repairs", *Cement and Concrete Research*, 29: 1067–1076.

Bartlett, F.M. and MacGregor, J.G. (1994) "Effect of Moisture Condition on Concrete Core Strengths", *ACI Materials Journal*, 91(3): 227–236.

Bazant, Z.P. (1979) "Physical Model for Steel Corrosion in Concrete Sea Structures", *Journal of the Structural Division*, ASCE, 105(6): 1137–1166.

Bazant, Z. P. (1992) "Improved Prediction Model for Time-Dependent Deformations of Concrete: Part 6 – Simplified Code-Type Formulation", *Materials and Structures*, 25(148): 219–223.

Bazant, Z.P. and Kaplan, M.F. (1996) *Concrete at High Temperatures, Material Properties and Mathematical Models*, Harlow: Longman Group Limited.

Beattie, A.G. (1983) "Acoustic Emission, Principal and Instrumentation", *Journal of Acoustic Emission*, 2(1/2): 95–128.

Beddoe, R. and Setzer, M.J. (1988) "A Low Temperature DSC Investigation of Hardened Cement Paste Subjected to Chloride Action", *Cement and Concrete Research*, 18: 249–251.

Beddoe, R. and Setzer, M.J. (1990) "Phase Transformations of Water in Hardened Cement Paste: A Low Temperature DCS Investigation", *Cement and Concrete Research*, 20: 236–242.

Bennett, E.W. and Loat, D.R. (1970) "Shrinkage and Creep of Concrete as Affected by the Fineness of Portland Cement", *Magazine of Concrete Research*, 22(71): 69–78.

Bentur, A., Diamond, S. and Berke, N.S. (1997) *Steel Corrosion in Concrete: Fundamentals and Civil Engineering Practice*, London: E & FN Spon.

Berke, N.S., Pfeifer, D.W. and Weil, T.G. (1988) "Protection against Chloride-Induced Corrosion", *Concrete International Design & Construction*, 10(12): 45–55.

Bertolini, L., Elsener, B., Pedeferri, P. and Polder, R. (2004) *Corrosion of Steel in Concrete: Prevention, Diagnosis and Repair*, Berlin: Wiley-VCH Verlag GmbH & Co. KGaA.

Bijen, J. (2003) *Durability of Engineering Structures: Design, Repair and Maintenance*, Cambridge: Woodhead Publishing Limited.

Bilger, W., Murphy, W.E. and Reidinger, J. (1986) *Inspection and Maintenance of Reinforced and Prestressed Concrete Structures*, FIP Guide to Good Practice, London: Thomas Telford,

Bray, D.E. and McBride, D. (1992) *Nondestructive Testing Techniques*, New York, Wiley.

Bray, D.E. and Stanley, R.K. (1997) *Nondestructive Evaluation: A Tool in Design, Manufacturing and Service*, Boca Raton, FL: CRC Press

Brendenberg, H. (ed.) (1980) *Proceedings of International Seminar on the Application of Stress-Wave Theory on Piles*, Stockholm, June.

Broomfield, J.P. (1997) *Corrosion of Steel in Concrete: Understanding, Investigation and Repair*, London: E & FN Spon.

Broomfield, J.P., Rodriguez, J., Ortega, L.M. and Garcia, A.M. (1995) "Field Measurement of the Corrosion Rate of Steel in Concrete Using a Microprocessor Controlled Unit with a Monitored Guard Ring for Signal Confinement", in *Proceedings of Techniques to Assess the Corrosion Activity of Steel-Reinforced Concrete Structures*, ASTM, STP 1276, Philadelphia.

BS1881: Part 120 (1983) *Method for Determination of the Compressive Strength of Concrete Cores*, London: British Standards Institute.

Bungey, J.H. and Millard, S.G. (1996) *Testing of Concrete in Structures*, 3rd edn, London: Blackie Academic & Professional.

CAC Cement Association of Canada (2006), <http://www.cement.ca>.

Calleja, J. (1980) "Durability", in *Proceedings*, 7th International Congress on the Chemistry of Cement, Paris, 1, pp. 2/1–2/48.

Campbell-Allen, D. and Roper, H. (1991) *Concrete Structures: Materials, Maintenance, and Repair*, Harlow: Longman Scientific & Technical

Carino, N.J., Sansalone, M. and Hsu, N.N. (1986) "A Point Source-Point Receiver Technique for Flaw Detection in Concrete", *Journal of American Concrete Institute*, 83(2): 199.

Champion, S. (1961) *Failure and Repair of Concrete Structures*, New York: John Wiley & Sons Inc.

Chandler, I. (1991) *Repair and Renovation of Modern Buildings*, New York: McGraw-Hill Inc.

Chandler, K.A. and Bayliss, D.A. (1985) *Corrosion Protection of Steel Structures*, Oxford: Elsevier Science Publishers.

Chen, W.F. and Duan, L. (2000) *Bridge Engineering Handbook*, Cambridge: CRC Press.

Chen, W.F. and Scawthorn, C. (2003) *Earthquake Engineering Handbook*, ICBO, SEA and CRC Press.

Cherry, B.W. (1989) "Cathodic Protection of Underground Reinforced Concrete Structures", paper presented at Institution of Corrosion Scientists and Technologists Second International Conference on Cathodic Protection Theory and Practice, June.

Chew, M.Y.L. (1993) "Effect of Heat Exposure Duration on the Thermoluminescence of Concrete", *ACI Materials Journal*, 90(4): 319–322.

Clear, K.C. (1992) "Effectiveness of Epoxy-Coated Reinforcing Steel", *Concrete International*, 14(5): 58–64.

Clifton, J.R., Beeghley, H.F. and Mathey, R.G. (1975) "Nonmetallic Coatings for Concrete Reinforcing Bars", *Building Science Series*, 65, US Department of Commerce, National Bureau of Standards, August.

Colotti, V. and Spadea, G. (2001) "Shear Strength of RC Beams Strengthened with Bonded Steel or FRP Plates", *Journal of Structural Engineering*, ASCE, 127(4): 367–373.

Concrete Society (1979) *Specification for Sprayed Concrete*, London: The Concrete Society.

Concrete Society (1980) *Code of Practice for Sprayed Concrete*, London: The Concrete Society.

Concrete Society (1981) *Guidance Notes on the Method of Measurement for Sprayed Concrete*, London: The Concrete Society.

Concrete Society (1984) *Repair of Concrete Damaged by Reinforcement Corrosion*, London: The Concrete Society.

Concrete Society (1987) *Concrete Core Testing for Strength*, London: The Concrete Society.

CRA (2006) see Concrete Repair Association Home Page, http://www.concreterepair.org.uk

Crom, T.R. (1986) "Shotcrete Nozzleman Certification", *Concrete International*, 8(2): 27–30.

CSA (1989) "Report of Technical Committee on Repair of Concrete Buildings", Draft 1.

Davidson, N., Padaratz, I. and Forde, M. (1995) "Quantification of Bridge Scour Using Impulse Radar", in G. Schickert and H. Wiggerhanser (eds), *International Symposium on Non-Destructive Testing in Civil Engineering*, Sept. 26–28, Berlin, pp. 61–68.

Davis, H.E. (1940) "Autogenous Volume Change of Concrete", Proceedings, ASTM, 40, pp. 1103–1110.

Dawson, J.L., John, D.G., Jafar, M.I., Hladky, K. and Sherwood, L. (1990) "Electrochemical Methods for Inspection and Monitoring of Corrosion of Reinforcing Steel in Concrete", in *Proceedings of Corrosion of Reinforcement in Concrete*, Oxford: Elsevier Applied Science, pp. 358–371.

Dowrick, D. (2003) *Earthquake Risk Reduction*, New York: Wiley.

Durand, B. and Gravel, C. (1995) "Evaluation of the Suitability of Lithium Hydroxide to Inhibit Expansion due to AAR both in New and Old Concrete Structures", in *Proceedings of the Second International Conference on AAR in Hydroelectric Plants and Dams*, Tennessee, Oct., pp. 507–528.

Emmons, P.H. (1994) *Concrete Repair and Maintenance Illustrated*, New York: R.S. Means Company.

Emmons, P.H. (2006) "Concrete: Understanding When and Why Repairs are Necessary." *Buildings*. http://www.buildings.com.

ENV 1992-1-2 (1995a) *Design of Concrete Structures – Part 1–2: General Rules – Structural Fire Design*, Brussels: European Committee for Standardization.

ENV 1993-1-2 (1995b) *Design of Steel structures – Part 1–2: General Rules – Structural Fire Design*, Brussels: European Committee for Standardization.

Felicetti, R., Gambarova, P.G. and Meda, A. (2004) "Guidelines for the Structural Design of Concrete Buildings Exposed to Fire", Workshop *fib Task Group 4.3.2, Fire Design of Concrete Structures*, Politecnico di Milano, Dec. 2–4.

FHWA (1990) *Hydraulic Engineering Circular (HEC) 18: Evaluating Scour at Bridges*, Washington, DC: U.S. Department of Transportation,

FHWA (1999) *Material and Procedures for Rapid Repair of Partial-Depth Spalls in Concrete Pavements, Manual of Practice*, FHWA Report No. FHWA-RD-99–152, Washington, DC: U.S. Federal Highway Administration.

FIB (2001) "Technical Report on Externally Bonded FRP Reinforcement for RC

Structures", *Bulletin No. 14*, International Federation for Structural Concrete, Lausanne, Switzerland.

FIP (1986) *Inspection and Maintenance of Reinforced and Prestressed Concrete Structures*, London: Thomas Telford.

FIP (1991) *FIP Guide to Good Practice of Repair and Strengthening of Concrete Structures*, London: Thomas Telford.

Folliard, K.J., Thomas, M.D.A. and Kurtis, K.E. (2003) *Guidelines for the Use of Lithium to Mitigate or Prevent Alkali-Silica Reaction (ASR)*, FHWA Report No. FHWA-RD-03–047.

Glass, G.K., Page, C.L. and Short, N.R. (1991) "Factors Affecting the Corrosion Rate of Steel in Carbonated Mortars", *Corrosion Science*, 32: 1283–1294.

Godfrey, J.R. (1984) "New Tools Help Find Flaws", *Civil Engineering*, 54(9): 34–41.

Gouda, V. K. (1970) "Corrosion and Corrosion Inhibition of Reinforcing Steel, I: Immersed in Alkaline Solutions", *British Corrosion Journal*, 5(9): 198–203.

Greene, G.W. (1954) "Test Hammer Provides New Method of Evaluating Hardened Concrete", *ACI J. Proc.*, 51(3): 249.

Gutenberg, B. and Richter, C.F. (1956) "Magnitude and Energy of Earthquakes", *Ann. Geofis.*, 9: 1–15.

Halicka, A. and Krol, M. (1999) "Evaluation and Testing of Bond Strength Between Ordinary and Expansive Concrete", in R.K. Dhir and M.J. McLarthy (eds), *Concrete Durability and Repair Technology*, London: Thomas Telford, pp. 493–501

Hall, J.F. (1999) "Discussion on 'the Role of Damping in Seismic Isolation' ", *Earthquake Engineering and Structural Dynamics*, 28: 1717–1720.

Hall, J.F. and Ryan, K.L. (2000) "Isolated Buildings and the 1997 UBC Near-Source Factors", *Earthquake Spectra*, 16: 393–411.

Halpin, J.C. (1984) *Primer on Composite Materials: Analysis*, Technomic Co.

Hanks, T.C. and Kanamori, H. (1979) "A Moment Magnitude Scale", *J. Geophys. Res.*, 84: 2348–2350.

Hartt, W.H. and Rosenberg, A.M. (1989) "Influence of $Ca(NO_2)2$ on Seawater Corrosion of Reinforcing Steel in Concrete", *American Concrete Institute*, Detroit, SP 65–33, 609–622.

Hausman, D. A. (1967) "Steel Corrosion in Concrete", *Materials Protection*, 6(11): 19–22.

Herakovich, C.T. (1998) *Mechanics of Fibrous Composites*, New York: John Wiley & Sons.

Hisano, M., Nagashino, I. and Kawamura, S. (1990) "Earthquake Observation of a Base-Isolated Building with Slide Bearings", in *Proc. 8th JEES*.

Holl, C.H. and O'Connor, S.A. (1997) "Cleaning and Preparing Concrete Before Repair", *Concrete International*, March, 60–63.

Holland, T.C. (1983) *Abrasion-Erosion Evaluation of Concrete Mixtures for Stilling Basin Repairs, Kinzua Dam, Pennsylvania*, Miscellaneous Paper No. SL-83–16 US Army Engineers Waterways Experiment Station, Vicksburg.

Hope, B.B., Neville, A.M. and Guruswami, A. (1967) "Influence of Admixtures on Creep of Concrete Containing Normal Weight Aggregate", in *RILEM International Symposium on Admixtures for Mortar and Concrete*, Brussels, Sept., pp. 17–32.

Hu, Y-X., Liu, S-C. and Dong, W. (1996) *Earthquake Engineering*, London: E & FN Spon.

Huang, X., Birman, V., Nanni, A. and Tunis, G. (2004) "Properties and Potential for Application of Steel Reinforced Polymer (SRP) and Steel Reinforced Grout (SRG) Composites", *Composites*, Part B, 36, 1, 73–82.

ICC (2002) *2003 International Building Code*, Country Club Hills, IL: International Code Council

ICC (2006) *2006 International Existing Building Code*, Country Club Hills, IL: International Code Council.

ICRI (1997) *Guide for Selecting and Specifying Materials for Repair of Concrete Structures*, International Concrete Repair Institute, Technical Guideline No. 03733.

ICRI (2002) *Guide for Surface Preparation for the Repair of Deteriorated Concrete Resulting from Reinforcing Steel Corrosion*, International Concrete Repair Institute, Technical Guideline No. 03730.

ICRI (2006) International Concrete Repair Institute website <http://www.icri.org>

Jessop, E.L., Ward, M.A. and Neville, A.M. (1967) "Influence of Water Reducing and Set-Retarding Admixtures on Creep of Lightweight Aggregate Concrete", in *RILEM International Symposium on Admixtures for Mortar and Concrete*, Brussels, Sept., pp. 35–46.

Jones, R. (1962) *Non-Destructive Testing of Concrete*, Cambridge: Cambridge University Press.

Jones, R.M. (1998) *Mechanics of Composite Materials*, 2nd edn, London: Taylor and Francis.

JSCE (2001) *Recommendations for Upgrading of Concrete Structures with Use of Continuous Fiber Sheets*, Tokyo: Research Committee on Upgrading of Concrete Structures with Use of Continuous Fiber Sheets, Japan Society of Civil Engineers.

Kahn, L.F. (2008) "ACI Code Requirements for Repair of Buildings", *Concrete International*, April, 51–54.

Kanamori, H. (1977) "The Energy Release in Great Earthquake", *J. Geophys. Res.*, 82: 2981–2987.

Kay, T. (1992) *Assessment and Renovation of Concrete Structures*, New York: Longman Scientific & Technical.

Keller, T. (2001) *Use of Fibre Reinforced Polymers in Bridge Construction, State-of-the-Art Report with Application and Research Recommendations*, EPFL, Switzerland.

Kelly, J.M. (1993) "The Role of Damping in Seismic Isolation", *Earthquake Engineering and Structural Dynamics*, 28: 3–20.

Kelly, J.M. (1997) *Earthquake-Resistant Design with Rubber*, New York: Springer-Verlag.

Kesler, C.E. and Higuchi, Y. (1953) "Determination of Compressive Strength of Concrete by Using its Sonic Properties", *Proc. ASTM*, 53: 1044.

Kesler, C.E. and Higuchi, Y. (1954) "Problems in the Sonic Testing of Plain Concrete", in *Proc. Int. Symp. on Nondestructive Testing of Materials and Structures*, 1, RILEM, Paris, p. 45.

Keyser, J.H. (1980) *Durability of Materials and Construction*, ASTM STP691, 38.

Khalifa, A., Gold, W.J., Nanni, A. and Aziz, A. (1998) "Contribution of Externally Bonded FRP to Shear Capacity of RC Flexural Members", *ASCE Journal of Composites for Construction*, 2(4): 195–203.

Klieger, P., Anderson, A.R., Bloem, D.L., Howard, E.L. and Schlintz, H. (1954)

"Discussion of 'Test Hammer Provides New Method of Evaluating Hardened Concrete,' " by Gordon W. Greene, *ACI J. Proc.*, 51(3): 256–251.

Knab, L.I. (1988) *Factors Related to the Performance of Concrete Repair Materials*, Technical Report REMR-CS- 12,, Washington, DC: U.S. Army Corps of Engineers.

Knab, L.I. and Spring, C. B. (1989) "Evaluation of Test Methods for Measuring the Bond Strength of Portland Cement Based Repair Material to Concrete", *Cement, Concrete and Aggregate*, 11: 3–14.

Kolek, J. (1958) "An Appreciation of the Schmidt Rebound Hammer", *Magazine of Concrete Research*, 10(28): 27.

Kolek, J. (1969) "Nondestructive Testing of Concrete by Hardness Methods", in *Proceedings of Symposium on NDT of Concrete and Timber*, Institute of Civil Engineers, London, p.15.

Kramer, S.L. (1996) *Geotechnical Earthquake Engineering*, Englewood Cliffs, NJ: Prentice Hall.

Kunieda, M., Kurihara, N., Uchida, Y. *et al.* (2000) "Application of Tension Softening Diagrams to Evaluation of Bond Properties at Concrete Interfaces", *Engineering Fracture Mechanics*, 65: 299–315.

Landis, E.N. and Shah, S.P. (1993) "Recovery of Microcrack Parameters in Mortar Using Quantitative Acoustic Emission", *Journal of Nondestructive Evaluation*, 12(4): 219–232.

Lea, F.M. (1970) *The Chemistry of Cement and Concrete*, London: Arnold.

Lee, J.S., Xi, Y. and Willam, K. (2008) "Properties of Concrete after High Temperature Heating and Cooling", *Journal of Materials*, ACI, July–Aug. 334–341.

Leung, C.K.Y. (2001) "Delamination Failure in Concrete Beams Retrofitted with a Bonded Plate", *ASCE Journal of Materials in Civil Engineering*, 13(2): 106–113.

Leung, C.K.Y. (2006) "FRP Debonding from a Concrete Substrate: Some Recent Findings against Conventional Belief", *Cement and Concrete Composites*, 28(8): 742–748.

Leung, C.K.Y., Chen, Z., Lee, S., Ng, M., Tang, J. and Xu, M. (2007) "Effect of Member Size on the Shear Failure of FRP Strengthened Reinforced Concrete Members", *ASCE Journal of Composites in Construction*, 1(5): 487–496.

L'Hermite (1960) *Proceedings of the IVth International Symposium on the Chemistry of* Cement, Washington, DC, 2, pp. 659–694.

Li, Z. (1996) "Microcrack Characterization in Concrete under Uniaxial Tension", *Magazine of Concrete Research*, 48(176): 219–228.

Li, Z. and Li, W. (2002) "Contactless, Transformer-Based Measurement of the Resistivity of Materials", United States Patent 6639401.

Li, Z., Li, F.M., Li, X.S. and Yang W.L. (2000) "P-Wave Arrival Determination and AE Characterization of Concrete", *Journal of Engineering Mechanics*, 126(2): 194–200.

Li, Z., Li, F.M., Zdunek, A., Landis, E. and Shah, S.P. (1998) "Application of Acoustic Emission Technique to Detection of Reinforcing Steel Corrosion in Concrete", *ACI Materials Journal*, 95(1): 68–76.

Li, Z., Mu, B. and Peng, J. (1999) "The Combined Influence of Chemical and Mineral Admixtures Upon the Alkali-Silica Reaction", *Magazine of Concrete Research*, 51(3): 163–169.

Li, Z., Mu, B. and Peng, J. (2000) "Aklali-Silica Reaction of Concrete with

Admixtures- Experiment and Prediction", *Engineering Mechanics*, ASCE, 126(3): 243–249.

Li, Z., Peng, J. and Ma, B. (1999) "Investigation of Chloride Diffusion for High-Performance Concrete Containing Fly Ash, Microsilica, and Chemical admixtures", *ACI Materials Journal*, 96(3): 391–396.

Li, Z., Rossow, E.C. and Shah, S.P. (1989) "Sinusoidal Forced Vibration of Sliding Masonry System", *Journal of Structural Engineering*, ASCE, 115(7): 1741–1755.

Li, Z. and Shah, S.P. (1994) "Localization of Microcracking in Concrete Under Uniaxial Tensile", *ACI Materials Journal*, July: 372–381.

Li, Z., Wei, X. and Li, W. (2003) "Preliminary Interpretation of Portland Cement Hydration Process using Resistivity Measurements", *ACI Materials Journal*, 100(3): 253–257.

Li, Z., Yao, W., Lee, S., Lee, C.H. and Yang, Z.Y. (2000) "Application of Infrared Thermography Technique in Building Finish Evaluation", *Journal of Nondestructive Evaluation*, 19(1): 11–19.

Lin, J.M. and Sansalone, M. (1997) "A Procedure for Determining P-wave Speed in Concrete for Use in Impact-Echo Testing Using a Rayleigh Wave Speed Measurement Technique", *Innovations in Nondestructive Testing*, SP-168, Detroit, MI: American Concrete Institute.

Lin, J.M., Sansalone, M. and Streett, W.B. (1996) "A Procedure for Determining P-Wave Speed in Concrete for Use in Impact-Echo Testing Using a Direct P-wave Speed Measurement Technique", *ACI Material Journal*, 93.

Liu, J., Xie, H., Xiong, G. *et al.* (1998) "Using Ultra-Sonic Methods to Judge Quality the New-Old Concrete Interface", in *Proceedings of the 4th Academic Conference on National Building Examination and Reinforcement in China*, Kunming, pp. 152–158 (in Chinese).

Long, A.E. and Murray, A. (1984) *The Pull-Off Partially Destructive Test for Concrete*, ACI SP-82, pp. 327–350.

Ludirdja, D., Berger, R.L. and Young, J.F. (1989) "Simple Method for Measuring Water Permeability of Concrete", *ACI Materials Journal*, 86(5): 433–439.

Mailvaganam, N.P. (1992) *Repair and Protection of Concrete Structures*, Boca Raton, FL: CRC Press.

Mailvaganam N.P., Pye, G.B. and Arnott, M.R. (1998) "Surface Preparation of the Concrete Substrate", Institute for Research in Construction, <http://irc.nrc-cnrc.gc.ca>

Maji, A.K., Ouyang, C. and Shah, S.P. (1990) "Fracture Mechanism of Quasi-Brittle Material Based on Acoustic Emission", *Journal of Material Research*, 5(1): 206–217.

Maji, A.K. and Shah, S.P. (1988) "Process Zone and Acoustic Emission Measurement in Concrete", *Experimental Mechanics*, 28, 27–33.

Maji, A.K. and Shah, S.P. (1989) "Application of Acoustic Emission and Laser Holography to Study Micro-Fracture in Concrete", in *Nondestructive Testing of Concrete*, SP-112, Detroit: American Concrete Institute, pp. 83–109.

Malek, A.M. and Patel, K. (2002) "Flexural Strengthening of Reinforced Concrete Flanged Beams with Composite Laminates", *ASCE Journal of Composites for Construction*, 6(2): 97–103.

Malhotra, H.L. (1956) "The Effect of Temperature on the Compressive Strength of Concrete", *Magazine of Concrete Research*, 8: 85.

Malhotra, V.M. and Carino, N.J. (2004) *Handbook of Nondestructive Testing of Concrete*, 2nd edn, Boca Raton, FL: CRC Press.

Mandar, J.B., Priestley, M.J.N. and Park, R. (1988) "Theoretical Stress-Strain Model for Confined Concrete", *ASCE Journal of Structural Engineering*, 114(8): 1804–1826.

Mangat, P.S. and Limbachiya, M.C. (1997) "Repair Material Properties for Effective Structural Application", *Cement and Concrete Research*, 27: 601–617.

Marosszeky, M., Yu, J.G. and Ng, C.M. (1991) "Prediction of Creep, Shrinkage, and Temperature Effects in Concrete Structures", ACI Publication SP126–70 Durability of Concrete, Second International Conference, 2: 1331–1334.

Mathey, R.G. and Knab, L.I. (1991) "Uniaxial Tensile Tested to Measure the Bond of In-Situ Concrete Overlays", NISTIR 4648.

Mays, G.C. (1985) "Structural Applications of Adhesives in Civil Engineering", *Materials Science and Technology*, 1(11): 937–943.

McKenna, J.K. and Erki, M.A. (1994) "Strengthening of Reinforced Concrete Flexural Members Using Externally Applied Steel Plates and Fibre Composite Sheets – A Survey", *Canadian Journal of Civil Engineering*, 21(1): 16–24.

McLeish, A. (1993) "Standard Tests for Repair Materials and Coatings for Concrete, Part I: Pull-Off Tests", Technical Note 139, CIRIA.

Mehta, P.K. (1980) "Durability of Concrete in Marine Environment – A Review", ACI SP-65, pp. 1–20.

Mehta, P.K. (1988a) "Sulfate Resistance of Blended Cements", in W.G. Ryan (ed.) *Papers, Concrete 88 Workshop Concrete Institute of Australia*, pp. 337–351.

Mehta, P.K. (1988b) "Durability of Concrete Exposed to Marine Environment – A Fresh Look", in V.M. Malhotra (ed.), *Concrete in Marine Environment*, ACI SP-109, pp. 1–29.

Mehta, P.K. and Gerwick, B.C. (1982) "Cracking-Corrosion Interaction in Concrete Exposed to Marine Environment", *Concete International*, 4(10): 45–51.

Mehta, P.K. and Monteiro, P.J.M. (2006) *Concrete: Microstructure, Properties and Materials*, 3rd edn, Maidenhead: McGraw-Hill.

Meinheit, D.F. and Monon, J.F. (1984) "Parking Garage Repaired Using Thin Polymer Concrete Overlay Concrete", *Concrete International*, 6(7): 7–13.

Mindess, S., Young, J.F, and Darwin, D. (2003) *Concrete*, 2nd edn, New Jersey: Pearson Education.

Monday, J.G.L. and Dhir, R.K. (1984) "Assessment of In-situ Concrete Quality by Core Testing", Special Publication, SP 82–20, Detroit: American Concrete Institute, pp. 393–410.

Nasser, K.W. and Al-Manaseer, A.A. (1987) "New Nondestructive Test", *ACI Concrete International*, 41.

National Research Council (1992) *Techniques for Concrete Removal and Bar Cleaning on Bridge Rehabilitation Projects*, Strategic Highway Research Program, SHRP-S-336, Washington, DC.

National Research Council (1993a) *Innovative Materials Development and Testing*, Vol. 5: *Partial Depth Spall Repair*, Strategic Highway Research Program, SHRP-H-356, Washington, DC.

National Research Council (1993b) *Concrete Bride Protection, Repair, and Rehabilitation Relative to Reinforcement Corrosion: A Methods Application Manual*, Strategic Highway Research Program, SHRP-S-360, Washington DC, pp. 47–98.

NCPP (National Center for Pavement Preservation) (2004) *Colorado DOT Distress Manual for HMA and PCC Pavements*, Michigan State University, 2857 Jolly Road, Okemos, MI 48864.

Netcomposites (2006) <http://www.netcomposites.com>

Neville, A.M. (ed.) (1975) "Fiber Reinforced Cement and Concrete", in *RILEM Symposium*, London: Construction Press.

Neville, A.M. (1996) *Properties of Concrete*, 4th edn., Harlow: Longman Group Limited.

Neville, A.M., Dilger, W.H. and Brooks, J.J. (1983) *Creep of Plain and Structural Concrete*, London: Construction Press.

Newman, A. (2001) *Structural Renovation of Buildings: Methods, Details, and Design Examples*, New York, McGraw-Hill

Nielsen, C.V., Pearce, C.J., and Bicanic, N. (2004) "Improved phenomenological modeling of transient thermal strains for concrete at high temperature", *Computers and Concrete*, March, 1(2): 189–209.

Obert, L. and Duvall, W.I. (1941) "Discussions of Dynamic Methods of Testing Concrete with Suggestions for Standardization", *Proc. ASTM*, 41, 1053.

Olson. L.D. and Wright, C.C. (1990) "Seismic, Sonic, and Vibration Methods for Quality Assurance and Forensic Investigation of Geotechnical, Pavement, and Structural Systems", in H.L.M. dos Reis (ed.), *Prof. Conf. on Nondestructive Testing and Evaluation for Manufacturing and Construction*, Hemisphere, p. 263.

Ouyang C.S., Landis, E. and Shah, S.P. (1991) "Damage Assessment in Concrete Using Quantitative Acoustic Emission", *Journal of Engineering Mechanics*, ASCE, 117(11): 2681–2698.

Page, C.L. (1992) "Nature and Properties of Concrete in Relation to Reinforcement Corrosion", in *Corrosion of Steel in Concrete*, Aachen, Feb. 17–19.

Page, C.L. and Treadway, K.W.J. (1982) "Aspects of the Electrochemistry of Steel in Concrete", *Nature*, 297: 109.

Parsons Brinckerhoff (1993) *Bridge Inspection and Rehabilitation: A Practical Guide*, ed. L.G. Silano, New York: John Wiley & Sons, Inc.

PCA, Portland Cement Association Home Page. 28<http://www.cement.org/index.asp>.

Pellegrino, C. and Modena, C. (2002) "Fiber Reinforced Polymer Shear Strengthening of Reinforced Concrete Beams with Transverse Steel Reinforcement", *ASCE Journal of Composites for Construction*, 6(2): 104–111.

Perkins, P.H. (1986) *Repair, Protection and Waterproofing of Concrete Structures*, London: Elsevier Applied Scientific Publishers Ltd.

Phan, L.T. (2004) "Codes and Standards for Fire Safety Design of Concrete Structures in the U.S.", in Workshop *fib Task Group 4.3.2*, Fire Design of Concrete Structures, Politecnico di Milano, Dec. 2–4.

Placido, F. (1980) "Thermoluminescence Test for Fire Damaged Concrete", *Magazine of Concrete Research*, 32(111): 112–116.

Popovics, J.S., Song, W., Achenbach, J.D., Lee, J.H. and Andre, R.F. (1998) "One-sided Stress Wave Velocity Measurement in Concrete", *ASCE Journal of Engineering Mechanics*, 124: 1346–1353.

Popovics, S. (1987) "Chemical Resistance of Portland Cement Mortar and Concrete", in W. L. Sheppard (ed.) *Corrosion and Chemical Resistance Masonry Materials Handbook*, New Jersey: Noyes Publications.

Powers, T.C., Copeland, L.E., Hayes, J.C. and Mann, H.M. (1954) "Permeability of Portland Cement Paste", *ACI Journal Proceedings*, 51(3): 285–298.

Pullar-Strecker, P. (1987) *Corrosion Damage Concrete: Assessment and Repair*, London: CIRIA Butterworths.

Raoof, M. and Hassanen, M.A.H. (2000) "Peeling Failure of Reinforced Concrete Beams with Fibre-Reinforced Plastics or Steel Plates Glued to Their Soffits", in *Proceedings of the Institution of Civil Engineers: Structures and Buildings*, 140: 291–305.

Rhode Island State Building Code (1997) 7th edn, Providence, R.I.

Richer, C.F. (1935) "An Instrumental Earthquake Scale", *Bulletin of the Seismological Society of America*, 25: 1–32.

Rizzo, E.M. and Sobelman, M. (1989) "Selection Criteria for Concrete Repair Materials", *Concrete International*, 11(9): 46–49.

Roberts, T.M. (1989) "Approximate Analysis of Shear and Normal Stress Concentrations in the Adhesive Layer of Plated RC Beams", *The Structural Engineer*, 67(12): 229–233.

Robins, P.J. and Austin, S.A. (1995) "A Unified Failure Envelope from the Evaluation of Concrete Repair Bond Tests", *Magazine of Concrete Research*, 47: 57–68.

Saccani, A. and Magnaghi, V. (1999) "Durability of Epoxy Resin-Based Materials for the Repair of Damaged Cementitious Composites", *Cement and Concrete Research*, 29: 95–98.

Sadegzadeh, M. and Kettle, R.J. (1988) "Abrasion Resistance of Surface Treated Concrete", *Cement, Concrete & Aggregates*, 10(1): 20–28.

Sansalone, M. (1997) "Impact-Echo: The Complete Story", *ACI Structural Journal*, 6.

Sansalone, M. and Carino, N.J. (1988) "Impact-Echo Method: Detecting Honeycombing, the Depth of Surface-Opening Cracks, and Ungrouted Ducts", *Concrete International*, 10(4): 38–46.

Sansalone, M. and Carino, N.J. (1989) "Detecting Delaminations in Concrete Slabs with and without Asphalt Concrete Overlays Using the Impact-Echo Method", *ACI Materials Journal*, 86(2): 175–184.

Sansalone, M. and Carino, N.J. (1990) "Finite Element Studies of the Impact-Echo Response of Layered Plates Containing Flaws", in W. McGonnagle (ed.), *International Advances in Nondestructive Testing*, 15th edn, New York: Gordon & Breach Science Publishers, pp. 313–3336.

Satake, J., Kamakura, M., Shirakawa, K., Mikami, N. and Swamy, R.N. (1983) "Long Term Resistance of Epoxy-Coated Reinforcing Bars", in A. P. Crane (ed.), *Corrosion of Reinforcement in Concrete Construction*, New York: The Society of Chemical Industry/Ellis Horwood Ltd, pp. 357–377.

Sebastian, W.M. (2001) "Significance of Midspan Debonding Failure in FRP-Plated Concrete Beams", *ASCE Journal of Structural Engineering*, 127(7): 792–798.

Seehra, S.S., Gupta, S. and Kumar, S. (1993) "Rapid Setting Magnesium Phosphate Cement for Quick Repair of Concrete Pavement", *Cement and Concrete Research*, 23: 254–266.

SEI/ASCE 11–99 (2000) *ASCE Standard – Guideline for Structural Condition Assessment of Existing Buildings*, Reston, Virginia: ASCE.

Shiu, K.N. and Stanish, K. (2008) "Extending the Service Life of Parking Structures", *Concrete International*, April, 43–49.

Smith, L.L., Kessler, R.J. and Powers, R.G. (1993) "Corrosion of Epoxy Coated

Rebar in a Marine Environment", *Transportation Research Circular*, 403, Transportation Research Board, National Research Council, pp. 36–45.

Soares, J.B. and Tang, T. (1998) "Bimaterial Brazilian Specimen For Determining Interfacial Fracture Toughness", *Engineering Fracture Mechanics*, 59(1): 57–71.

Spoelstra, M.R. and Monti, G. (1999) "FRP-Confined Concrete Model", *ASCE Journal of Composites for Construction*, 3(2): 143–150.

Stehno, G. and Mall, G. (1977) "The Tear-Off Method: A New Way to Determine the Quality of Concrete in Structures on Site", in *Proceedings of RILEM International Symposium on Testing In-Situ of Concrete Structures*, Budapest, pp. 335–347.

Steinbach, J. and Vey, E. (1975) "Caisson Evaluation by Stress Wave Propagation Method", *Journal of Geotechnical Engineering*, ASCE, 101(4): 361.

Stokes, D.B. (2002) "The Role of Lithium to Mitigate ASR in Existing Concrete", paper presented at the Workshop for Use of Lithium to Mitigate Alkali-Silica Reactivity, Transportation Research Board, 81st Annual Meeting, January 13, Washington, DC.

Suaris, W. and van Mier, J.G.M. (1995) "Acoustic Emission Source Characterization in Concrete under Biaxial Loading", *Materials and Structures*, 28(182): 444–449.

Taylor, M.A. (1992) "Passive Acoustic Emission for Quantitative Evaluation of Freeze Thaw and Alkali Aggregate Reaction in Concretes", in Nondestructive Testing of Concrete Elements and Structures, Proceedings of Sessions sponsored by the Engineering Mechanics Division of the American Society of Civil Engineers in Conjunction with the Structures Congress, San Antonio, TX, 13–15 April.

Teng, J.G., Chen, J.F., Smith, S.T. and Lam, L. (2002) *FRP Strengthened RC Structures*. Chichester: John Wiley & Sons, Ltd.

Tovey, A.K. (1986) "Assessment and Repair of Fire-Damaged Concrete Structures-an Update", in T.Z. Harmathy (ed.) *Evaluation and Repair of Fire Damage to Concrete*, Detroit, MI: American Concrete Institute.

Trabanelli, G. (1986) "Corrosion Inhibitors", in F. Mansfeld (ed.), *Corrosion Mechanism*, New York: Marcel Dekker.

Triantafillou, T.C. (1998) "Shear Strengthening of Reinforced Concrete Beams Using Epoxy-Bonded FRP Composites", *ACI Structural Journal*, 95(2): 107–115.

Triantafillou, T.C. and Antonopoulos, C.P. (2000) "Design of Concrete Flexural Members Strengthened in Shear with FRP", *ASCE Journal of Composites for Construction*, 4(4): 198–205.

Trout, J. (1997) *Epoxy Injection in Construction*, Hanley-Wood, Inc.

Troxell, G.E, Davis, H.E, and Kelly, G.W. (1968) *Composition and Properties of Concrete*, 2nd edn, New York: McGraw-Hill.

Utah DOT Research News (1998) "Clear Penetrating Sealers– the Concrete Protection of Tomorrow", Number 98–2.

Valenta, O. (1969) "Kinetics of Water Penetration into Concrete as an Important Factor of Its Deterioration and of Reinforced Corrosion", *RILEM International Symposium on the Durability of Concrete*, Prague, Part I, pp. 177–193.

Vassie, P.R. (1978) "Evaluation of Techniques for Investigating the Corrosion of Steel in Concrete", Department of the Environment, Department of Transport, TRRL Report SR397, Crowthorne.

Verbeck, G.J. (1975) "Mechanism of Corrosion in Concrete", in *Corrosion of Metals in Concrete*, ACI SP-49.

Viktorov, I. (1967) *Rayleigh and Lamb Waves*, trans. W. P. Mason, New York: Plenum Press.

Vuorinen, J. (1985) "Applications of Diffusion Theory to Permeability Tests on Concrete, Part I: Depth of Water Penetration into Concrete and Coefficient of Permeability", *Magazine of Concrete Research*, 37(132): 145–152.

Wall, J.S. and Shrive, N.G. (1998) "Factors Affecting Bond Between New and Old Concrete", *ACI Material Journal*, 85: 117–125.

Warner, J. (1984) "Selection of Repair Materials", *Concrete Construction – World of Concrete*, 29(10): 865–871.

Warner, J. (1997) "Methods of Repairing and Retrofitting (Strengthening) Existing Buildings", paper presented at workshop on Earthquake-Resistant Reinforced Concrete Building Construction (ERCBC), University of California, Berkeley, Jul 11–15, 1977. Reprinted in *ACI Seminar Course Manual*, ACI Infrastructure Seminar – Rehabilitation of Concrete Structure (Undated), pp. 61–92.

Weil, S. (1995) "Non-Destructive Testing of Bridge, Highway and Airport Pavements", in G. Schickert, and H. Wiggerhanser (eds), *International Symposium Non-Destructive Testing in Civil Engineering*, Sept. 26–28, Berlin, pp. 467–474.

Wood, H.O. and Neumann, F. (1931) "Modified Mercalli Intensity Scale of 1931", *Bulletin of the Seismological Society of America*, 21: 277–283.

Wyss, M. and Brune, J. (1968) "Seismic Moment, Stress and Source Dimensions", *Journal of Geophys. Res.*, 73: 4681–4694.

Xanthakos, P.P. (1996) *Bridge Strengthening and Rehabilitation*, Upper Saddle River, NJ: Prentice Hall.

Xi, Y. and Li, Y. (2004) "Materials, Testing Methods, and Construction Practices for Fast Concrete Deck Repair", Internal Report CU/SESM/XI-2004/001 University of Colorado at Boulder, and Report to Colorado DOT.

Xi, Y., Shing, B., Abu-Hejleh, N., Asiz, A., Suwito, Xie, Z.H., Ababneh, A. (2003) *Assessment of the Cracking Problem in Newly Constructed Decks in Colorado*, Colorado Department of Transportation, Report No. CDOT-DTD-R-2003-3.

Xiong, G., Liu, J. and Li, G. (2002) "A Way of Improving Interfacial Transition Zone Between Concrete Substrate and Repair Materials", *Cement and Concrete Research*, 32: 1–5.

Yang, Q., Zhang, S. and Wu, X. (2000) "Deicer-Scaling Resistance of Phosphate Cement-Based Binder for Rapid Repair of Concrete", *Cement and Concrete Research*, 30: 1807–1813.

Yang, Q., Zhang, S. and Wu, X. (2002) "Deicer-Scaling Resistance of Phosphate Cement-Based Binder for Rapid Repair of Concrete", *Cement and Concrete Research*, 32: 165–168.

Yang, Q., Zhu, B., Zhang, S. and Wu, X. (2000) "Properties and Application of Magnesia-Phosphate Cement Mortar for Rapid Repair of Concrete", *Cement and Concrete Research*, 30: 1807–1813.

Yue, L.L. and Taerwe, L. (1992) "Creep Recovery of Plain Concrete and Its Mathematical Modelling", *Magazine of Concrete Research*, 44(161): 281–290.

Zhang, J., Qin L. and Li, Z. (2008) "Hydration Monitoring of Cement-Based Materials with Resistivity and Ultrasonic Methods", accepted for publication, *Materials and Structures*.

Zoldners, N.G. (1957) "Calibration and Use of Impact Test Hammer", *ACI J. Proc.*, 54(2): 161.

Index

Note: *f* indicates a figure and *t* a table